Preservation of Food by Ionizing Radiation

Volume III

Editors

Dr. Edward S. Josephson

Department of Nutrition and Food Science
Massachusetts Institute of Technology
Cambridge, Massachusetts

Dr. Martin S. Peterson

Editorial Consultant
Science and Technology
Natick, Massachusetts

CRC Press
Taylor & Francis Group
Boca Raton London New York

CRC Press is an imprint of the
Taylor & Francis Group, an informa business

First published 1983 by CRC Press
Taylor & Francis Group
6000 Broken Sound Parkway NW, Suite 300
Boca Raton, FL 33487-2742

Reissued 2018 by CRC Press

Library of Congress Cataloging in Publication Data
Main entry under title:

Preservation of food by ionizing radiation.
Bibliography: p.
Includes index.
1. Radiation preservation of food.
I. Josephson, Edward S. II. Peterson,
Martin S.
TP371.8.P74 664'.0288 81-18118
ISBN 0-8493-5323-8 (v. 1) AACR2
ISBN 0-8493-5324-6 (v. 2)
ISBN 0-8493-5325-4 (v. 3)

A Library of Congress record exists under LC control number: 81018118

Publisher's Note
The publisher has gone to great lengths to ensure the quality of this reprint but points out that some imperfections in the original copies may be apparent.

Disclaimer
The publisher has made every effort to trace copyright holders and welcomes correspondence from those they have been unable to contact.

ISBN 13: 978-1-315-89689-2 (hbk)
ISBN 13: 978-1-351-07599-2 (ebk)

Visit the Taylor & Francis Web site at http://www.taylorandfrancis.com and the
CRC Press Web site at http://www.crcpress.com

If you give a man a fish, he will have a meal.
If you teach him to fish, he will have a living.
If you are thinking a year ahead, sow seed.
If you are thinking ten years ahead, plant a tree.
If you are thinking one hundred years ahead, educate the people.
By sowing seed once, you will harvest once.
By planting a tree, you will harvest tenfold.
By educating the people, you will harvest one hundredfold.

Anonymous Chinese Poet
420 B.C.

FOREWORD

I first became acquainted with the concept of preserving food by exposing it to ionizing radiation when this revolutionary food preservation process had just emerged from its status as a speculative research venture and was attracting attention as a feasible preservation process. This was in the early 1950s. To many, however, the concept at that time still belonged to the realm of theory.

Many of us, however, and I am proud to be counted among them, recognized the advent of an exciting new preservation method with the potential of bringing about substantial economic, nutritional, and public health advantages. Its applications were apparent even at that relatively early stage of development. It had been demonstrated that:

1. Ionizing radiations at a comparatively low radiation level could destroy the microorganisms, insects, and parasites in food causing spoilage and disease; could inhibit losses caused by sprouting during the storage of potatoes and onions; and could extend the shelf life of many food products in a fresh-like condition.
2. Ionizing radiations applied at the level required for sterilization could permit long-term storage without refrigeration of several meat, poultry, and seafood products although with some compromise in taste, odor, and texture. This left, to be sure, a challenging area of research to discover the why's and wherefore's in order to obtain the product improvement necessary for acceptance by the consumer.

It was recognized very quickly that proof of safety for consumption ("wholesomeness") had to be demonstrated in order for irradiated foods to receive universal acceptance by consumers.

Research begun in those early days was intensive. The effort applied, which was of the highest professional quality, indicated the confidence of those seeking to lay the foundations for an ultramodern food preservation process. It was an impressive program.

The going was not always smooth. Skepticism, reluctance to provide the funding required, and the failure on the part of many food scientists and technologists to recognize the values of the process, dogged the path of progress. But those of us who were close enough to the project to foresee its value persisted in our efforts to gain industrial and governmental support and on a few crucial occasions we kept the program alive. We could see how each advance engendered further advances—the story of science in its essence—and that a body of knowledge was developing that would eventually solve the seemingly unsolvable problems standing in the way of harnessing nuclear energy for another peaceful purpose.

The applications of nuclear energy for peaceful purposes are many. In my view, its use in preserving food and reducing the incidence of food-borne diseases can provide great benefits to mankind. Nuclear radiation effectiveness at low dosage to prevent waste due to insect infestations and sprouting of tubers and bulbs during storage has been demonstrated. Its application at sterilization levels can reduce the need for expensive energy-consuming refrigeration facilities and their maintenance. The foregoing are only a few of its many values.

It is a pleasure to report, on the basis of the analysis which this book provides, that the early promise of the radiation preservation of food is now approaching realization. We are on the home stretch.

Melvin Price
Member, House of Representatives
Congress of the United States
1981

PREFACE

Preservation of Food by Ionizing Radiation presents a comprehensive account of the underlying science and technology of this newest of food preservation processes. The distinguished scientists, drawn from all quarters of the globe, who submitted chapters to this treatise are responsible in their own right for important research and development advances in radiation preservation and are fully cognizant of the fundamental knowledge that has accumulated in three decades of intense study. The main aspects of this remarkable preservation process are covered — the process; its chemical, physical, and biological effects on food; the problems; the limitations; the successful applications; and the growing list of products cleared for consumption. The references that document each chapter are extensive and most if not all of the scores of workers who have contributed to the advance of radiation preservation are listed therein. It can be said without exaggeration that this monumental work will provide the technically informed reader with the current status of the preservation of food by ionizing radiation.

The authors, it should be noted, are at work in numerous laboratories, not only those in the U.S., but those in other parts of the world.

But the work of this widely dispersed group is highly integrated — first by such prestigious organizations as the United Nations' International Atomic Energy Agency, Food and Agriculture Organization, and World Health Organization and second by major symposia, conferences, and panel meetings held in various parts of the world, e.g., Bombay, Karlsruhe, Vienna, Natick, Rome, Geneva, Boston, Wageningen, Tokyo, and Colombo.

Both the scientific talent and the research facilities have increased greatly over the years. The large number of scientists at work in food irradiation is attested by the hundreds of authors cited in the reference sections of the three volumes.

In short, human resources and research facilities for studying the enormously complex phenomena associated with the preservation of food by ionizing radiation and for applying the results to the design of viable processes have grown impressively since the inception of the program.

The preservation of food by ionizing radiation is a very young branch of food processing but it has already found a number of applications — the inhibition of sprouting in potatoes and onions (now a substantial commercial operation in Japan), the disinfestation of insects in grain, and the reduction of spoilage organisms in spices coming first to mind. The list of products approved for human consumption in various nations of the world is constantly growing. Included in the list are products made free of pathogens.

There are roadblocks, of course, but advances in the sciences involved in radiation preservation research have overcome and will continue to overcome many of the barriers that stand in the path that leads to the ultimate objective, a successful industrial process for preserving food by ionizing radiation.

In addition to its potential as an industrial process serving the civilian market, irradiated foods, particularly irradiation sterilized meats, would fulfill critical human needs in areas remote from conventional sources of supply. Substantial savings in money and energy would accrue since these products require no refrigeration. It was these valuable military applications as first perceived by Dr. Ralph G. H. Siu that stimulated him to initiate and obtain continuing support for the U.S. Army's program and to bring forth the first treatise in the U.S. on the subject in 1957. American astronauts who ate irradiation sterilized ham on the moon and who with Soviet cosmonauts ate irradiation sterilized meats in outer space have attested to their excellent organoleptic quality.

The irradiation process has great usefulness in the elimination of food-borne pathogens such as salmonella and that dreaded organism, *Clostridium botulinum,* which develops a toxin that can be fatal if ingested. Irradiation is a promising alternative to reduce dependence upon ethylene oxide and other chemical food additives which are presently used to control pathogens but which are under scrutiny by health authorities for carcinogenicity.

Food scientists the world over should keep abreast of advances in knowledge and techniques in this developing new food process. The place to start is with these three volumes, which are, without question, the most comprehensive and the most authoritative source of information on the basic science and technology yet published on food preservation by the application of ionizing radiation.

Dale H. Sieling
Technical Director (Retired)
U.S. Army Natick Research
and Development Laboratories
1981

ACKNOWLEDGMENTS

The editors are grateful to their many friends and associates who have assisted in the preparation of *Preservation of Food by Ionizing Radiation*. For his initial encouragement and interest in producing a definitive treatment of this most unique of food processing methods, the editors are greatly indebted to Dr. Ralph G. H. Siu, formerly Technical Director, Office of the Quartermaster General, Washington, D.C. Special thanks are due Professor Marcus Karel, of the Department of Nutrition and Food Science at the Massachusetts Institute of Technology (MIT) for his continuous encouragement and his generosity in making the facilities of the Department available for executing the various tasks of manuscript processing. Especial thanks are due Ms. Patricia B. Olstead, Librarian, U.S. Army Natick Research and Development Laboratories, (NLABS), for obtaining copies of publications hard to come by (a very sizeable task), and to Ms. Stephanie Tosches, Barbara Farrell Parsons and Ruth Fortner at NLABS, and Molly Crowley and Janet Copley at MIT for secretarial assistance of various kinds, but for the most part tasks related to manuscript processing.

The editors are deeply indebted to their wives, Blanche Josephson and Wilma L. Peterson, for their understanding and support. In numerous ways they lightened the burden of a heavy task.

Presenting, definitively, the history, nature and effect of ionizing radiations on food has been a cooperative effort on the parts of authors, secretaries, editors, including the able editors of CRC Press, and the many people working diligently behind the scenes.

Edward S. Josephson
Martin S. Peterson
1981

GLOSSARY OF TERMS

Becquerel (Bq) — The unit of radioactivity. The amount of radioactivity which decays at a rate of one nuclear transformation per sec. This unit supersedes the Curie in the SI units.

Beta particle — A high-speed electron ejected from an atomic nucleus in certain types of radioactive disintegration. These electrons were described as beta particles before their identity as electrons had been established.

Curie (Ci) — The unit of radioactivity. A curie was originally defined as the number of nuclear disintegrations occurring each second in 1 g of radium. It is now defined as that amount of radioactivity which involves the disintegration of 3.7×10^{10} atomic nuclei in 1 sec.

Electron — A particle possessing a negative electric charge which is an integral part of every neutral atom. It has a mass of 1/1840th of that of the hydrogen atom. Free electrons can be produced in various ways, collected into beams in a vacuum and accelerated in electrical machines to a high energy potential. Electrons and beta particles are identical in nature, the term beta particle being reserved to describe electrons originating from a disintegrating nucleus in radioactive decay.

Electron Volt (eV) — A unit of energy equivalent to the amount of energy gained by an electron when accelerated by a potential of 1 V. One electron volt = 1.6×10^{12} erg. Larger multiple units are used to describe the energy level of electrons and X-rays from electrical machine sources, or beta particles and gamma rays from radioactive sources, viz: keV = 1000 eV; MeV = 1 million eV.

Gamma radiation — Electromagnetic radiation of a very short wavelength produced by the spontaneous disintegration of the atomic nucleus of certain radioactive nuclides. Gamma rays are of the same nature as X-rays of short wavelengths.

Gray (Gy) — The unit of energy absorbed from ionizing radiation by the matter through which the radiation passes. A radiation dose of 1 Gy involves the absorption of 1 J of energy by each kilogram of matter through which the radiation passes. Large multiple units are frequently used to express the radiation dose in food irradiation, viz., 1000 Gy = 1 kGy. The Gy is equivalent to 100 rad.

Rad — The unit of energy absorbed from ionizing radiation by the matter through which the radiation passes. A radiation dose of 1 rad involves the liberation of 100 erg of energy into each gram of matter through which the radiation passes. Larger multiple units are frequently used to express the radiation dose in food irradiation, viz: krad = 1000 rad; Mrad = million rad. The rad is being superseded by the Gray.

Radappertization — Exposure to ionizing radiation of food in hermetically sealed packaging at doses necessary to kill all organisms of food spoilage or public health significance. Doses used are greater than 1 Mrad. Analogous to thermal sterilization (canning).

Radicidation — Exposure of food to ionizing radiation at doses necessary to kill all nonsporeforming pathogens. Doses used are generally below 1 Mrad. Analogous to pasteurization.

Radurization — Exposure of food to ionizing radiation to reduce populations of organisms in order to delay onset of spoilage. Doses used are generally below 1 Mrad. Analogous to pasteurization.

Rem — A special unit of dose equivalent. The dose-equivalent in rems is equal numerically to the absorbed dose in rads multiplied by the quality factor, the distribution factor, and any other necessary modifying factors. The rem represents the quantity of radiation that is equivalent, in biological damage of a specified sort, to 1 rad of 250 kV X-rays.

Rep(r) — Acronymn for Roentgen Equivalent Physical. An obsolete unit of absorbed dose of any ionizing radiation with a magnitude of 93 ergs/g. Superseded by rad, being superseded by Gray. Larger multiple units are frequently used to express the radiation dose in food irradiation, viz: kGy = 1000 Gy. Supersedure of terms: 1 rep = 0.93 rad; 1 rad = 1.075 rep (1.075 r); 1 Gray (Gy) = 100 rad; 100 rad = 107.5 rep (107.5 r).

X-rays — Electromagnetic radiation of a wide variety of short wavelengths usually produced by a machine in which a beam of fast electrons in a high vacuum bombards a metallic target. X-rays of short wavelengths are of the same nature as gamma rays except for their origin. X-rays are sometimes termed Roentgen rays.

THE EDITORS

Edward S. Josephson, Ph.D., is Senior Lecturer in the Department of Nutrition and Food Science at the Massachusetts Institute of Technology, Cambridge, Mass. He received the A.B. degree in Mathematics from Harvard University in 1936 and the Ph.D. degree (with Distinction) in Biochemistry from the Massachusetts Institute of Technology in 1940.

Dr. Josephson spent his postdoctoral years as the Williams-Waterman Research Fellow in Nutritional Biochemistry at the Massachusetts Institute of Technology (1940-1942). Subsequently, he served until 1946 as Biochemist at the National Institutes of Health's World War II Malaria Drug Testing Laboratory in the U.S. Penitentiary Hospital in Atlanta, Ga. During the post World War II years and the Korean War, he continued his research on chemotherapy of malaria and other tropical diseases as Senior Biochemist at the National Institutes of Health's Laboratory of Tropical Diseases in Bethesda, Md. (1946-1952).

After serving as Principal Biochemist in the U.S. Army's Office of the Chief Chemical Officer in Washington, D.C. (1952-1954), Dr. Josephson transferred to the Army's Research and Development Laboratories at Natick, Mass., where he held positions with increasing responsibilities until his retirement from government service in 1975. From 1961 until retirement he directed and administered the Natick Laboratories' Food Irradiation Program, holding the positions of Associate Director for Food Irradiation until 1972 and Deputy Technical Director for the Food Service Systems and Pollution Abatement Programs (1972-1975).

Dr. Josephson has served as Science Advisor to the Speaker of the New Hampshire House of Representatives in Concord, N.H. and advisor and consultant on preserving food by ionizing radiation to the international Atomic Energy Agency, Food and Agricultural Organization, Inter-American Nuclear Energy Commission of the Organization of American States, the Governments of India, Iran, and Israel and commercial companies.

Dr. Josephson's extensive travels have taken him to the Arctic, Europe, Japan, and developing countries in Latin America, the Near, Middle, and Far East, and Central, South, and Southeast Asia. In addition to teaching at the Massachusetts Institute of Technology, he has taught at Oglethorpe University and Emory University Medical School, Atlanta, Ga., American University, Washington, D.C.; and Harvard University, Cambridge, Mass. He has also lectured at a number of universities and research centers in the U.S. and abroad. Dr. Josephson has more than 100 publications, primarily on food preservation by ionizing radiations and chemotherapy of malaria.

Martin S. Peterson, Ph.D., now retired, was formerly editor of scientific, technological, and engineering reports on subjects within the scope of the mission of the U.S. Army Natick Research and Development Laboratories(NLABS), Natick, Mass. One such area was the preservation of food by ionizing radiations, a project in which the Army was interested because irradiated food had several logistical benefits; namely, extended shelf life without refrigeration, simplicity of preparation, reduction of kitchen labor, and minimal changes in flavor, texture, odor, and color.

Since retirement Dr. Peterson has continued to be active in editorial tasks — as co-editor with A. H. Johnson of the *Encyclopedia of Food Technology*, 1974, and the *Encyclopedia of Food Science*, 1978, edited by M. S. Peterson and A. H. Johnson. For the past several years he has participated in editing the annual *Posture Reports* of the Natick Laboratories and the *Annual Historical Review* of these Laboratories. He has had a number of other consulting assignments.

Dr. Peterson's many years in university teaching were often coordinated with an editorial function. At the University of Nebraska, where he taught composition and literature, he was Associate Editor of the *Prairie Schooner,* a literary quarterly sponsored by the University. During the war years (1940 to 1945) he was Acting Editor of the College of Agriculture technical reports. On leaving Nebraska and accepting a post with the Quartermaster Food and Container Institute, Chicago, he was editor of the *Activities Report,* 1948 to 1958, Research and Development Associates for Military Food and Packaging Systems, Inc. From 1952 to 1962, he was editor of two professional journals, *Food Technology* and *Food Research,* published by the Institute of Food Technologists.

Dr. Peterson received an A.B. from Reed College in 1920, an M.A. from the University of Nebraska in 1928, and a Ph.D. in American Literature from the same institution in 1932.

CONTRIBUTORS

Ernest K. Akamine
Emeritus Plant Physiologist
Emeritus Professor of Plant Physiology
College of Tropical Agriculture and Human Resources
Department of Botany
University of Hawaii at Manoa
Honolulu, Hawaii

John H. Brower
Research Entomologist
Biological Control Research Unit
Stored-Products Insects Research Laboratory
U.S. Department of Agriculture
Savannah, Georgia

Ari Brynjolfsson
Science and Advanced Technology
U.S. Army Natick Laboratories
Natick, Massachusetts
Department of Food Science and Nutrition
Massachusetts Institute of Technology
Cambridge, Massachusetts

Arthur K. Burditt, Jr.
Laboratory Director
U.S. Department of Agriculture
Yakima Agriculture Research Laboratory
Yakima, Washington

J. F. Diehl
Director and Professor
Federal Research Center for Nutrition
Karlsruhe, West Germany

J. Farkas
International Facility for Food Irradiation Technology
Pilot Plant for Food Irradiation
Wageningen, The Netherlands

Frank M. Fraser
General Manager
Industrial Products
Atomic Energy of Canada, Ltd.
Commercial Products
Ottawa, Ontario
Canada

Harry E. Goresline*
Kaumberg, Austria

Nicholas Grecz
Scientist I and Head
Cancer and Radiation Biology Research Department
Cancer Therapy Institute
King Faisal Specialist Hospital and Research Centre
Riyadh, Kingdom of Saudi Arabia

Robert D. Jarrett, Sr.
Director
Radiological Safety Staff
U.S. Department of Agriculture
Beltsville, Maryland

Edward S. Josephson
Senior Lecturer
Department of Nutrition and Food Science
Massachusetts Institute of Technology
Cambridge, Massachusetts

John J. Killoran
Polymer Chemist
Food Engineering Laboratory
U.S. Army Natick Research and Development Laboratories
Natick, Massachusetts

Betty L. King
Associate Professor of Biology
Northern Virginia Community College
Alexandria, Virginia

Joseph J. Licciardello
Supervisory Research Food Technologist
Northeast Fisheries Center
National Marine Fisheries Service
U.S. Department of Commerce
Gloucester, Massachusetts

Paisan Loaharanu
Food Preservation Section
Joint FAO/IAEA Division of Isotope and Radiation Applications of Atomic Energy for Food and Agricultural Development
Vienna, Austria

Thomas G. Martin, III
Director
Safety and Radiation Protection
U.S. Army Natick Research and Development Laboratories
Natick, Massachusetts

Akira Matsuyama
Leader of Japanese Expert Team for AP4 Project
Japan International Cooperation Agency
Faculty of Agricultural Technology
Bogor Agricultural University
Bogor, Indonesia
Emeritus Radiobiologist
Institute of Physical and Chemical Research
Saitama, Japan

William L. McLaughlin
Radiation Physics Division
Center for Radiation Research
National Bureau of Standards
Washington, D.C.

James H. Moy
Professor of Food Engineering
Chairman, Graduate Faculty in Food Science
Department of Food Science and Human Nutrition
University of Hawaii at Manoa
Honolulu, Hawaii

W. W. Nawar
Professor
Department of Food Science and Human Nutrition
University of Massachusetts
Amherst, Massachusetts

John T. R. Nickerson**
Professor Emeritus
Massachusetts Institute of Technology
Emeritus Member and Fellow of the Institute of Food Technologist

Thomas A. Olejnik
Johnson and Johnson
East Windsor, New Jersey

E. C. Pollard
Hargitt Research Fellow
Zoology Department
Duke University
Durham, North Carolina

Warren J. Ramler
Senior Vice President and General Manager
RPC Industries
Plainfield, Illinois

Louis J. Ronsivalli*
Laboratory Director
Northeast Fisheries Center
National Marine Fisheries Service
U.S. Department of Commerce
Gloucester, Massachusetts

Durwood B. Rowley
Head
Food Microbiology Group
Science and Advanced Technology Laboratory
U.S. Army Natick Research and Development Laboratories
Natick, Massachusetts

Michael G. Simic
Pulse Radiolysis Group Leader, CRR
National Bureau of Standards
U.S. Department of Commerce
Washington, D.C.,

Irwin A. Taub
Chief, Plant Products
Food Technology Division
Food Engineering Laboratory
U.S. Army Natick Research and
Development Laboratories
Natick, Massachusetts

Elvin W. Tilton
Research Entomologist
Chemical Control Research Unit
Stored-Products Insects Research and
Development Laboratory
U.S. Department of Agriculture
Savannah, Georgia

K. Umeda
Chief Advisor
Instituto Technologico Pesquero del
Peru
Ministerio de Pesqueria
Lima, Peru

Walter M. Urbain*
Professor Emeritus
Michigan State University
Sun City, Arizona

Karoly Vas**
Director
Central Food Research Institute
Budapest, Hungary

Chia Ping Wang
Research Physicist
Food Systems Equipment Division
Food Engineering Laboratory
U.S. Army Natick Research and
Development Laboratories
Natick, Massachusetts

* Retired
** Deceased

PRESERVATION OF FOOD BY IONIZING RADIATION

Edward S. Josephson and Martin S. Peterson

Volume I

Volume II

Volume III

TABLE OF CONTENTS

Volume III

Chapter 1

RADURIZATION AND RADICIDATION: MEAT AND POULTRY

Walter M. Urbain

TABLE OF CONTENTS

I. RADURIZATION

The purpose of radurization of meats and poultry is to extend the product life. Partly because of considerations of microbiological safety and partly for reasons of product quality, in the case of these foods radurization must be used with refrigeration. Radurization is a low-dose process. Its specific objective is to reduce the initial microbial population to such an extent that a longer than usual time period is needed for the development of ultimate microbial spoilage. Radurization does not prevent ultimate microbial spoilage; it serves only to delay it, and the delay constitutes the product life extension.

There are several types of red meats, each derived from different animal species. There are special differences which affect end results. Red meats of greatest interest are beef, veal, pork, and lamb. With poultry, interest in radurization is limited largely to chicken. With all meats, the principal interest in radurization is its application to fresh meats. A secondary possibility is the radurization of those processed meats distributed as perishable items, including sliced cold cuts and fresh, cooked, smoked, and/or cured sausages.

Except for fresh sausage, perishable processed meats are generally distributed in vacuum packages. This type of package protects the product against quality loss caused by action of atmospheric oxygen and also extends product life by delaying microbial spoilage. Vacuum packaging combined with good manufacturing practices easily provides normal lives for these products of 30 days or more. Such stability is adequate for centralized processing and packaging into retail units and for distribution over large market areas.

Athough there have been some investigations of the use of radiation to extend the life of processed meats,[1-3] there seems to be little need for radurization. Justification of the irradiation cost would appear difficult.

There appear to be, however, excellent opportunities for the radurization of fresh meats and poultry. Even with the availability of good refrigeration, the perishability of these products imposes severe restrictions on their distribution. In the U.S. almost all fresh meats are sold from a display case as prepackaged items. This also is the practice in certain other countries and is one that is growing. With good refrigeration, the display case life of a retail cut is a maximum of about 3 days. This short life requires that final cutting and packaging be done in the retail store. Except possibly for some local situations, there is not sufficient time for the preparation and transport of retail cuts of fresh meat from a centralized facility to retail stores. In the U.S. distribution time from a centralized facility will vary but could be as long as 3 weeks.[4]

The present procedure for distribution of fresh meats involves the delay of final cutting and packaging until the time the retail cut is offered for sale. Transport of fresh meats from the meat packing establishment to the retail store is accomplished by shipping the meat as sides, quarters, or wholesale cuts. This maintains a minimum ratio of surface to mass, thereby minimizing microbial contamination as well as exposure to atmospheric oxygen. A recent distribution improvement has been the vacuum packaging of wholesale cuts or larger pieces. Other improvements have involved reduction of the initial microbial population through various bactericidal measures (e.g., chlorinated water wash). Efforts to market frozen retail cuts generally have not been successful. Vacuum packaging excludes oxygen and darkens the meat to an unacceptable degree.

There are a number of reasons to seek ways to centralize final cutting and packaging of meats and poultry. Although the present system of fresh meat distribution is effective, particularly in terms of meat quality, it does incur some costs that could be reduced by centralization. Cost reduction possibilities through centralization are

1. Labor savings through improved efficiency and utilization, i.e., increased productivity
2. Reduced shipping costs, since only what is sold is shipped — trimmings, bones, etc. are retained at the point of origin
3. Better utilization of by-product trimmings, etc. due to concentration of volume at a central location
4. Reduced capital requirements for facilities — one central unit can produce the volume of a number of store units, thereby saving on equipment and building space
5. Opportunities for better quality control exist in a centralized production system
6. Increased merchandising flexibility (Because the retail cuts will be available, a given store needs to stock only what its customers want. This avoids forced selling at that store of all the cuts in a side or other wholesale unit.)

There is ample evidence of the effectiveness of irradiation in accomplishing the delay of microbial spoilage of fresh meats and poultry.[5-8] Changes in meats occur in storage, however, that are not related to microbial spoilage and which adversely affect consumer acceptance.[9,10] These changes include

1. Oxidation by atmospheric oxygen of the pigment myoglobin to yield brown or gray discolorations
2. The formation of a serum-like liquid exudate, termed "weep" or "drip" from the cut surface of the lean
3. Oxidation by atmospheric oxygen of the meat lipids which causes off-flavors

In the short time period involved in the present procedure for handling retail cuts, 3 days or less, none of these deteriorative changes occur sufficiently to affect consumer acceptance. If the microbial spoilage was delayed so as to permit a 3-week instead of 3-day life, then these other spoilage pathways can become important. Hence irradiation alone would not result in acceptable life extension. Ways to prevent discoloration, drip, and lipid oxidation must also be used.[9]

A procedure has been developed which is designed to control all four pathways of deterioration of fresh meats.[11,12] It entails the following steps:

1. To the cut of meat is added not more than 0.5% of a condensed phosphate, (e.g., sodium tripolyphosphate [TPP]). This may be done by dipping the meat in an aqueous solution or by other appropriate means.
2. Each cut is wrapped in an oxygen-permeable, moisture-impermeable flexible film.
3. A number of wrapped individual cuts are placed in a bulk container, which is vacuumized.
4. The bulk-packaged meats are irradiated with gamma radiation, using doses in the range of 1 to 2 kGy at product temperatures between 0 and 10°C.
5. The bulk package is stored and transported at temperatures between 0 and 5°C.
6. At the retail store, the individual wrapped cuts are removed from the bulk package about ½ hr prior to displaying for sale.
7. The retail cuts are displayed at temperatures in the range 0 to 5°C and are sold within 3 days.

In this procedure irradiation controls microbial spoilage. Phosphate greatly reduces drip and preserves the red color of meat. The packaging system provides an anaerobic environment for all but the last 3 days of market life. It protects color, prevents lipid oxidation, and has an effect on the ultimate microbial outgrowth pattern.

Table 1
EFFECT OF VARIOUS DOSES OF GAMMA RADIATION ON THE TOTAL PLATE COUNT OF VACUUM PACKAGED STORED BEEF STEAKS

Dose (kGy)	Days stored at 4°C 0 (TPC/g)	7 (TPC/g)	14 (TPC/g)	21 (TPC/g)
0	1.6×10^6	1.0×10^7	5.6×10^7	7.0×10^8
0.5	1.8×10^4	6.2×10^4	7.7×10^6	3.0×10^7
1.0	8.0×10^3	3.3×10^5	3.0×10^6	9.0×10^6
2.5	1.0×10^2	2.8×10^2	8.6×10^3	9.8×10^3
5.0	60	1.0×10^2	2.0×10^2	2.2×10^2
10.0	<10	<10	<10	<10

From Urbain, W. M., *Proceedings of the Bombay Symposim on Radiation Preservation of Food*, IAEA, Vienna, 1973, 505. With permission.

Table 2
THE THRESHOLD DOSE FOR THE COMMON RED MEATS AND CHICKEN FOR AN IDENTIFIABLE "IRRADIATION" FLAVOR

Food	Threshold dose (kGy)
Pork	1.75
Beef	2.5
Chicken	2.5
Lamb	6.25

From Sudarmadji, S. and Urbain, W. M., *J. Food Sci.*, 37, 671, 1972. With permission.

Table 1 shows the effect of various doses of gamma radiation on the total plate count (TPC) of stored beef steaks. A TPC of about 10^8 organisms per gram can be regarded as indicating a condition of incipient microbial spoilage. Although these data show the advantage of higher doses, there are limitations on the amount of radiation that can be used. In addition to killing spoilage microorganisms, irradiation can cause undesired side effects. With meats, the principal such effect is the development of a characteristic "irradiated" flavor. This resembles scorching and generally is regarded as objectionable. The dose at which a detectable irradiated flavor occurs varies with the species. Sudarmadji and Urbain[13] determined the threshold dose for the irradiated flavor for a number of raw protein foods irradiated at 5 to 10°C. Their values for the common red meats and chicken are shown in Table 2. The flavor was determined after cooking the meats.

The use of subfreezing temperatures during irradiation reduces the amount of irradiation flavor.[14,15] Irradiation at subfreezing temperatures is a very necessary part of the radappertization of meats, because the high dose requirements of this process generate a strong irradiated flavor if the irradiation is carried out at temperatures above freezing.[16] Theoretically this low-temperature technique can be used in radurization of meats to allow use of doses above the threshold levels indicated in Table 2. The extra costs for freezing and subsequent thawing and the possible associated product quality

Table 3
EFFECT OF PHOSPHATE PRETREATMENT, VACUUM PACKAGING, RADIATION, AND SUBSEQUENT EXPOSURE TO THE ATMOSPHERE ON COLOR AND METMYOGLOBIN DEVELOPMENT ON FRESH BEEF DURING STORAGE AT 5°C

		21 Days in vacuum		21 Days in vacuum plus 1 day in air	
Dose (kGy)	Phosphate	Color (visual)	% Metmyoglobin	Color (visual)	% Metmyoglobin
0	No	Brown	11.5	Brown	28.2
	Yes	Purple	0.0	Red	2.0
0.5	No	Brown	9.0	Brown	24.7
	Yes	Purple	0.0	Red	3.2
1.0	No	Brown	5.0	Brown	18.0
	Yes	Purple	0.0	Red	7.0
2.0	No	Brown	11.5	Brown	28.0
	Yes	Purple	0.0	Red	3.0
3.0	No	Brown	10.3	Brown	27.5
	Yes	Purple	0.0	Red	9.5
4.0	No	Brown	21.2	Brown	43.0
	Yes	Purplish brown	9.0	Reddish brown	11.5
5.0	No	Brown	47.0	Brown	66.0
	Yes	Purplish brown	10.0	Reddish brown	20.2

From Urbain, W. M., *Proceedings of the Bombay Symposium on Radiation Preservation of Food,* IAEA, Vienna, 1973, 505. With permission.

loss make this approach seem impractical. Mercuri et al.[17] reported that chicken irradiated in the frozen state and stored thawed developed spoilage earlier than chicken irradiated unfrozen. At the present time the only feasible route to acceptable flavor appears to be to keep the dose below the threshold value.

The radurization dose for meats, within limits, is a process variable whose value in part is governed by the level of the initial microbial load and in part by the product-life extension desired. In practice, a dose giving the product performance desired would be used. It would have to be, however, below that which results in a detectable irradiation flavor.

A second limitation on dose arises from the fact that too much irradiation will discolor meat. Under some conditions this darkening occurs immediately upon irradiation. A high dose rate can increase the amount of discoloration.[18] Discoloration of phosphate-treated vacuum-stored beef occurs at about 4 kGy. As stated, TPP protects the color of beef. Table 3 shows this protective action and also the damaging effect of higher doses of radiation.

Dipping thin cuts (e.g., steaks, chops) for about 30 sec in a 10% aqueous solution of TPP results in the incorporation of about 0.5% of the phosphate in the meat. Lesser amounts can be secured by lowering the concentration of the dipping solution. Over a 3-week storage period at 5°C, the drip loss is less than 1% for meats containing 0.5% TPP.

Lipid oxidation is not affected by 1 kGy of radiation nor by the presence of TPP. Figure 1 shows the thiobarbituric acid (TBA) values for beef obtained during storage at 5°C. Exclusion of oxygen prevents oxidation. When the meat is exposed to air, oxidation of the lipid occurs. The TBA value of about 2 obtained after 3 days' exposure to air, however, indicates a level of oxidation too low to be detected as an off-flavor.

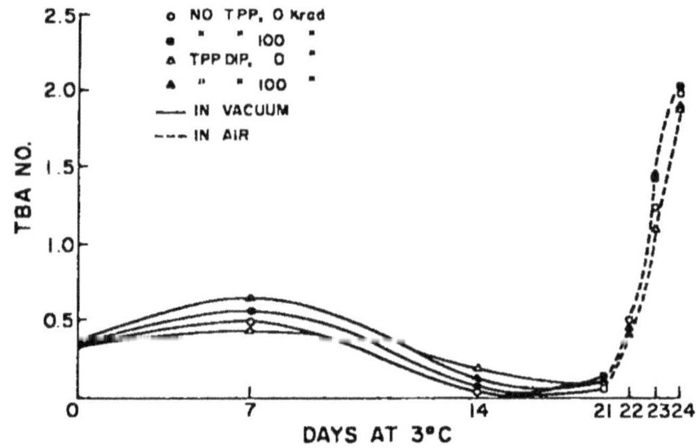

FIGURE 1. Lipid oxidation changes in irradiated and nonirradiated fresh beef as measured by TBA values (milligram of malonaldehyde per kilogram of meat) during storage at 5°C in vacuum for 21 days plus 3 days in air. (From Urbain, W. M., *Proceedings of the Bombay Symposium of Radiation Preservation of Food*, IAEA, Vienna, 1973, 505. With permission.)

The above-cited data demonstrate the effectiveness of the process in controlling all four identified pathways of meat deterioration on storage. When held for the maximum period considered necessary to enable centralized cutting and packaging (21 days) plus 3 additional days under display-case conditions, beef has an appearance comparable with freshly cut meat. Presumably it would be acceptable to a purchaser. There is, however, the need to determine also the acceptability of the meat in terms of its eating quality. By using consumer-type panels and a 9-point hedonic rating scale, the data shown in Table 4 were obtained on four indexes of quality: odor, flavor, juiciness, and overall quality. For testing, the beef (steaks) was cooked to a medium-well condition by heating for 30 min in air oven operating at 157°C. These panel data indicate good acceptance of the beef.

With radurization not all microorganisms present are killed. As the food is held, the survivors may multiply. For this outgrowth there are at least two concerns: (1) the possibility of a new health hazard being introduced and (2) the alteration of the character of the ultimate spoilage in a way that the consumer may not recognize as spoilage. Groesbeck[19] investigated the spoilage microflora of beef dipped in a 10% TPP solution for 30 sec, irradiated with 1 kGy, and stored for 21 days at 5°C in vacuum plus 5 additional days in air. Microbiological analyses were made at 0, 10, 21, and 26 days of storage. They consisted of (1) determination of TPCs and (2) characterization of representative isolates from total-count spread-plates in terms of percentages of:

A. Gram positive organisms
  Cocci
    Catalase positive and negative rods (nonsporulating)
    Catalase positive
    Lactobacillus
B. Gram negative organisms
  Rods
C. Yeasts
D. Unknown

Table 4
AVERAGE SCORES (9-POINT HEDONIC SCALE) BY 24 PANELISTS FOR BEEF

	0 Days		24 Days	
	Not treated	Treated[a]	Not treated[b]	Treated[a,c]
Odor	6.3	6.2	6.3	6.3
Flavor	6.3	6.5	5.5	5.8
Juiciness	6.4	6.0	5.8	6.3
Overall quality	6.2	6.2	5.8	6.2

[a] Phosphated and irradiated.
[b] Frozen 24 days.
[c] Stored 21 days in vacuum plus 3 in air at 5°C.

From Urbain, W. M., *Proceedings of the Bombay Symposium of Radiation Preservation of Food*, IAEA, Vienna, 1973, 505. With permission.

and (3) determination of the presence of the folowing organisms: *Salmonella*, coliforms and fecal coliforms, coagulase-positive *Staphylococcus aureus*, and *Clostridium perfringens*.

The findings may be summarized as follows:

1. The irradiated samples had counts of the order of 10^7 organisms per gram after 21 days of vacuum storage and 10^8 organisms after an additional 5 of air storage at 5°C. These were about 1 log less than the nonirradiated. The total counts were about equally divided between mesophiles and psychrophiles. The classification of the organisms found in the total count spread plates revealed the following:
   - A. In general the irradiated samples tend to contain Gram-positive organisms, whereas Gram-negative rods predominate in nonirradiated samples.
   - B. The organisms detected in all samples were largely anaerobes or facultative anaerobes. Yeasts were seldom found.
   - C. No Gram-positive catalase-negative cocci were found.
   - D. Gram-positive catalase-positive cocci were found in all types of samples.
   - E. No spore-forming Gram-positive rods were found.
   - F. Gram-positive rods, catalase-positive (nonsporulating), were found in all types of samples.
   - G. *Lactobacilli* were found in all types of samples. They predominated at the end of the storage period in the irradiated samples.
   - H. All Gram-negative organisms were rods. They occurred in all samples, but predominated in the nonirradiated samples at the end of the storage.
   - I. No Gram-negative cocci were found. This finding excludes the presence of *Moraxella* and *Herellea*.
2. Of coliform bacteria, none were found in the irradiated samples. Many (about 10^7/g) were found in the stored nonirradiated beef.
3. Of fecal coliforms, essentially none (in one sample only 43 organisms) were found in the irradiated samples.
4. Of *C. perfringens*, none (<5) were found in any sample.
5. Of Salmonellae, none were found in any sample.
6. Of coagulase-positive *S. aureus*, none (>5>1000) were found in any sample.

In interpreting the above findings, the following items are noted:

1. The procedure involves an extended period of anaerobic holding. Because of this, growth is limited to anaerobes and facultative anaerobes. This also limits the growth of yeasts.
2. Catalase-positive cocci include *Staphylococci, Micrococci,* and *Sarcina.* These occurred in small numbers, even in the irradiated samples. Except for *Staphylococcus aureus,* these have no health significance, but can contribute to microbial spoilage.
3. Catalase-positive Gram-positive rods, nonsporulating, generally are not putrefactive. They can, however, contribute to spoilage. Some are lactic-acid producers.
4. *Lactobacilli* have no health-hazard significance. They can reduce the pH which can have implications concerning retarding the growth of other organisms, e.g., *Salmonellae* or *Staphylococcus aureus.*
5. The Gram-negative rods include the *Pseudomonas, Achromobacter, Flavobacterium,* Enterobacteriaceae, and others. These are common meat contaminants and usually account for the typical spoilage of fresh meat. The meats treated in the contemplated manner are essentially free of them.

Based on these findings, meats treated as contemplated contain surviving organisms that in time cause microbial spoilage. This spoilage can be recognized through sensory changes. The dominant outgrowth group of organisms is the *Lactobacilli.* The outgrowth differs from the usual *Pseudomonas* and *Achromobacter.* The findings of this study revealed no new health hazard. In fact, the dominant outgrowth of the *Lactobacilli* could be regarded as lessening any existing potential hazard.

The sensory spoilage that occurs ultimately is easily recognizable by the consumer as spoilage. This fact has importance in that at the spoilage stage the consumer will not be misled in judging the suitability of the product for consumption.

In consideration of the fact that surviving microorganisms in radurized foods have been exposed to ionizing radiation, there is the possibility that mutations may have occurred. Both Idziak[20] and Ingram[21] have reported that there is no evidence that such mutations constitute a health problem with radurized foods.

Nonspore-forming bacteria with a very high resistance to radiation have been isolated from beef. These include *Micrococcus radiodurans*[22] and *Morexella acinetobacter.*[23] These organisms have D-values such that they survive radurization. At the present time the significance, if any, of these organisms in the spoilage of meats or in consumer health is unknown.

In the development of the above-described process, most of the experimental work was done with beef. This meat was chosen partly because of its popularity as a fresh meat, but mostly because it seems to involve the greatest problems of preservation. Beef is more highly pigmented than the other meats and is most sensitive to discoloration. It was observed that some beef muscles have a greater tendency to discolor than others. Drip from beef is more of a problem because it carries the meat pigment in higher concentration than that from other meats. Beef is one of the meats with a relatively low threshold dose for the formation of a detectable irradiation flavor. Other meats, notably pork and chicken, may be amenable to a simpler treatment than was used with beef. Phosphates, for example, may not be needed with these meats.

As judged by the laboratory studies, in the total picture the reported process provides the means to extend the market life of fresh red meats and poultry sufficiently to meet a requirement as long as 3 weeks for the centralization of final cutting and packaging operations. The next logical step would be to extend the work to a larger scale in order to determine process parameters more precisely, to engineer the various process steps including the packaging system, to determine market acceptance of the meats, and to establish costs and benefits. In the U.S. and in most countries this cannot be done at present. In the U.S. this step will require regulations of the U.S. Food and

Drug Administration and of the U.S. Department of Agriculture. Since 1974 in Canada there has been approval for the market testing of irradiated chicken and the FAO/IAEA/WHO Joint Expert Committee on the Wholesomeness of Irradiated Foods recommended approval of radurized and radicidized chicken in 1976. Recently the Netherlands also has approved irradiation of chicken. The studies of the U.S. Army on radappertized beef, pork, ham, and chicken, aimed at determining the wholesomeness of these products, may provide a basis for the U.S. regulations on low-dose treatment of meats and poultry. This will require acceptance of the principle that results of wholesomeness studies performed with one dose of radiation can be extrapolated to another dose. Taub et al.[24] have indicated that such an extrapolation is valid.

II. RADICIDATION

A. Bacteria

In the case of raw red meats and poultry, the pathogen of greatest concern and the one which has been given the greatest attention for control by irradiation is the bacterium *Salmonella*. The epidemiology of salmonellosis as related to the role of food in its transmission to man has been reviewed by Granville[25] and Hobbs.[26] Red meats and poultry frequently carry *Salmonellae* and it has been proposed that irradition be used to accomplish their inactivation. In order to free the meats of the organism a reduction factor of 10^7 has been regarded as necessary. Licciardello et al.[27] suggested a dose of 4.75 kGy for a 10^7 reduction for poultry meat. Presumably a comparable dose would be needed for the red meats. Quinn et al.[28] proposed doses of 5 to 8 kGy for securing a 10^7 reduction of nonspore-forming microorganisms associated with seafoods and which are capable of causing infection or poisoning. In addition to *Salmonella*, these organisms included *Shigella, Neisseria, Microbacterium, Escherichia, Proteus, Streptococcus*, and *Staphylococcus*.

The doses indicated for radicidation are high and exceed the threshold doses for irradiated flavor development in meats (with the possible exception of lamb) when the irradiation is carried out above 0°C. Irradiation at subfreezing temperatures would overcome this difficulty. In view of the rather large costs that would be involved in such processing including freezing, irradiation, and thawing, it seems unlikely that a meat processor would undertake it voluntarily. While government regulation could impose radicidation, the risk-benefit balance for the consumer would need to be considered. In terms of cost to the consumer it appears to this author to be an unwarranted application of irradiation. This statement is offered with the realization that others may have different views.

B. Helminths

In reviewing the irradiation of foods in connection with helminths, Wharton[29] listed references as early as 1904. Over the years there has been serious interest in inactivating meat-borne parasites by irradiation. Table 5 lists doses reported for the principal meat-borne helminths. If immediate destruction of the organism is not required, lower doses may be used. Van Kooy and Robijns[32] reported that a dose of 3 kGy plus 7 days of holding of beef at 2°C are fatal to *Cysticercus bovis*.

In the U.S. the meat supply is relatively free of helminths and treatment of all meats by radicidation for this purpose would seem to impose an unwarranted cost. In other countries in which parasitic infestation is more severe, such cost may be justified. More important than cost, however, is the effect of the high doses involved upon flavor and other sensory properties. As with radicidation to inactivate pathogenic bacteria, irradiation at subfreezing temperatures would be helpful in limiting sensory changes, but here, too, there would be an added cost. Freezing alone is one method of inactivating helminths.

Table 5
DOSES FOR IMMEDIATE INACTIVATION OF MEAT-BORNE HELMINTHS

Organism	Dose (kGy)	Ref.
Trichinella spiralis	6.3	30
Cysticercus bovis	4—5	31
C. cellulosae	4—5	31
C. pisiformis	4—5	31

Gibbs et al.[33] reported that 0.02 to 0.03 kGy of ionizing radiation prevent maturation of *Trichinella spiralis.* This is a substantially smaller dose than is required for immediate kill, and would prevent reproduction of the organism in the intestine of the host. While obviating that phase of trichinosis associated with migration of the helminth from the intestine into muscle tissue and other parts of the body, it would not prevent the initial phase of the disease associated with the presence of the ingested organisms in the intestine of the host. For this reason such a low-dose treatment appears to be inadequate as a public health measure.

C. Viruses

Although meat and poultry may be contaminated with viruses potentially pathogenic to man, exact information is lacking on the role, if any, of foods in transmitting viral diseases. There is, however, a strong interest in the virus of the foot-and-mouth disease (FMD). This virus causes serious disease in meat animals and is highly contagious. Raw meats from countries known to have the FMD virus cannot be imported into the U.S. which along with Canada and Central American countries is free of the virus. The FMD virus is especially small, being of the order of 10 to 12 nm in diameter. Undoubtedly for reason of its small size, a large dose is needed to inactivate it. Massa[34] reported a D-value of 4.81 kGy for the liquid virus and 6.26 kGy for the dried virus. He recommended a dose of at least 20 kGy for treating meat. Because of the high-dose requirement the use of irradiation to inactivate the FMD virus has generated little interest.

REFERENCES

1. **Billon, J.,** Experimental radurization of sliced ham vacuum packed in cellulose or plastic film, *Food Irradiat.*, (European Information Centre for Food Irradiation), 11(1,2), 4, 1970.
2. **Cagna, D., Campanini, M., and Vicini, E.,** Radiopasteurization of vacuum-packed smoked frankfurters, *Ind. Conserve*, 49, 7, 1974.
3. **Watanabe, H., Kume, T., Aoki, S., Ito, H., and Sato, T.,** Shallow-dose irradiation and prevention of slime production or putrefaction in Vienna sausages by electrons, *Food Irradiat.*, (Shokuhin-Sosha), 9, 55, 1974.
4. **Anon.,** Cost-Benefit Analysis of Radiation Pasteurized Meat under Centralized Cutting Operations, NYO 3831-1, U.S. Atomic Energy Commission, Washington, D.C., 1968.
5. **Urbain, W. M.,** The Effect of Sub-Sterilizing Doses of Cathode Rays and Gamma Rays on the Keeping Quality of Beef, Rep. No. 1, Contract QM R & D (Natick) No. 2 Period December 6, 1954 to November 1, 1955, U.S. Army Natick Research and Development Laboratories, Natick, Mass., 1955.
6. **Wolin, E. F., Evans, J. B., and Niven, C. F., Jr.,** The microbiology of fresh and irradiated beef, *Food Res.*, 22, 683, 1957.

7. **Coleby, B.**, The effects of irradiation on the quality of meat and poultry, *Int. J. Appl. Radiat. Isot.*, 6, 115, 1959.
8. **Ingram, M. and Thornley, M. J.**, Changes in spoilage pattern of chicken meat as a result of irradiation, *Int. J. Appl. Radiat. Isot.*, 6, 122, 1959.
9. **Urbain, W. M.**, Radiation preservation of meats and poultry, in *Radiation Preservation of Foods*, National Academy of Sciences Publ. No. 1273, Washington, D.C., 1965, 87.
10. **Rhodes, D. N. and Shepherd, H. J.**, The treatment of meats with ionizing radiations. XIII. Pastuerisation of beef and lamb, *J. Sci. Food Agric.*, 17, 180, 1966.
11. **Urbain, W. M. and Giddings, G. G.**, Factors related to market life extension of low dose irradiated fresh meat and poultry, *Radiat. Res. Rev.*, 3, 389, 1972.
12. **Urbain, W. M.**, The low-dose radiation preservation of retail cuts of meat, in *Proceedings of the Bombay Symposium on Radiation Preservation of Food*, International Atomic Energy Agency (IAEA), Vienna, 1973, 505.
13. **Sudarmadji, S. and Urbain, W. M.**, Flavor sensitivity of selected animal protein foods to gamma radiation, *J. Food Sci.*, 37, 671, 1972.
14. **Hannan, R. S.**, Scientific and Technological Problems Involved in Using Ionizing Radiations for the Preservation of Food, Special Rep. No. 61, Department of Scientific and Industrial Research, Her Majesty's Stationery Office, London, 1955.
15. **Brasch, A. and Huber, W.**, Reduction of undesirable by-effects in products treated by radiation, *Science*, 108, 536, 1948.
16. **Wierbicki, E., Brynjolfsson, A., Johnson, H. C., and Rowley, D. B.**, Preservation of meats by ionizing radiation, Rapporteurs' Papers — Paper No. 14, Eur. Meet. Meat Res. Workers, Berne, Switzerland, 1975.
17. **Mercuri, A. J., Kotula, A. W., and Sanders, D. H.**, Low-dose ionizing irradiation of tray-packed cut-up fryer chickens, *Poulry Sci.*, 45, 1105, 1966.
18. **Ballantyne, W. W.**, Color Stability of Fresh Beef as Affected by Ionizing Dose Rate and Added Inorganic Phosphate, Ph.D. thesis, Michigan State University, E. Lansing, 1970.
19. **Groesbeck, C. P.**, The Spoilage Microflora of Irradiated Beef, M. S. thesis, Michigan State University, E. Lansing, 1973.
20. **Idziak, E. S.**, Radiation sensitivity of microorganisms. Effect of radiation on microorganisms, *Int. J. Radiat. Sterilizat.*, 1, 45, 1973.
21. **Ingram, M.**, Microbiology of foods pasteurized by ionising radiation, Tech. Rep. IFIP-R33, International Project in the Field of Food Irradiation, Karlsruhe, 1975.
22. **Duggan, D. E., Anderson, A. W., and Elliker, P. R.**, Inactivation of radiation-resistant spoilage bacterium *Micrococcus radiodurans, Appl. Microbiol.*, 11, 398, 1963.
23. **Welch, G. B. and Maxcy, R. B.**, Characterization of radiation-resistant vegetative bacteria in beef, *Appl. Microbiol.*, 30, 242, 1975.
24. **Taub, I. A., Angelini, P., and Merritt, C., Jr.**, Irradiated food: validity of extrapolating wholesomeness data, *J. Food Sci.*, 41, 942, 1976.
25. **Granville, A.**, The epidemiology of salmonellosis in relation to its transmission by food and feed products, in *Radiation Control of Salmonellae in Food and Feed Products*, Tech. Rep. Ser. No. 22, IAEA, Vienna, 1963, 9.
26. **Hobbs, B. C.**, Salmonellae in foods and animal feeding stuffs, in *Radiation Control of Salmonellae in Food and Feed Products*, Tech. Rep. Ser. No. 22, IAEA, Vienna, 1963, 35.
27. **Licciardello, J. J., Nickerson, J. T. R., and Goldblith, S. A.**, Elimination of *Salmonella* in poultry with ionizing radiation, in *Elimination of Harmful Organisms from Food and Feed by Irradiation*, IAEA, Vienna, 1968, 1.
28. **Quinn, D. J., Anderson, A. W., and Dyer, J. F.**, The inactivation of infection and intoxication microorganisms by irradiation in seafood, in *Microbiological Problems in Food Preservation by Irradiation*, IAEA, Vienna, 1967, 1.
29. **Wharton, D. R. A.**, Action of ionizing radiations on helminths, in Radiation Preservation of Food, Quartermaster Corps, U.S. Army, U.S. Government Printing Office, Washington, D. C., 1957, 235.
30. **Gould, S. E., Van Dyke, J. G., and Gomberg, H. J.**, Effect of X rays on trichina larvae, *Am. J. Pathol.*, 29, 323, 1953.
31. **Pawel, O.**, Devitalization of cysticerci by gamma radiation, in *Elimination of Harmful Organisms from Food and Feed by Irradiation*, IAEA, Vienna, 1968, 91.
32. **Van Kooy, J. G. and Robijns, K. G.**, Gamma irradiation elimination of *Cysticercus bovis* in meat, in *Elimination of Harmful Organisms from Food and Feed by Irradiation*, IAEA, Vienna, 1968, 81.
33. **Gibbs, H. C., MacQueen, K. F., and Pullin, J. W.**, Studies on the effects of gamma radiation on pork infected with *Trichinella spiralis, Can. J. Pub. Health*, 55, 191, 1964.
34. **Massa, D.**, Radiation inactivation of foot-and-mouth disease virus in the blood, lymphatic glands and bone marrow of the carcasses of infected animals, in *Proceedings of the Karlsruhe Symposium on Food Irradiation*, IAEA, Vienna, 1966, 329.

Chapter 2

RADURIZATION AND RADICIDATION: FISH AND SHELLFISH

J. T. R. Nickerson, J. J. Licciardello, and L. J. Ronsivalli

TABLE OF CONTENTS

I. INTRODUCTION

Soon after the discovery of X-rays in 1895, studies were initiated to determine their biological effects. Although some early patents were granted on the use of ionizing radiation to preserve food, the first work of any real importance in this area was reported in 1943.[1]

In 1950 Nickerson et al.[2] reported on the treatment of mackerel fillets with a cathode ray dose of 1.5 million rep. This appears to be the first literature report on the treatment of fish with ionizing radiation. Although the main purpose of this study was to eliminate microorganisms on one of a pair of fillets so as to ascertain the cause of spoilage in mackerel fillets, that is, autolytic or microbial, the authors did observe through organoleptic examination that the irradiated fillets had a longer storage life on ice compared to the controls.

The earliest report that could be found in the literature dealing with low-dose irradiation of fish followed by refrigerated storage for shelf-life extension was by Nickerson et al. in 1954.[3] In this study haddock fillets were irradiated with cathode ray doses ranging from 4×10^5 to 6×10^5 rep with subsequent storage at 36 to 40°F (2.2 to 4.4°C).

During the early 1950s food irradiation research in the U.S. was sponsored principally by the U.S. Atomic Energy Commission (AEC), the U.S. Army Quartermaster Corps, and the U.S. Naval Supply Research and Development Facility.[4] The thrust was mainly on effects of irradiation, particularly sterilizing doses, on a wide variety of foods and foodstuffs, and also on microbial inactivation. As it became apparent that sterilizing doses in most instances produced undesirable side effects in many foods, the research emphasis began to shift to low-dose irradiation and subsequent refrigerated storage. Consequently in 1960 through the joint effort of the Interdepartmental Committee on Food Irradiation and the Division of Isotope Development (AEC) it was decided that the AEC would sponsor a comprehensive research program to determine the feasibility of low-dose treatment of certain food items (particularly fish) for the benefit of a civilian market, whereas the Department of Defense would support research to develop a sterilization process basically for military feeding. Soon after this decision Proctor et al.[5] on a contract with the AEC surveyed the literature and published a report on the technical, economic, and practical feasibility of radiation preservation of fish. This was followed by a contract report by Nickerson et al.[6] which contained an outline of projects and experimental protocol for determining the feasibility of radiation preservation of marine products. With this background information providing guidance the AEC launched a national program on the low-dose irradiation of seafoods drawing on the combined research efforts of universities, industry, and the technological laboratories of the Bureau of Commercial Fisheries.*

During the early years of seafood irradiation studies the types of ionizing radiation sources generally employed included mechanical, electron accelerators such as the Van de Graaff generator and Capacitron®, radioisotope sources such as cobalt-60, and spent fuel rods from a nuclear reactor. Each of these types of sources was not without some form of limitation for use in large-scale experiments requiring large amounts of food to be irradiated. For effective penetration with electron accelerators thick samples had to be frozen and then irradiated on both sides. The most common cobalt-60 source in the U.S. at the time consisted of a shielded, encapsulated pipe of the isotope with an inside diameter approximately 2 in. (5 cm). Samples to be irradiated were usually contained in a capped test tube or sealed vial. The spent fuel rod sources such as the Material Testing Reactor at Arco, Idaho, allowed for the irradiation of samples sealed in large cans; however, samples usually had to be shipped to and from the remote irradiation site in the frozen state to preserve quality. Consequently when the low-dose irradiation program was inaugurated in 1960, the AEC elected to provide some of its contractors with on-site irradiation facilities designed principally for research leading eventually to pilot plant studies. The units designated Research Irradiator were located at the Massachusetts Institute of Technology (MIT) (Cambridge), University of California (Davis), University of Washington (Seattle), University of Florida (Gainesville), and University of Hawaii (Honolulu). They were a pool-type gamma source of about 30 kCi of cobalt-60 and with a capacity for irradiating about 75 lb (34 kg) of food per hour at a dose of 1 Mrad.[7] Two rectangular plaques of cobalt-60 sources were arranged vertically at the bottom of a 12-ft (3.7-m) deep, water-filled, stainless-steel tank. Three stainless-steel watertight containers could be automatically lowered into the irradiation field between the two plaques or on one side for a prescribed time and then raised by

* Now, National Marine Fisheries Service (NMFS).

a motor-driven elevator that was controlled by a timing device. With this unit it was possible to irradiate under controlled temperature and controlled gaseous environment.

The Marine Products Development Irradiator (MPDI), which is located at the Gloucester Laboratory of the National Marine Fisheries Service (NMFS) in close proximity to many fresh-fish processing plants, is a semicommercial-scale production facility with an irradiation capability of 1000 lb (454 kg)/hr at a dose of 0.5 Mrad. This unit was designed for conducting storage studies, shipping tests, and consumer tests on a pilot-plant scale. The source plaque, which initially contained 250,000 Ci of cobalt-60, is normally stored at the bottom of a pool of water. For irradiation purposes the source is raised out of the water on a track to a horizontal position sandwiched between two parallel conveyors. The product to be irradiated is packed in 20- or 30-lb (9.1 or 13.6 kg) fish tins which enter the irradiation chamber by conveyor through a labyrinth, make one pass beneath the source plaque on the lower conveyor, and then another pass above it on the upper conveyor.

The Shipboard Irradiator was designed and built for irradiation of fish at sea.[8] In 1966 the unit was installed aboard the NMFS (then Bureau of Commercial Fisheries) research vessel, *M.V. Delaware*. The irradiator measured 8 ft 10 in. (2.6 m) in height and 47 × 55 in. (1.2 × 1.4 m) at the base. It weighed 17.5 short tons (15.9 metric tons). The unit was located in one of the holds of the ship bolted to four steel beams set in a concrete bed. The radiation source was a cobalt-60 plaque of 27,000 Ci contained within a lead-walled cask. For irradiation the sample was placed in a stainless steel box which was then loaded into a shuttle device that lowered the sample box into the cask in juxtaposition with the source plaque.

Other irradiation facilities existed in the U.S. However, these were not designed and constructed principally for studying radiation preservation of seafoods and thus will not be described.

Two objectives are possible with radiation treatment of foods. One is to produce a shelf-stable product at room temperature and this has been usually referred to as radiosterilization or radiation sterilization; the other is to reduce the number of commensal microorganisms and extend the shelf life. This treatment has been designated radiopasteurization or radiation pasteurization. It was argued by certain individuals that some ambiguity was associated with these terms. For example, in practice radiation sterilization is never achieved since very high doses may be required to completely inactivate certain resistant spores or viruses. The conditions actually existing in radiation sterilized foods are akin to commercial sterility in heat-processed foods. With low-dose irradiation one can either selectively eliminate all pathogenic nonspore-forming microorganisms or one can indiscriminately reduce the numbers of the total microflora. Consequently, an international group of experts proposed the following terminology be adopted to properly identify the three different treatments.[9] *Radappertization* would refer to the radiation sterilization treatment that was analagous to heat sterilization (appertization). *Radicidation* would describe the irradiation treatment for eliminating nonspore-forming pathogens and *radurization* would apply to the process of nonselective reduction in the total microbial count to extend shelf life similar to heat pasteurization. This proposal has slowly received broad acceptance. One of the reasons for slowness of acceptance is that these terms do not appear in standard English dictionaries, and this discourages their use by most writers who recognize the value of the courtesy of using only words that can be found in standard English dictionaries. However, the major deterrent probably arose from the labeling requirements stipulated by the Food and Drug Administration (FDA) for irradiated foods when approved.[10,11]

II. ADVANTAGES AND DISADVANTAGES OF THE RADIATION TREATMENT OF SEAFOODS

The rationale for preserving foods with ionizing radiations is based on the fact that this form of energy is an effective bactericide and contrary to heat energy used in conventional heat processing and heat pasteurization methods, there is only a small temperature rise during radiation.[12] As with heat energy, the higher the input of radiation energy, the greater the bactericidal effect and the greater the organoleptic changes in the product.[13] However, the radiation-induced organoleptic changes, in lean fish especially, are objectionable as compared to heat-induced organoleptic changes which are generally acceptable. Consequently, except for smoked products, most unfrozen fish and fish products cannot be irradiated at doses that exceed 300 krad. Fatty fish (e.g., mackerel) and shellfish which have relatively strong tastes that tend to mask the off-flavors which develop from irradiation can, however, be exposed to doses that exceed 300 krad. Although seafoods can be preserved by other means, radiation treatment, at sufficiently low levels ($\sim$200 krad) has the least adverse effect on their organoleptic qualities. There are certain obvious advantages to preserving seafoods by any method, including radiation, and there are some possible disadvantages to the use of radiation. It might be of interest to examine these as well as certain other aspects of the process that are not so easily categorized. (It should be kept in mind that low-dose irradiation is being considered only for unfrozen fish that must be refrigerated following the radiation treatment.)

A. Advantages

It is generally agreed that the low-dose radiation preservation treatment of seafoods would have four major advantages: (1) expansion of the fresh-seafood market, (2) stabilization of the supply-and-demand of fresh seafoods, (3) greater use of the seafood resource, and (4) stabilization of quality without any appreciable organoleptic change. A common aspect of the first three advantages cited above is that they are an expected consequence of radiation preservation or of preservation by other processes. On the other hand, radiation is unique among processes in that it alters the physical and organoleptic properties of the product the least.

1. Expansion of Market

Although fresh, unfrozen fish is more acceptable to many consumers than is frozen fish or fish that has been otherwise processed, its availability is largely restricted to areas that are within a few hundred miles from the fishing port from which it originates. Under the best holding conditions, the bulk of the unfrozen fish that lands in the U.S. has a shelf life of about 1 week. Considering the inadequacy of the quality control that unfrozen fish may encounter in the distribution chain, it is not surprising that high-quality, fresh fish is not easily attainable in the inland areas of the country, except for fish caught in nearby rivers, ponds, and lakes. The treatment of fish with about 100 krad would extend its shelf life by 1 week or more at refrigerator temperatures above freezing. Such an extension would be sufficient to permit distribution of treated products to any major city in the country, thus increasing the market for fresh fish substantially.

2. Stabilization of Supply and Demand

In most cases, retailers of fresh fish tend to buy less than they expect to sell in order to eliminate or at least minimize the economic loss that would otherwise result from the spoilage of the unsold portion of their purchases. This is especially true when the wholesale price of fish is high (the wholesale prices of fish fluctuate greatly and often).

This usually results in a demand that is not completely met — a situation that has an adverse impact on the demand-related growth of seafood consumption. The availability of fresh fish at retail depends on a number of variables (e.g., weather, size, and location of the stocks) which create frequent and large fluctuations in the supply, which in turn are reflected in frequent and large fluctuations in price. When conditions are favorable, the supply of fresh fish usually exceeds the demand and, since it is known that the fish will not retain their good quality for long, the catch cannot be sold entirely and there is a glut which causes the value to drop sharply. The glut portion of the catch, which can be large, must finally be sold at a low price to producers of fish meal. The use of irradiation to preserve fish would tend to stabilize the supply by allowing processors to buy the large catches and to hold, on short inventory, the unsold portion so as not to glut the market. The inventory would serve a second important purpose; it could be used to fill any unexpected surge in the demand. Under these conditions, fishermen could use the size of the processor's inventory as a guide to the amount of fish that they might be able to sell. Also, since they could sell large catches at a reasonable price, it would not be necessary for them to practically give away the gluts only to have to return soon to sea, even in bad weather, simply to meet their income requirements as well as to keep the demand satisfied.

3. Greater Use of the Product

Because of the relatively short shelf life of unfrozen seafoods, a significant proportion of these undergo degradation of quality to the point that they are just barely acceptable, and quite often not acceptable, by the time that they are purchased. Fish of borderline or poor quality are often discarded after they have been purchased. Much poor quality fish does not even reach the retail counter. Instead it is sold to fish meal manufacturers at a much lower price than it could demand as human food. The preservation of fish by irradiation would extend the period during which its quality remains acceptable, thereby providing more time to sell it and thus eliminating or sharply reducing the amount of fish lost to spoilage.

4. Stabilization of Quality

When one examines the immediate effect of any of the processes that can be used in the U.S. to preserve seafoods, only refrigeration at temperatures above 0°C and low-dose radiation ($\sim$ 200 krad) can produce a significant effect without creating changes in the physical and organoleptic characteristics of the products. Theoretically, refrigeration at temperatures below 0°C (freezing) can also be effective without creating significant organoleptic changes, but, in practice, frozen seafoods do not have the same high-quality image of unfrozen seafoods. During freezing, seafoods can become tough, rancid, and dehydrated, and some flavor components may be lost to the "drip water". The effectiveness of either refrigeration (at temperatures above 0°C) or radiation (at dose levels $\sim$200 krad) is limited, but when the radiation treatment is followed by refrigeration, the maximum benefit is obtained.

B. Disadvantages

Consideration of the use of radiation to preserve food has surfaced at least two possible disadvantages of the process: (1) fear of radiation and (2) abuse handling. However, unlike the advantages cited above which clearly will accrue as a consequence of using the process, the disadvantages which have been identified are speculative and may not necessarily follow.

1. Fear of Radiation

It has been speculated that the word "radiation" carries with it fearful connotations to both the processor and the consumer. The heavy shielding around radiation sources and the strict precautions associated with their operation are bound to create an impression of possible extreme hazard. To many, the word "radiation" may connote "radioactivity", "fallout", and other hazardous, nuclear-related phenomena. Because of these possible public responses, it is further speculated that an expensive education program would be required to place the process in proper perspective before the public. However, whether or not the fear theory is correct remains to be seen. In at least two pertinent reports, the theory is generally contradicted, and the public is reportedly not afraid of the process nor of irradiated foods.[14,15] (Perhaps this attitude reflects more the consumer's confidence in the government than in the process itself.)

2. Abuse Handling

It has been theorized that there exists the potential for a relaxation of effort on the part of handlers throughout the distribution chain who will depend on the preservative effect of radiation to control the quality of the product. While it is true that such abuse is probable, there is no way to predict its significance. It should be apparent that the advantages through shelf-life extension afforded by radiation could easily be neutralized by a breakdown of quality control practices and it therefore requires that we emphasize that they be strictly maintained.

C. Other Aspects of the Process

Although certain advantages and possible disadvantages associated with the preservation of foods with radiation are readily categorized as such, there are certain aspects of the process that are not. Two of these, "wholesomeness" and "safety" are covered elsewhere in this chapter. In this section we will examine: (1) organoleptic effects of low-dose radiation and (2) treatment costs.

1. Organoleptic Effects

It is known that the radiation treatment of seafoods may result in the development of off-odors and off-flavors. However, the effect depends on the dose level and numerous existing data indicate that the levels intended for use in the radiation treatment of seafoods are too low to produce any significant organoleptic effects.[16,17]

2. Treatment Costs

Estimates of the cost to treat seafoods with low doses of radiation range from $0.01 to $0.02/lb, but that excludes the cost for packaging the product which is an essential step if the maximum preservation effect is to be achieved.[14] Packaging prevents recontamination with bacteria and, if made from gas-impermeable material, reduces oxygen availability to oxidative reactions involved in the spoilage of refrigerated, unfrozen seafoods. Packaging, depending on the type of container desired, could add about $0.02 to $0.05/lb.

Although it was once feared that the addition of a few pennies to the price of a pound of fish might have an adverse effect on sales, there is no reason to believe that there is any current basis for this fear. First of all, it is difficult to perceive how the addition of a few pennies to the retail price of 1 lb (454 g) of haddock fillets is at all significant when that price undergoes weekly fluctuations to any point in the range $1.50 to $3.00/lb.[18] Second, and, quite important, it has been reported that some consumers are willing to pay as much as $1.00/lb more for high quality fillets than for fillets of unknown quality and that many consumers will pay at least $0.20/lb extra for the premium quality product.[18]

From the standpoint of cost, there is at least one other consideration. While frozen fish may not compete with unfrozen fish and vice versa, one does have to consider the rising costs of electrical energy which have more than doubled over the past decade against the declining costs of atomic energy which are now about 50% less than they were a decade ago. This changing price structure might allow radiation treated seafoods to make inroads on the market for frozen seafoods.

III. IMPEDIMENTS TO RADIATION TREATMENT OF SEAFOODS

It has been stated that "Never in the history of man has any method of food processing invoked so much comment, induced so great a controversy, nor required such an expenditure of funds before its use could be permitted even on a limited basis as has the use of ionizing energy".[19] Defined in the U.S. as a food additive by the controversial Food Additives Amendment of 1958, radiation energy has been intensely investigated by numerous teams in many countries for more than 2 decades. The objectives of these studies included (1) evaluations of the potential for applications, (2) wholesomeness of irradiated foods, and (3) the toxigenic effects of radiation on foods. Not all of the results earned credibility and, in particular, some of the wholesomeness studies were termed inadequate. The results of several studies tended to discredit the radiation process, and while most of the discredits were refuted, they made a significant adverse impact on its chances for approval by the FDA. Those studies that demonstrated the economic and technological feasibilities of ionizing energy for preserving foods have succeeded in generating a world-wide interest in the process. Those studies that were directed at the negative aspects of the process have succeeded in identifying problem areas that have dulled some of the interest. It is, therefore, important to examine these suggested problems with the objective of arriving at the appropriate perspectives and to identify whatever further study is genuinely required. To maintain an appropriate perspective it is imperative that two basic facts be remembered. First, ionizing radiation is a form of energy that is lethal to microorganisms, through a sequence of drastic physical attacks at the atomic level of their constituents, which results in a variety of serious alterations to their life processes (e.g., cell permeability and reproduction). As such, it must be conceded by proponents of the process that we are dealing with a system which is capable of producing effects in the food components that should be evaluated for the possible destruction of nutrients upon which the body depends or for the possible creation of toxic agents. Second, other forms of energy that are equally lethal to microorganisms are also responsible for serious physical and chemical alterations of food components which also may result in the destruction of nutrients and in the production of toxic agents. Accordingly, it must be conceded by opponents of the process that while radiation is unique, in some respects, among other forms of energy, it is not so unique in its effects. Therefore, while it is appropriate to prove the safety of ionizing radiation for preserving foods, the process should be studied by the same set of rules by which other processes should be similarly studied. In addition to the foregoing, the use of low levels of ionizing energy needs to be investigated for its possible implication in botulism, as will be described in more detail below. Again, there is much evidence to substantiate that radiation is not unique in this respect, and any process that does not result in commercial sterility may also be implicated in botulism, at least for some foods. In consideration of the foregoing discussion, the potential problems that require analysis are (1) induced radioactivity, (2) effects on nutrients, (3) toxicity and carcinogenicity effects, and (4) effects on microbial composition.

A. Induced Radioactivity

When matter is irradiated at high energy levels (e.g., >15 meV) certain of its components can become radioactive and this radioactivity is detectable by currently available monitoring instruments. This phenomenon is the basis for some conjecture that the preservation of food with ionizing radiation may make it radioactive. However, it has been shown that when matter is irradiated at energy levels of 12 MeV or less, no induced radioactivity can be detected in the target material with the current instrumentation. In a test to evaluate the potential hazard for induced radioactivity in foods, energy levels as high as 24 MeV electrons were used, and in such instances, the longest-lived and among the most important nuclides in foods, ^{22}Na, was activated to a level of only 0.5% of the maximum allowable concentration set by the National Committee on Radiation Protection.[20] Other nuclides commonly found in foods are either very short-lived or not activated to detectable levels. The naturally occurring radionuclides in foods, ^{40}K and ^{14}C, have higher measurable radioactivity than can be produced by sterilizing foods with doses of ionizing radiation at energy levels of 24 MeV.[21] Given the above facts, it can be concluded that the use of energy levels less than 12 MeV electrons will not induce any detectable radioactivity in irradiated foods. To add a safety factor, it has become common practice to irradiate foods with electrons at energy levels that do not exceed 10 MeV. Among isotope sources, ^{60}Co and ^{137}Cs have maximum energies well below the threshold for inducing measurable radioactivity in food.

B. Effect of Ionizing Energy on Food Nutrients

It has been shown that ionizing radiations can produce oxidation and reduction products from carotenoids, lipids, and alcohols, and decarboxylation and dehydrogenation products from organic acids. It oxidizes and hydrolyzes sugars in the presence of water, it fragments and aggregates proteins, it deaminates free amino acids, and it dephosphorylates nucleic acids. From the above, it can be seen that this form of energy should be expected to affect food nutrients, and it does. Vitamin A is sensitive to ionizing radiation, and significant amounts are destroyed when foods containing it are irradiated at high dose levels.[22] Although vitamin D is produced in the body as a consequence of exposure to the radiation energy from the sun and although it is produced by UV radiation of precursor sterols such as 7-dehydrocholesterol for the fortification of milk, it is still susceptible to some destruction by radiation.[23] Vitamin E also has been found to be destroyed by sterilizing doses of ionizing energy.[24] Vitamin K is very unstable to ionizing energy and it is largely destroyed by sterilizing doses.[22] Vitamin C, in solution, is also unstable to irradiation, although the instability is inversely proportional to the concentration of the vitamin.[25] Of the B vitamins, it has been reported that thiamine (B_1), riboflavin (B_2), pantothenic acid, pyridoxine (B_6), and folic acid were destroyed by less than 1 Mrad, but niacin was not.[26] However, as with most experiments the nutrients were not as unstable *in situ* as in vitro, and the apparent protection of the food depended upon the nature of the food itself. Cyanocobalamin (B_{12}) is reportedly also unstable to irradiation, being even more radiosensitive than vitamin C.[26] Choline demonstrated various susceptibilities to irradiation, depending on which of its analogs was tested.[27] In testing the radiation sensitivity of cholesterol, early experiments indicated it to have a low radioresistance, until it was established that it was not the radiation that destroyed the sterol but rather it was destroyed by the decomposition products of the solvents used to extract it. In consideration of the destructive effects of solvents used in extracting nutrients and of the protective effects of food components, results of in vitro tests cannot be accepted at face value. Once thought to be carcinogenic, irradiated cholesterol showed no signs of carcinogenicity in subsequent experiments.[28] Antibiotics, hormones, steroids, multivitamin preparations, anticoagulants, alkaloids, sulfonamides, and other pharmaceutical preparations were found to be stable at sterilizing doses.[29] It was concluded that the irradiation effects on proteins and fats are unimportant.[30-32]

From the foregoing discussion, we have seen that irradiation at sterilizing and near-sterilizing doses does degrade a variety of nutrients. However, the overall effect is not significant since radiation is not unique in this respect, and various studies have shown that the sterilization of foods with thermal energy has equally destructive effects on their nutrient components.[33,34] To put it another way, the use of radiation for preserving foods does not degrade their nutritional quality more than it is degraded by preserving them with heat.[33,35-37]

C. Toxicity Effects of Ionizing Energy

While consideration has been given to the probability of the production of toxic agents as a result of ionizing energy on various food components (e.g., the formation of toxic levels of hydrogen peroxide) no real hazard from such toxic agents has been demonstrated.[38] Because of the complexity of foods, it is highly improbable under current capabilities that one can make a comprehensive study of the details of all chemical events resulting from any of the various treatments (including radiation) which foods may undergo. It should therefore not be expected that the problem of possible toxicity can be resolved without a systematic and exhaustive investigation of radiation chemistry. The best single method now available to the investigator for resolving the potential problem of toxicity is that of animal feeding. It is the only technique that can demonstrate the integrated effects of radiation on the development of toxic compounds. Even though the effects on animals may not be accurately extrapolated to effects on humans, it is still the most accurate indicator of effects on humans when reasonable judgment is used in evaluating the results. Since animal feeding studies can be used to detect toxigenic and carcinogenic effects simultaneously, their results can therefore be examined to determine the effect of irradiation on both the toxicity and the carcinogenicity in foods. Neither carcinogenicity nor toxicity was reported in the results of feeding tests involving irradiated cod,[39-43a] ocean perch,[40-42a] tuna,[44,45] and shrimp.[46,47]

Whereas most wholesomeness feeding experiments were conducted with seafoods that had been irradiated at relatively high-dose levels (megarads), it would appear reasonable to assume that the results of these tests should be extrapolatable to seafoods irradiated at relatively low doses (hundreds of kilorads). Analyses of the radiation chemistry occurring in cod muscle irradiated at either high or low doses support the thesis that radiolytic products are similar in both cases but that their amounts are proportionately larger in substrates receiving higher doses of radiation.[48,49] They do not support the thesis that wholesomeness data obtained with food irradiated at relatively high doses cannot be extrapolated to food irradiated at relatively low doses because of the possibility that an intermediate hazardous radiolytic product might be formed in foods irradiated at low doses and remain as a hazard, whereas they may be transformed to harmless compounds in foods irradiated at high doses.

Insofar as the mutagenic properties of ionizing radiations have been demonstrated in model systems, the significance of these findings is questionable. As with radiomimetic agents such as alcohol and caffeine which are so classified because of their mutagenic properties which resemble those of radiation, the intimated hazard is not real. It is not real for two reasons. The first reason is that a physiologically sophisticated organism such as man has the ability to deal with low levels of toxic materials or otherwise to exclude them from his system. The second reason is that analysts testing the safety of materials usually use high concentrations of the material which are unrelated to the actual use amounts that would normally be ingested by man.

D. Effects of Radiation on Microbial Composition

Because of the specific lethal effects of radiation on microorganisms, it has been theorized that microbial shifts under low levels of radiation might give rise to botulism. Also because of the mutagenic effects of radiation, there may emerge radioresistant microbes.

1. Botulism Hazard

Few words in the food processing industry, in food regulatory agencies, and to the consumer carry the strong connotations of the word "botulism". It is therefore understandable that any process which might be suspected of fostering this disease requires a rigid examination in that regard. It has been theorized, and it is believed by some, that low-dose irradiation of seafoods represents a botulism hazard. The rationale for this theory is based on a few facts: pseudomonads are the normal spoilage microorganisms in seafoods;[50,51] these bacteria have a low radiation D-value (less than 10 krad);[52] *Clostridium botulinum* Type E, a psychrotrophic member of this species, grows at temperatures as low as 38°F (3.3°C);[53] and its radiation D-value is about 130 krad.[54] Based on the foregoing facts, it is believed that a radiation dose of 150 krad will eliminate the pseudomonads (reducing their number through about 15 log cycles), but it will effect only a 1-log reduction in the numbers of Type E *C. botulinum*. With the competitive pseudomonads eliminated it is believed that the growth of the *C. botulinum* will be accelerated resulting in the production of toxin, the seafoods will not undergo typical malodorous spoilage, and the seafoods may become toxic before they become objectionable to the consumer. In this discussion, it should be remembered that the pseudomonads are aerobes and *C. botulinum* Type E is an anaerobe. The above facts take on added significance when it is considered that domestic refrigerators are seldom colder than 38°F (3.3°C) and that irradiated seafoods are normally enclosed in a package (to prevent recontamination), and sometimes, the package may be of a gas-impermeable material which is responsible for a lowering of the oxygen tension within the package.

Thus far we have examined the facts that support the theory. It is just as important to note the facts that do not support the theory. Although the qualitative incidence of *C. botulinum* is relatively high, the quantitative incidence is low. For example, the highest contamination level found by one research team was 0.17 spore per gram of fish.[55] The fear that irradiated fish will not spoil by normal patterns and that they will become unacceptable is without basis, and enough data have been produced to demonstrate that they do reach a state of spoilage where they become organoleptically unacceptable before they become toxic.[56]

At this point it is well to examine the rationale for both points of view. It is true that the spoilage bacteria normally present in seafoods are probably eliminated even at low levels of radiation and this would appear to be sufficient justification for assuming that the seafoods will not be subjected to typical spoilage reactions. However, the data invalidate this assumption, and it could be assumed that while radiation readily destroys the spoilage bacteria, it does not have a significant effect on any of the enzymes that have already been produced by the bacteria. It is well known that enzymes are generally resistant to radiation. Thus, whereas the concentration of enzyme is kept at some low level by eliminating its source, it nevertheless is able to degrade the seafood, albeit considerably more slowly.

Keeping in mind that irradiated seafoods do spoil, do become organoleptically unacceptable, and that the natural quantitative incidence of *C. botulinum* Type E is much less than a spore per gram, the question that remains to be answered is, will seafoods become recognizably spoiled before botulinum toxin can be produced? Experiments were conducted to determine the times, at different storage temperatures, that it takes radiation-treated (0 to 200 krad) seafoods to become organoleptically unacceptable.[56] Table 1 is a typical example of the data obtained. It should be pointed out that the range of values obtained was large, especially at the lower temperatures where there is little significance because botulinum is unable to grow at <38°F (3.3°C). There are at least three reasons responsible for the wide range in values. The first reason is that the radiation dose cannot be applied uniformly, as is well known by radiation scientists;

Table 1
TIMES (IN DAYS) AT WHICH RADIATION TREATED HADDOCK FILLETS BECAME ORGANOLEPTICALLY UNACCEPTABLE (AVERAGES OF SIX EXPERIMENTS)

Radiation dose (krads)	Storage temperature °F(°C)								
	33(0.6)	38(3.3)	40(4.4)	42(5.6)	44(6.7)	46(7.8)	50(10.0)	55(12.8)	60(15.6)
0	36	14	16	11	7	10	6	5	4
50	34	19	19	16	14	10	9	—	7
100	44	24	31	22	15	14	13	10	7
200	57	35	41	30	18	20	16	12	10

the second reason is that the starting quality of the product was not uniform; and the third reason is due to inherent error in the test panel arising partly from the expected imprecision of the human response and partly from the variation in the composition of the test panel from test to test. However, while some of the ranges of values were quite high, it may be of no significance if the upper limits of the ranges are far below toxin-outbreak times. Figure 1, a plot of the data in Table 1, is included in order to observe the relationship between the holding temperature and the postirradiation holding times. It is obvious that the product shelf life is increased as the dose is increased, but the effect of the dose is generally greatest in the lower range of temperatures where the botulism hazard does not exist. At temperatures above 38°F (3.3°C) where botulinum growth may occur, the longest shelf life was 35 days. This time is considerably less than that found for the growth of *C. botulinum* Type E under optimum conditions.[53] Varying toxin outbreak times have been reported, and Table 2 illustrates the type of data that have been produced.[57] There are significant variations in the data, some of which have been attributed to species differences as appears to be borne out in the table. It has been theorized that the production of botulinum toxin is dependent on the fat content of the substrate and some data support this relationship. However, it has been shown that the presence of an antibiotic polypeptide in haddock inhibits the growth of *C. botulinum* Type E,[58] and it may be that the presence of this antibiotic is by coincidence in some way related to the fat content. The presence of such a polypeptide in haddock might account for its longer shelf life in Table 2.

There are other reasons for the variations in the literature on toxin-outbreak times and these are covered elsewhere.[59] Prominent among these are faulty inoculum techniques which could account for errors of two orders of magnitude and which can make the botulism hazard appear more probable than it actually is. This discussion has not covered *C. botulinum* Type F, another psychotrophic member of this species. However, from toxin-outbreak times reported for this organism it appears that the probability of a Type F hazard is similar to a Type E hazard and requires no further elaboration here.[57] The probability of a Type E hazard has been discussed and a method for assessing this probability has been proposed.[59]

2. *Mutagenic Radioresistant Microbes*

It has been theorized that multiple sublethal exposures of a substrate to ionizing radiation might produce radioresistant mutants among its microbial inhabitants. The consequence of such an occurrence is obvious, especially when the microbes involved are pathogens such as salmonellae. There is evidence that the radioresistance of microbes can be increased with repeated doses. However, our earlier work indicates that in order to increase the radioresistance of salmonellae, it was necessary to select survivors and to subject them to ten or more exposures to radiation.[60] But this situation

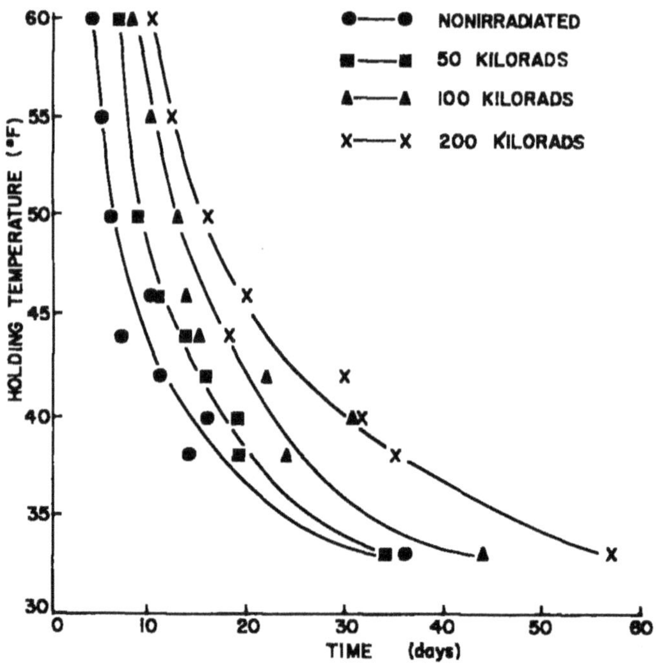

FIGURE 1. Times at which haddock fillets (irradiated and nonirradiated) become organoleptically unacceptable (averages of six experiments).

could hardly be encountered in practice, where the probability of giving a substrate more than one treatment of ionizing radiation would be small and inadvertent. As far as the probability of producing mutant pathogens from nonpathogens is concerned, there is no evidence that this has occurred.

IV. PROCESSING CONSIDERATIONS

A. Criteria for Selecting Optimum Dose

There are three fundamental criteria which must be considered when selecting the optimum dose for low-dose irradiated seafood. The first condition which must be satisfied is that the dose level selected does not produce any detectable organoleptic change. When testing for flavor change it should be pointed out that the method of preparing the food for tasting can change the threshold level at which flavor changes are evident. Steaming seafood is the most sensitive culinary method of preparation for detecting irradiation off-flavor whereas deep-frying tends to mask off-flavors. The second criterion is the shelf-life desired at a particular storage temperature, and this too is a function of the dose. This dose, however, must not exceed the threshold dose for organoleptic change. The third criterion is the degree of safety desired from a botulism hazard. With increasing irradiation dose there is a concomitant increase in the shelf life of the product which at first hand may seem desirable. However, it is possible to extend greatly the shelf life by irradiation to the point that *C. botulinum* Type E, if present, could proliferate under favorable conditions and produce toxin before the product has become recognizably spoiled. Unless one could absolutely guarantee that the irradiated product would be held under conditions which precluded outgrowth of *C. botulinum*, it would be deemed essential to employ a dose that would ensure normal spoilage to precede toxin production.

Table 2
SOME TOXIN OUTBREAK TIMES IN IRRADIATED HADDOCK AND PETRALE SOLE FILLETS PACKED IN OXYGEN-IMPERMEABLE FILMS AND STORED AT 42°F (5.6°C)

Irradiation dose (krad)	Toxin-outbreak times (days)	Inoculum (spores/g)	Substrate
0	< 55	10^4	Haddock
100	< 55	10^4	Haddock
200	< 55	10^4	Haddock
0	< 34	10^2	Petrale sole
50	< 34	10^2	Petrale sole
100	32	10^2	Petrale sole
200	21	10^2	Petrale sole

From Eklund, M. W., Poysky, F. T., and Wieler, D. I., Final Summary report by National Marine Fisheries Service to U.S. Atomic Energy Commission, Contract No. AT(49-7)-2442, Seattle, Wash., November 1967.

B. Double-Dose Application

It has been reported that the stage of growth of bacterial cultures markedly affects radiation sensitivity.[61] *Escherichia coli* was found to be most sensitive during the logarithmic growth phase and least sensitive while in the lag or stationary phase. This information suggested to Liston and Matches[62] that irradiation of fresh-caught fish with a low dose and with subsequent storage at 34°F (1.1°C) for about a week to ensure that surviving microorganisms would be in the log phase, followed by a second low-dose treatment, might result in a greater shelf-life extension compared to the two separate doses being administered as a single dose. With strictly fresh fillets the double-dose treatment was shown to be effective when the primary and secondary dose were low such as 50 + 50 krad. The double-dose treatment lost its advantage when higher doses such as 100 + 100 krad were employed. The advantage also diminished as the initial quality of the fillets decreased. These researchers also showed that irradiation of whole fish immediately after capture with 50 krad followed by a treatment with 100 krad to the fillets prepared from these fish resulted in an extension of storage life. Hannesson and Dagbjartsson[63] also demonstrated that reirradiation (150 krad) of fillets from previously irradiated (50 krad) whole eviscerated cod stored on ice was effective in extending the storage life.

One might surmise that through double-dose irradiation the possibility arises that unsold irradiated fish could be reclaimed from retail markets, given an additional low-dose treatment, and then put back into distribution. This procedure would increase the risk of a botulism hazard. To prevent this practice the FDA has specified in the labeling requirement for low dose irradiated foods that wholesale packages and invoices of bulk shipment bear the statement "Treated with ionizing radiation — do not irradiate again."

C. Shipboard Irradiation

Reports in the scientific literature indicated that the higher the quality of fish at the time of irradiation the longer the shelf-life extension at a given dose.[62,64,65] A logical extension of this finding then was to irradiate fish at sea immediately after capture. Two independent trials were conducted by the NMFS aboard vessels equipped with irradiators, the *M. V. Delaware* on the east coast and the *M. V. Miller Freeman* on the west coast.[8,66] The conclusion of both studies was that fish irradiated immediately after being caught while still in prerigor had the longest storage life and best quality. It was also demonstrated by researchers at Louisiana State University that the shelf life of shrimp could be extended 10 to 20 days by irradiation at sea with low doses ranging from 100 to 300 krad.[67]

Shipboard irradiation offers several advantages. Not only does it allow for the landing of top quality fish, but it also permits the vessel to remain at sea longer or travel to more distant fishing grounds thus ensuring a larger catch. Also, by irradiating fish when the bacterial content is minimal, a lower dose such as 50 krad can be employed to achieve the desired effect. The use of a lower dose would also provide for an increased rate of throughput in the (isotope) irradiator. However, there are some disadvantages associated with shipboard irradiation and these include increased space requirements and more importantly, inefficient use of the radiation source which most likely will not operate continually. This could be obviated by installing the irradiator on a processing "mother" ship which would be supplied with fish by satellite fishing vessels.

A limitation of storing fish on board ship in refrigerated sea water is that eventually there is a buildup of bacteria in the sea water. In an experiment with whole eviscerated haddock Nickerson and Licciardello[68] demonstrated that by continuous recirculation of the refrigerated sea water through a gamma radiation field, the rate of bacterial proliferation could be decreased. A shipboard irradiator would offer the potential for irradiating refrigerated sea water in a vessel equipped with that system of transporting fresh-caught fish.

D. Combination Treatments

Various processes or additives have been combined with irradiation either concurrently or as a pretreatment either to effect an additional increase in shelf-life extension or to lessen or mask the development of irradiation off-flavors and odors. Some investigators have applied a heat treatment prior to irradiation not only for inactivating enzymes but also for reducing the initial bacterial load which lowers the dose requirement for a given shelf-life extension. Some of the products which have been treated to advantage with a combined heating and low-dose irradiation include lobster and shrimp.[69,70] It has, however, been demonstrated that with regard to the sequence of heating and irradiation, the most effective combination for inactivating microbial cells occurred when the two were applied simultaneously.[71-73]

Goldblith and Nickerson[74] reported the irradiation of cooked lobster meat or cooked blue crab meat at 140°F (60°C) with 125-krad (48 min) or 250-krad (96 min) gamma rays while packed in metal cans either in air or vacuum. They concluded that with low plate count lobster or crab meat, irradiation with 125 krad at 32°F (0°C) was adequate for extending storage life at 36 to 40°F (2.2 to 4.4°C) whereas with moderately contaminated meats irradiation with 125 krad at 140°F was the preferred method. These investigators also irradiated haddock fillets at 32, 100, 110, or 120°F (0, 37, 8, 43.3, or 48.9°C).[75] With subsequent storage at 32°F (0°C) irradiation at 110 or 120°F provided a longer storage life compared to irradiation at 32°F, whereas with storage at 43 to 45°F (6.1 to 7.2°C) the most effective treatment was irradiation at 120°F with 350 krad. An important factor to consider in simultaneous irradiation and heating is the dose rate of the source because this governs the length of time the sample is exposed to the elevated temperature during irradiation. Some products may deteriorate in quality if held too long at the elevated temperature due to thermal effects alone. Also, some method such as microwave energy for rapidly heating the product for irradiation and cooling following irradiation would be required.

The success of a low-dose irradiation treatment is contingent upon subsequent storage at refrigerator temperatures to retard outgrowth of surviving microorganisms. One might expect that the presence of a bactericidal agent at the time of irradiation would further extend the storage life by inhibiting the outgrowth of survivors particularly those with radiation damage. In this context various chemical preservatives including sodium benzoate (0.1%), potassium sorbate (0.1%), sodium salts of methyl and propyl esters of parahydroxy benzoic acid (0.1%), and furylfuramide (0.002%) when combined with low-dose irradiation effectively extended the storage life of Dover sole, whale meat, or horse mackerel flesh.[76,77] It was shown, for example, that Dover sole irradiated with 100 or 300 krad and stored at 42 to 44°F (5.6 to 6.7°C) attained a plate count of 10^6/g after 6 or 10 days, whereas samples similarly treated but containing 0.1% sodium benzoate reached the same count after 9 or 19 days, respectively.[78] Tomiyama et al.[77] demonstrated that it was more efficient to incorporate the antimicrobial agent prior to irradiation than following it. The use of potassium sorbate is recommended when it is desired to suppress outgrowth of yeasts in irradiated products.

Antibiotics such as chlortetracycline (5 to 10 ppm) and tylosin added to various seafoods have been shown to exert a complementary effect when combined with low-dose irradiation.[77,79,80] Awad et al.[81] reported that raw shrimp irradiated with 500 krad remained acceptable up to 5 weeks at 38°F (3.3°C), whereas the irradiated product containing 5 ppm chlortetracycline was still in good condition after 8 to 10 weeks. As with chemical preservatives the antibiotic provides a greater lethal effect if added prior to irradiation.

There are several processing techniques for minimizing off-flavor development resulting from radiation treatment. These include elimination of atmospheric oxygen, irradiation in the frozen state, and addition of free-radical scavengers. An alternate procedure is to incorporate seasonings or flavoring agents which not only create a flavorful product but also tend to mask the off-flavor. Roy and Kaylor[82] found that raw shrimp treated with a spice (turmeric) and irradiated with 100 to 300 krad had a longer storage life at 33°F (0.6°C) compared to the same product receiving comparable irradiation alone. Spinelli and Miyauchi[83] demonstrated that the addition of 1 μmol/g of 5′ inosine monophosphate increased the acceptability of irradiated petrale sole or Pacific ocean perch fillets during storage.

V. PACKAGING

The role of packaging in protecting radiopasteurized seafoods would be the same as for the nonirradiated counterpart but with the following additional requirements: (1) the package must protect the product from microbial recontamination, (2) it should not impart any off-odor or off-flavor to the product, (3) it should be resistant to physical or chemical change in properties caused by irradiation, (4) it should withstand exposure to wet products or melting ice for periods in excess of 4 weeks, (5) it must prevent loss of water, and (6) the packaging material must be readily available in the required package sizes and forms and be reasonable in cost.[84,85] Radiopasteurized seafoods can be bulk packed in 10-, 20-, or 30-lb rectangular, metal containers with tight fitting lids. However, this type of closure may not prevent the ingress of contaminating microorganisms. Metal cans with hermetic seals fulfill the preceding requirements and have been used successfully in studies with irradiated fish. However, this would not be an ideal container for retail marketing of fresh fish because of lack of visibility and also because of its association with heat-sterilized foods which may unwittingly lead to the product being stored at room temperature. Irradiated fish packaged in hardboard trays with a film overwrap developed a cardboard flavor.[84] Tinker et al.[86] evaluated 18 commonly used flexible films for their suitability as packaging materials in the form of sealed pouches. Cellophane and rubber hydrochloride were considered unsuitable because of the high rate of failure of the seals. Polyethylene and polypropylene were not recommended because of the high rate of gas transmission. The suitable films included those with low gas permeability such as aluminum foil laminates, nylon, and saran types.

When the principal objective of radiopasteurization is shelf-life extension at chill temperatures, packaging seafood under vacuum in an oxygen-impermeable film is the preferred mode. Two advantages result from this treatment. First, the development of oxidative rancidity, particularly in fatty species, is suppressed.[87-90] The addition of an antioxidant in conjunction with vacuum packaging was found to enhance further the flavor stability of stored, irradiated, fatty fish.[91] Second, and perhaps most important, an atmosphere of low-oxygen tension prolongs the lag time for outgrowth of microorganisms surviving irradiation and also causes a shift in the spoilage microflora to species that are less active biochemically than those normally responsible for spoilage of refrigerated fish.[50,92-94]

VI. EFFECT OF STORAGE TEMPERATURE

The storage life of radiopasteurized seafoods depends on a number of factors which include the initial concentration of microorganisms at the time of irradiation, the qualitative and quantitative composition of the microbial flora, packaging method, absorbed dose, the presence or absence of microbial inhibitors, and the storage temperature. The last factor is extremely important. It is well documented that the rate of spoilage of fresh fish is much less over an increment of temperature in the vicinity of the freezing point compared to the same increment at higher temperatures.[95,96] There are much published data which indicate that the shelf life of low-dose irradiated seafoods was approximately doubled when the storage temperature was decreased 9°F (5°C) from 42°F (5.6°C) to 33°F (0.6°C).[88,92,97-100] The effect of lowering the holding temperature is to increase the length of the lag phase and the generation time for psychrophiles.[101] Irradiation has a similar effect on the lag phase.[102] Perhaps the two treatments applied concurrently act synergistically since it is acknowledged that the postirradiation temperature has an influence on the survival of injured bacterial cells.[103]

VII. BIOLOGICAL CONTAMINANTS

A. Nonpathogenic Bacteria

The microbial flora of fresh-caught marine fish taken from cold waters consists principally of *Pseudomonas, Achromobacter (Acinetobacter* and *Moraxella), Micrococcus,* and *Flavobacterium — Cytophaga* species with a minor proportion of *Cornyebacterium, Vibrio, Bacillus, Proteus,* and yeast.[104-106] These microorganisms are present on the skin, the slime layer, and gills. The muscle tissue has been generally found to be sterile. The initial flora consists predominantly of Gram-negative psychrophiles, but the net result of handling aboard the fishing vessels and in the filleting plants is an increase not only in the total population but also in the numbers of Gram-positive mesophiles.[106,107] Eventually, spoilage at refrigerator temperatures is caused by psychrophiles of the genera pseudomonas and achromobacter.[108] In contrast the floral composition of warm-water marine fish is predominated by Gram-negative mesophiles; however, since these genera do not grow much below 50°F (10°C), pseudomonads and achromobacter usually prevail and are responsible for spoilage at chill temperature storage.[106,109] The spoilage pattern of fresh-water fish has been reported to be similar to that of marine fish.[110]

The extension in shelf life of seafoods with low doses of irradiation is due to the reduction of the pseudomonad population to an insignificant level. The pseudomonads are among the most radiosensitive of microbial species.[52,111] A dose range of 7 to 23 krad has been reported for 99% inactivation of a pseudomonas species.[111] In general Gram-negative species are more radiosensitive than Gram-positives, and rod forms are more radiosensitive than coccoid types. With the application of a radiopasteurizing dose of ionizing radiation there is usually a shift from a mixed Gram negative-Gram positive to a predominantly Gram-positive microbial flora.[112] A definitive floral composition following a low-dose treatment cannot be given because this depends on the qualitative and quantitative composition of the initial flora. This can vary from one lot of fish to another because of season, fishing grounds, method of capture, postmortem age, holding temperature, and postharvest contamination.[106,113] Since a variation in radiation resistance (yeast >micrococcus >achromobacter >flavobacterium >pseudomonas) was found to exist among the different genera of microorganisms isolated from seafoods, the absorbed dose also plays a role in the quantitative and qualitative aspects of the microflora after irradiation.[102] In most reported studies involving seafoods the microflora following a 100- to 300-krad treatment has been predominantly achromobacter.[50,79,114-116] The mode of packaging exerts a deciding influence on the spoilage flora that develops on seafoods after low-dose irradiation treatment. At storage temperatures below 50°F (10°C) the spoilage microflora of low-dose (100 to 300 krad) irradiated, air-packed seafoods comprised mainly achromobacter.[79,113,114,116,117] With samples irradiated at higher dose levels ultimate spoilage may be caused by yeasts; however, this can be obviated by vacuum-packing.[114,118] Packaging under low-oxygen tension favors the selection and growth of lactobacilli in irradiated seafoods and eventually these organisms constitute the major portion, if not all, of the spoilage flora.[50,94,119]

B. Nonspore-Forming Pathogens

Fish and shellfish can be contaminated with certain nonspore-forming pathogens including *Salmonella, Shigella,* enteropathogenic *E. coli,* enteric *Streptococci, Staphylococcus aureus, Vibrio parahaemolyticus,* and virus.[120] This can occur when the seafood is taken from polluted waters or as a result of postharvest contamination. The presence of these pathogens in seafoods is more problematic in developing countries than in the industrialized nations.[63] It has been demonstrated that some of these pathogens *(Salmonella* and *Staphylococcus)* do not grow in perishable foods at or below 42°F (5.6°C).[121,122] Although it is recommended that radiopasteurized seafoods be held below this temperature, the presence of these pathogens in irradiated foods still should be reduced to an insignificant level to decrease the potential hazard of food poisoning.

In general, viruses are relatively radioresistant and it is not expected that a radiopasteurizing dose will effect any significant degree of reduction in their numbers. In contrast, *V. parahaemolyticus,* which has been considered the leading causative agent of bacterial gastroenteritis in Japan from eating fish, is relatively radiosensitive. For various strains inoculated in crabmeat a dose of 25 krad caused a 2- to 5-log cycle reduction in numbers.[123]

Quinn et al.[124] and Anderson[125] determined radiation survival curves for various pathogens or nonpathogenic members of various genera containing pathogenic species. The organisms were more resistant in a food substrate including oysters, crabmeat, shrimp, and salmon than in Hartsell's broth. The survival curves in the food substrate, particularly shrimp and crab meat, exhibited a tailing off which was absent or less apparent in the broth. For comparative purposes the ranges of D_{10} values for the various microorganisms in the different food substrates were as follows:

Microorganism	D_{10} (krad)
Escherichia coli	15—35
Proteus vulgaris	10—20
Shigella (3 serotypes)	25—40
Salmonella (7 serotypes)	50—100
Streptococcus faecalis	75—100
S. pyogenes	50—100
Staphylococcus aureus	80—190

In view of the nonlinear nature of the survival curves, the preceding D_{10} values cannot be used for computing a dose for complete elimination. Also, the relative radiation sensitivity of *E. coli* and probably coliforms would restrict the use of these types as indicator organisms in irradiated foods.

Survival and growth of pathogenic microorganisms on irradiated foods depend not only on the dose and substrate, but also on the storage temperature and numbers of competitive saprophytes present. At 53.6°F (12°C) *S. aureus* grew well in sterile crabmeat but was inhibited in irradiated (100 to 200 kilorad) crabmeat because of competition from other surviving saprophytic bacteria.[126] The organism failed to grow at 46.4°F (8°C) even in sterile crabmeat. Salmonella grew competitively in irradiated (100 krad) crabmeat at 71.6°F (22°C) but not at 51.8°F (11°C).[122] On English sole fillets irradiated with 50 krad growth occurred at 46.4°F (8°C) but not at 42.8°F (6°C). The effect of increasing the irradiation dose is to reduce the number of competitive survivors and thus enhance the opportunity for growth of those pathogens which are poor competitors. Salmonella can be eradicated from certain fish products with moderately low-dose levels. Treatment of fish meal with 500 krad reduced the salmonella count by 6 log cycles.[127] The same dose was also adequate for eliminating salmonella from smoked lake whitefish fillets which had been soaked in a suspension of 2 billion salmonella cells per milliliter.[128]

C. Spore-Forming Pathogens

The spore-forming pathogens of concern in irradiated fishery products would be members of the genus *Clostridium*, in particular, *C. botulinum*. Of the seven serological types discovered to date (A, B, C, D, E, F, and G), types A, B, C, E, and F have been isolated from fish or shellfish, and/or from marine sediments. However, types C and D are rarely, if ever, implicated in human botulism. The organisms, if present, are usually located on the skin or gills or in the viscera if the fish have been feeding. The serotype most commonly encountered on fish and shellfish is type E. Like other clostridia this serotype is basically a soil microorganism but has been recovered from fresh-water lakes and from marine environments particularly in estuarial regions. For a more detailed account of this entire subject matter, the reader is referred to the comprehensive review article by Hobbs.[129] The spores of type E are less radioresistant than spores of types A or B.[54] Decimal reduction doses of 132 krad in beef stew and 120 to 220 krad in haddock homogenate have been reported for type E.[54,130]

The initial segment of the radiation survival curve for clostridial spores, however, usually contains a shoulder and under this circumstance a significant reduction in numbers is not to be expected from a pasteurizing dose of radiation.[131,132] Thus, control of botulism in low-dose irradiated fish or shellfish would have to be through temperature control or the presence of a chemical inhibitor. This would not be too difficult in the case of the proteolytic strains of types A and B since they do not germinate and grow at temperatures below 50°F (10°C); however, nonproteolytic strains of types B, E, and F could be problematical since they have been shown to grow and produce toxin at a temperature as low as 38°F (3.3°C).[53,133,134]

The potential hazard attributed to low-dose irradiation of marine products is that with the extended shelf life resulting from a reduction of the natural microflora, particularly the spoilage producing pseudomonads, spores of *C. botulinum*, if present, would have ample opportunity to grow and produce toxin without the product being recognizably spoiled. In assessing the potential botulism hazard the following variables have to be considered: (1) initial concentration and strain of spores, (2) species of fish, (3) packaging method, (4) irradiation dose, and (5) storage temperature.

The initial concentration of spores governs the length of time required for toxin production at a given temperature and also the minimal temperature for growth. To initiate toxin production by type E at 38°F an inoculum of 4 to 12 million spores per 35 g of substrate was required.[53] With regard to species, fatty fish such as herring are generally considered to be more botulinogenic than lean fish.[135] This phenomenon was originally thought to be due to differences in fat content; however, later studies seemed to implicate glucose or ribose as the critical nutrient.[134,136] The fact that irradiation retards the disappearance of these sugars in certain fish may explain why some researchers found irradiated fish to be a better substrate for toxin production than nonirradiated fish. The presence or absence of these sugars may also account for the variability in toxin outbreak time between different batches of the same species. Goldblith and Nickerson[55] reported a factor in haddock muscle or press juice which tended to inactivate spores of type E during standing.

Packaging in an oxygen-impermeable film or container either under vacuum or in an inert gas will probably accelerate rate of toxin formation; however, if the spores are situated in deep muscle tissue or between two tightly pressed fillets where reducing conditions prevail, and if the substrate is capable of supporting growth of *C. botulinum*, toxin production will probably occur regardless of the external atmosphere.

One effect of the irradiation dose is the selection of the spoilage microflora. At doses less than approximately 100 krad there is a very high probability that the spoilage microflora of irradiated, temperate-water fish will be predominantly pseudomonads. This group of microorganisms is usually responsible for the normal fishy, putrid, spoilage pattern of fish that is recognizable to most individuals. At dose levels in excess of approximately 100 krad the predominant spoilage microflora may be achromobacter if air-packed or lactobacilli if vacuum-packed. The former produce sweet, fruity odors at spoilage whereas the latter produce sour odors in fish. Neither of these two spoilage types produce changes as offensive as that produced by pseudomonads. As a consequence, if toxin production and spoilage occur simultaneously in stored irradiated fish, samples irradiated at about 100 krad or less may be rejected as inedible at spoilage whereas samples receiving a higher dose may not.

Another consequence of increasing the dose is to extend the refrigerated storage life and this increases the probability of toxin production occurring before recognizable spoilage if the storage temperature is conducive to growth of *C. botulinum*. The two most important factors which determine whether irradiated fish spoil before toxin production can occur are storage temperature and level of contamination. As the storage temperature and contamination level of spores increase, the safety margin between rejection due to spoilage and toxin development decreases.[137] To determine whether irradiated marine products present a greater botulism hazard compared to nonirradiated products, Ronsivalli et al.[59] proposed the following test procedure: (1) determine the spoilage time at various storage temperatures for irradiated and nonirradiated fish based on unanimous rejection by a nonexpert panel; (2) measure the time required for toxin to develop to the threshold level in inoculated packs of irradiated and nonirradiated product; and (3) compare the times observed in steps 1 and 2 to obtain a measure of safety margin. Studies of this type were carried out at MIT, Cambridge, Mass.; Torry Research Station, Aberdeen, Scotland; and the NMFS laboratory, Seattle,

Wash.[134,135,138,139] Inoculum levels ranged from about 10^2 to 10^6 spores per gram, and storage temperatures varied from 38 to 50°F (3.3 to 10°C).

In general, the results indicated that at the higher storage temperatures and with a high inoculum of spores there either was a very small or no margin of safety. Based on inoculated pack studies at their respective laboratories, Eklund and Poysky[134] at Seattle NMFS and Hobbs[137] at Torry Research Station concluded that low-dose irradiation of fish would present no botulism hazard if the product were maintained at a storage temperature below 41 to 42°F (5 to 5.6°C). The use of spore challenges at a concentration on the order of 10^6/g seems impractical in view of the fact that toxin was shown to be present in fillets shortly after inoculation at this level.[134] What little information is available on the natural quantitative level of contamination of seafoods with type E *C. botulinum* indicates the numbers to be low. In a survey of herring taken from Norwegian fishing grounds the incidence of type E spores was estimated as one spore per 16 g of fish or less.[140] Nickerson et al.[141] examined the intestines of haddock, cod, and flounder (blackback) caught in the Gulf of Maine. The highest level of type E cells was found to be 10/100 g.

In a survey to determine the level of contamination of commercially produced haddock fillets, 24% of samples taken from 5 different plants were positive for type E.[55] The highest concentration of cells found in any of the contaminated fillets was 17/100 g. More data are needed in this area as a guideline for establishing a logical spore challenge in determining the safety of radiopasteurized seafood. An adequate process for low-dose irradiation of seafood which has been approved will have some inherent safety factors. First, the concentration of botulinum spores used in the safety study will probably be several orders of magnitude greater than the natural spore level; second, toxin outbreak times are based on mouse deaths which require considerably less toxin for the same effect on humans (based on weight extrapolation); third, toxin development is determined by i.p. injection of mice and the LD_{50} by this route is much less compared to oral ingestion;[142] and fourth, the toxin of *C. botulinum* type E is heat labile and since fish is not usually consumed raw (except in some oriental countries), normal cooking would inactivate any toxin that may have developed.[142,143]

The use of ionizing radiation for control of botulism in smoked or cured fishery products appears promising. The smoked flavor allows for the use of higher irradiation doses by masking irradiation off-flavors. The relatively high salt content of cured fish in conjunction with ionizing radiation may lessen the requirement for nitrite which is normally used to prevent outgrowth of botulinum spores in cured products and the use of which is currently under investigation by FDA.

D. Insects and Parasites

A dose of 15 to 25 krad was adequate to prevent insect damage (by sterilization) in smoked or dried fish produced in tropical countries.[144-146] Treatment with chemicals, the only other effective method, is precluded because of toxic residues. Larvae of the nematode, *Anasakis marina,* in lightly salted (5%) herring fillets were only partially inactivated (44% survival) after irradiation with 600 krad.[147] A higher salt content or higher dose would probably have produced a greater degree of inactivation, but would also have caused undesirable flavor changes.

VIII. ASSESSMENT OF QUALITY OF IRRADIATED PRODUCTS

The quality of seafoods can be monitored during refrigerated storage by measuring changes which occur as a result of autolytic processes, microbial action, or chemical (oxidative) reactions. In a critical review of the numerous objective tests that have been devised for evaluating quality of fresh or frozen fish, Gould and Peters[148] stated that "no objective test has been agreed upon that consistently correlates with the subjective taste panel method." Shewan and Hobbs[149] were of the opinion that the best method for judging overall eating quality is through an evaluation of various attributes such as odor and texture of the raw product and appearance, odor, and flavor of the cooked product. After all, this is the method by which the consumer will judge the product. It should be borne in mind that the method of culinary preparation is important since, as was previously indicated, deep-fat frying tends to mask off-flavors as compared to steaming or baking.

Chemical tests for assessing quality are based on measuring decomposition products resulting from autolytic reactions such as nucleotide degradation or microbial growth. The principal spoilage organisms in fish are the pseudomonads. These microorganisms possess very strong metabolic capabilities such as proteolysis, lipolysis, reduction of trimethylamine oxide, and production of hydrogen sulfide. The decomposition products that accrue as a result of this metabolic activity can be measured by tests for total volatile bases, total volatile acids, trimethylamine, and volatile reducing substances. The feasibility of applying these tests to seafood depends on the irradiation dose, method of packaging, and storage temperature since all of these factors influence the dominant spoilage flora. The bacterial count of the product prior to irradiation is also important since with high initial count products a significant quantity of bacterial enzymes will remain following a low-dose treatment even though most of the cellular organisms have been inactivated. During storage these enzymes can form the decomposition products referred to previously.

Spinelli et al.[88,130,131] conducted studies to determine the applicability of various chemical tests for monitoring quality of stored, irradiated seafood. These researchers found that following low-dose irradiation and storage at refrigerator temperatures there is a rapid decline in organoleptic score due to autolysis. This is followed by a period of slow decline during which the microflora is beginning to proliferate but hasn't as yet produced sufficient decomposition products, and finally, a period of rapid decline in sensory score due to the accumulation of bacterial metabolic end products. The following tests were considered useful for measuring quality of irradiated seafoods during storage: (1) the decline in content of inosine monophosphate or accumulation of hypoxanthine; (2) total volatile bases (amount produced at spoilage will not be as much as in nonirradiated sample); (3) trimethylamine (the amount at spoilage decreases with increasing dose and the test is not recommended at doses exceeding 200 krad; (4) total volatile acids; and (5) decrease in free glucose. Tests with limited acceptability included histamine content, thiobarbituric acid number, low molecular weight carbonyls, and decrease in free ribose.

It should be emphasized that the preceding recommended tests cannot be applied across the board to all species of seafood because of variations in the initial quantities or variations in the rate of decomposition of the precursor compounds. Whereas the usefulness of the trimethylamine test has been corroborated for low-dose treatment, there still remains some controversy as to the usefulness of the total volatile acids and total volatile bases tests.[152] Some researchers have found these tests to correlate with sensory change and others have not.[85,100,153] Part of this discrepancy may be due to differences in the initial microflora of the fish because of origin and also differences in the magnitude of the radiation dose applied.

The total viable plate count (TPC) has not been considered to be too reliable as an index of quality of fresh fish, the principal reason being that the microflora is usually composed of many different species whereas only a few of these species are actually responsible for the eventual spoilage.[154] The predominant spoilage organisms may vary because of the origin of the fish, type of seafood, storage temperature, packaging method, and processing treatment, if any. In North Atlantic cod the chief spoilage organisms were found to be the Gram-negative pseudomonads whereas in fish taken from tropical waters *Proteus* and *Aeromonas* prevailed at spoilage.[155,156] In Pacific oysters spoilage was chiefly caused by growth of lactobacilli and in Dungeness crabmeat *Achromobacter* dominated the spoilage flora.[115] Processing treatments such as heat, irradiation, etc. depending on the dose, will influence the spoilage flora as will packaging methods that exclude oxygen. Although one cannot assign absolute numerical values to indicate spoilage in view of these variables cited, in general it has been found that for nonirradiated fish TPC at spoilage is usually on the order of 10^6/g whereas for low-dose irradiated product the corresponding value is approximately 10^8/g.[63,102,113,150]

IX. MARINE FISH SPECIES

Investigations concerning the application of ionizing radiation to fish and shellfish have been carried out mainly for purposes of extending the storage life of these products at refrigerator temperatures above freezing. The reason for limiting treatment to low doses is that when high-dose levels of ionizing radiation are applied to fish and shellfish, undesirable changes, especially in flavor, are produced in these foods. Whereas the treatment of numerous species of fish, crustaceans, and molluscs with ionizing radiation has been investigated, most of the research has been applied to those species which are relatively plentiful in the northwestern Atlantic and the northeastern Pacific oceans. Some species from other areas have been examined to determine whether or not they are suitable for this type of preservation treatment.

A. Anglerfish *(Lophius americanus)*

The only investigation concerning the treatment of anglerfish with ionizing radiation was reported by Carver et al.[157] In this instance the dressed fish were treated, prerigor, aboard ship with gamma ray dose levels of 150 krad and iced in the usual commercial manner. It was found that the radiation treatment did not affect the organoleptic qualities of this species. The results of this investigation indicated that treatment in this manner would extend the iced storage life of this fish and provide a product of better quality to consumers.

B. Bombay Duck *(Harpodon nehereus)*

In India, cleaned and filleted Bombay duck packed in polyethylene bags containing brine was irradiated at a level of 400 krad with gamma rays and held at 10 to 12°C (50 to 53.6°F). The results indicated that the irradiated product was acceptable with respect to odor, texture, and taste after storage for a period of 21 days. The nonirradiated controls spoiled in 3 days at the stated temperatures. Irradiated and vacuum-dried Bombay duck was found to be superior in quality to the nonirradiated sun-dried fish.[158] Mavinkurve et al.[159] observed that the spoilage pattern of Bombay duck which had been treated with ionizing radiation at levels of 250 to 1000 krad was different from that of the nonirradiated fish. Irradiation treatment at levels of 400 to 500 krad provided a six-fold extension of the storage life of this fish at 10 to 12°C. Typical putrid odors did not develop in the irradiated product. Sawant et al.[160] found that irradiation dose levels of 250, 500, and 1000 krad when applied to Bombay duck extended its storage life at 10 to 12°C (50 to 53.6°F) to 14, 21, and 46 days, respectively. The nonirradiated control, held at the same temperature, spoiled in 3 days. Aiyar[161] found that irradiation dose levels of 100 to 250 krad applied to Bombay duck increased its storage life to 3 to 4 times that of the nonirradiated fish.

C. Butterfish *(Peprilus triacanthus)*

Carver and Steinberg[162] found that the maximum permissible dose of ionizing radiation for butterfish was 460 krad and that the dose of maximum acceptability was 230 krad. When treated with 230 krad of ionizing radiation and stored at 1.7°C (35°F) butterfish had a storage life of 49 days. The nonirradiated controls spoiled in 23 days at the stated temperature. Generally, there was no appreciable difference in the odor, flavor, and appearance between the controls and the irradiation-treated fish. Shewan[163] stated that butterfish, when treated with 250 krad of ionizing radiation, had a storage life at 1.5°C (34.7°F) 10 to 20 days longer than that of the nonirradiated fish held at the same temperature. Coleby and Shewan[164] reported that the maximum permissible dose of ionizing radiation for butterfish is 500 to 700 krad. Carver et al.[157] found that the treatment of butterfish with 150 krad of ionizing radiation did not affect its organoleptic qualities. Brooke and Steinberg[13] irradiated butterfish at levels of 120 to 1900 krad. Product treated at levels of 450 krad or less had the greatest acceptability. When irradiated at 330 or 460 krad and held at 1.7°C (35°F) the storage life of butterfish was tripled compared to that of the untreated controls held at the same temperature.

D. Cod

1. Atlantic Cod (Gadus morhua morhua)

Shewan and Liston[165] found that the best results were obtained in processing cod with ionizing radiation by treatment in the frozen state at a level of 500 krad or in the unfrozen state at a level of 250 krad. Carver and Steinberg[162] reported that the maximum irradiation dose for cod fillets is 460 krad and that the dose of greatest acceptability is 120 krad. When cooked by steaming the frozen controls were preferred over the irradiated product, but when cooked by deep-fat frying the frozen controls and product irradiated at a level of 230 krad were not appreciably different. When irradiated at 230 krad, held for 49 days at 1.7°C (35°F), then cooked by steaming, cod fillets were judged to be acceptable. Nonirradiated cod fillets held for 23 days at 1.7°C (35°F) were judged to be spoiled.

Slavin and Ronsivalli[166] concluded that, because of economical reasons, the optimum irradiation dose for cod fillets is 150 krad. When irradiated at this level and held for 20 days at 0.6°C (33°F) this product was rated as "good", and when held for 30 days under these conditions, the rating was "fair to good". When held at 5.6°C (42°F), the irradiated cod fillets were rated as "good" after 20 days and spoiled after 30 days of storage. All taste tests were made on fillets which had been deep-fat fried. Rhodes[167] found no differences in odor, flavor, and appearance of nonirradiated cod fillets and in fillets of the same species treated with ionizing radiation at levels as high as 300 krad. Brooke and Steinberg[13] found that the treatment of cod fillets at a level of 330 krad of ionizing radiation tripled their storage life at 1.7°C (35°F).

Ronsivalli et al.[85] stated that when packed in cans, in air or under vacuum, the optimum irradiation dose for cod fillets is 150 krad. Steinberg[168] found that cod fillets packaged in hermetically sealed cans without vacuum had a storage life of 12 to 14 days at 0.6°C (33°F) while those which were treated at 150 krad of ionizing radiation after packaging in the same manner had a storage life of 30 days at this temperature.

Slavin et al.[169] found that cod fillets were acceptable after irradiation treatment at a dose level of 150 krad and holding for 28 days at 0.6°C (33°F). The postirradiation shelf life of fillets cut from cod 1 to 2 days out of water and held for more than 3 days at 5.6°C (42°F) was not increased significantly. Fillets from 1- to 2-day-old fish held for 1 day at 5.6°C (42°F) then irradiated had a shelf life of 8 to 13 days at 0.6°C (33°F). All taste tests were carried out on fish cooked by steaming. Coleby and Shewan[164] stated that the maximum permissible irradiation dose for cod is 800 to 1500 krad. Slavin et al.[17,170] verified the optimum irradiation dose for cod fillets at 150 krad and noted that the shelf-life extension beyond that of the control, in products held at 0.6°C, was 28 days.

Wittfagel[171] found that when cod fillets were packaged in plastic (air present) and held at 4°C (39.2°F) the storage life of untreated controls was 10 days while fillets irradiated at a level of 75 krad had a storage life of 21 days. Slavin et al.[172] irradiated eviscerated cod fish aboard ship pre and postrigor at a level of 100 krad, then held them on ice (about 0.6°C). Fillets cut from fish irradiated prerigor were "fair to good" after 28 days of storage while fillets cut from fish irradiated postrigor were only "fair" after 14 days of storage. Fillets from the prerigor irradiated fish were tested after cooking by both steaming and deep-fat frying. When cooked by steaming fillets from fish held 15 days at 0.6°C (33°F) were judged to have a "good" texture and "fair" odor and flavor. After 22 days of storage, the appearance, odor, flavor, and texture of this product were rated as "fair to borderline". When cooked by deep-fat frying the various organoleptic criteria were rated as "fair to good" after both 29 and 36 days of storage at 0.6°C (33°F). Similar results were obtained with cod fish irradiated aboard ship at a level of 200 krad and held at 0.6°C (33°F).

Ronsivalli et al.[173] packaged cod fillets in gas-permeable polyethylene pouches and in gas-impermeable, coated, polyester pouches. These products were irradiated at 50, 100, or 200 krad or not irradiated and held at 0.6, 4.4, or 7.8°C (33, 40, or 46°F). The products packaged in the gas-impermeable pouches had a slightly longer shelf life than those packaged in polyethylene pouches, but the difference was less marked at the higher storage temperatures. The storage life of cod fillets packaged by either method was extended by irradiation treatment. Expert panelists generally tended to indicate a shorter shelf-life time for cod fillets than did nonexpert panelists.

Metlitskii et al.[174] treated cod fillets with ionizing radiation at a level of 250 krad and found that they had a storage life of 35 to 37 days at 2°C (35.6°F). The storage life of the nonirradiated control fillets at this temperature was 7 to 9 days. Ampola et al.[90] stated that the optimum dose of ionizing radiation for cod fillets is 150 krad and that this treatment extended the storage life at 1°C (33.8°F) to 30 days. When irradiated fillets were deep-fat fried, taste panels found them to be as acceptable as the frozen controls (also deep-fat fried) stored for the same period at a temperature below −17.8°C (0°F). Fillets were packaged in air and under vacuum in hermetically sealed cans. Vacuum packing caused off odors to develop in the irradiated fillets.

Ronsivalli et al.[175] reported the results of organoleptic tests made by a consumer-type panel on cod fillets irradiated at 150 krad and on frozen control fillets. The irradiated samples were rated at approximately the same levels as the controls (see Table 3). It was stated that when the storage temperature for irradiated fish fillets was increased 10°C (18°F) above 0°C (32°F) the spoilage rate approximately tripled. Carver et al.[8] found that the storage life of eviscerated cod fish on ice could be doubled or tripled by treating with ionizing radiation, aboard ship, at a level of 100 krad. Prerigor fish appeared to be more suitable for irradiation treatment than did postrigor fish.

Ronsivalli et al.[176] reported on the storage life of nonirradiated cod fillets and of cod fillets irradiated at 75, 100, or 200 krad when held at 0.6, 3.3, 5.6, 8.9, and 12.8°C (33, 38, 42, 48, and 55°F). Quantities of 20 lb of product were packed in commercial containers prior to storage or to irradiation treatment and storage. Storage extension times obtained by irradiation treatment were reported (see Table 4). Miyauchi[177] reported that cod fillets were irradiated at Gloucester, Mass., and shipped by rail with the nonirradiated control samples to Seattle, Wash. The irradiation dose levels used were 100 and 200 krad. When received at Seattle the control fillets were spoiled whereas the irradiated fillets had an additional storage life of 7 to 14 days at 0.6°C (33°F).

2. Pacific Cod (Gadus macrocephalus)

Miyauchi[178] packed Pacific cod fillets in cans without vacuum, surrounded the cans with ice, and shipped them from Seattle, Wash., to Idaho Falls, Idaho, for treatment with ionizing radiation at levels of 100, 230, 460, or 700 krad. The products, including

Table 3
ACCEPTABILITY TESTS ON RADIATION-PASTEURIZED SEAFOOD (U.S. ARMY TEST AND EVALUATION COMMAND, FORT LEE, VA.)

Product	Panel members (no.)	Dose (krad)	Storage temperature (°C)	Storage period (days)	Average taste score: Irradiated	Average taste score: Nonirradiated frozen controls
Haddock fillets[a]	314	250	0.6	15	5.8	6.1
				29	5.8	6.2
Petrale sole fillets[a]	333	200	0—1.7	16	5.9	6.4
				28	6.5	6.6
Haddock fillets[b]	693	150	0.6	7	6.2	5.9
				14	6.0	6.2
Cod fillets[b]	588	150	0.6	17	6.6	6.5
				24	6.3	6.2

Note: Based on hedonic scale where 9 = like extremely; 8 = like very much; 7 = like moderately; 6 = like slightly; 5 = neither like nor dislike; 4 = dislike slightly; 3 = dislike moderately; 2 = dislike very much; 1 = dislike extremely.

[a] Packed in air, hermetically sealed cans.
[b] Packed in 30-lb (13.6-kg) fillet tins to approximate commercial conditions.

From Ronsivalli, L. J., Kaylor, J. D., and Slavin, J. W., *Freezing and Irradiation of Fish*, Kreuzer, R., Ed., Fishing News (Books), London, 1969, 482.

Table 4
THE MAXIMUM SHELF LIFE VALUE (IN DAYS) OF IRRADIATED AND NONIRRADIATED COMMERCIAL-QUALITY COD FILLETS (EXPERIMENTS NO. 1, 2, AND 3)

Radiation dose (krad)	Storage temperature °F (°C): 33 (0.6)	38 (3.3)	42 (5.6)	48 (8.8)	55 (12.8)
0	12—14	11	7—8	5—7	3—4
75	23—40	11—27	9—11	8—11	6—7
100	30—49	15—30	11—15	10—12	6—9
200	45—50	29—32	22—23	13—15	8—10

From Ronsivalli, L. J., King, F. J., Ampola, V. G., and Holston, J. A., Bureau of Commercial Fisheries Technological Laboratory, for U.S. Atomic Energy Commission Contract No. AT(49-11)-1889, Gloucester, Mass., 1970

nonirradiated controls, were reshipped to Seattle. The time required for shipment and treatment was 5 days. After the samples arrived in Seattle they were held at 0 to 1.7°C (32 to 35°F). The nonirradiated fillets were "very poor" after 14 days of storage, while the irradiated fillets were rated as "good" for a period of 21 days and "borderline" for 42 to 63 days, depending upon the dose level applied. The irradiated fillets were slightly darkened and tended to lose flavor during long storage. It was concluded that the optimum dose level for Pacific cod fillets is 460 krad or less.

Miyauchi et al.[179] vacuum packed Pacific cod fillets in polyester pouches and treated them with ionizing radiation at levels of 100 to 200 krad then held them, along with the nonirradiated controls, at 0.6°C (33°F). The control samples had a shelf life of 4 to 7 days, while the shelf life of the irradiated fillets was 2 to 3 times greater than that of the controls. Quantities of 12 or 25 lb of Pacific cod fillets in commercial metal containers or in paperboard cartons, lined with oxygen-impermeable plastic, were treated with ionizing radiation at levels of 100 or 200 krad and shipped to different cities in the U.S. During shipment temperatures of −2.2 to 5°C (28 to 41°F) were encountered. Upon arrival the fillets had an additional storage life of at least 14 days. Pacific cod fillets were also irradiated at 100 or 200 krad or not irradiated and held at 10, 7.8, or 5.6°C. (50, 46, or 42°F). The fillets were tested by an expert panel after cooking by steaming. Control samples had a storage life of 3 to 9 days depending on the storage temperatures. The storage life of the irradiated samples was 7 to 23 days depending upon the dose level applied and the storage temperature used. Consumer-type panelists tended to indicate longer storage life times for both nonirradiated and irradiated samples than did the expert panelists.

E. Flounder

1. Blackback or Winter Flounder (Pseudopleuronectes americanus)

Carver and Steinberg[162] stated that the maximum permissible dose of ionizing radiation as applied to blackback flounder is 930 krad and that the dose of greatest acceptability is 460 krad. Brooke and Steinberg[13] found that the most acceptable dose of ionizing radiation for blackback flounder was 450 krad or less. Ronsivalli et al.[175] treated blackback flounder with ionizing radiation at a low-dose level and shipped them along with nonirradiated controls, under commercial conditions, for distances of approximately 1500 to 3000 mi (2400 to 4800 km). It was found that the irradiated product had a shelf life of at least 1 week longer than the nonirradiated controls handled under the same conditions. The irradiated product also had a high level of quality for 1 week or more longer than that of the nonirradiated controls. The acceptability of the irradiated flounder had been previously established by consumer-type taste tests made at Fort Lee, Va.

2. Dover Sole (Microstomus pacificus)

Shiflett[76] determined the storage life of dover sole fillets by TPC (bacterial) after treating with ionizing radiation at levels of 100, 300, or 500 krad with or without the addition of 0.1% of sodium benzoate, potassium sorbate, or the methyl or propyl esters of parahydroxy benzoic acid. He found that the addition of the chemical microbial inhibitors doubled the refrigerated storage life of the irradiated fillets. Miyauchi[177] treated dover sole, aboard ship, with 50 or 100 krad of ionizing radiation. The treated fish were found to have better storage characteristics than the nonirradiated control fish. When treated with ionizing radiation in the prerigor state, dover sole had a longer storage life on ice than when irradiation treated in the postrigor condition. When dover sole was irradiated, at sea, at a level of 50 krad, held on ice for 14 days, then filleted, these fillets had a 4-day longer refrigerated storage life than did those cut from nonirradiated fish held under the same conditions for the same period of time. He found that irradiated dover sole held for longer than 14 days on ice prior to filleting may undergo a loss of quality caused by autolysis and discoloration of the flesh.

3. English Sole (Parophrys vetulus)

Miyauchi et al.[179] packed English sole fillets in hermetically sealed cans and treated them with ionizing radiation at levels of 100, 200, or 300 krad. The refrigerated storage life of the treated fillets was found to be as much as three times greater than that of the nonirradiated control fillets. Steinberg[168] stated that the storage life of English sole fillets packed in hermetically sealed cans and treated with ionizing radiation at a level of 200 krad then held at 0.6°C (33°F) is 25 days. The nonirradiated control fillets packed and held under the same conditions had a storage life of 4 to 6 days. Miyauchi et al.[180,181] packed 25-lb quantities of English sole fillets in Scotchpack®* or in polyethylene bags and irradiated some of each type of pack at a level of 200 krad. These products, along with nonirradiated controls packaged in the same manner, were held at 0.6°C (33°F) for periods of 7 and 14 days. After storage, the control and treated fillets were repacked under vacuum in Scotchpack® or on paperboard trays with a cellophane overwrap and again stored at 0.6°C (33°F). After repacking the control samples prestored for 7 days had an additional storage life of 5 days while the irradiated fillets prestored for 6 days had an additional storage life of 9 days. There was no difference in fillets prepackaged in Scotchpack® and those prepackaged in polyethylene. At the end of 14 days of prepackaged storage the nonirradiated fillets were spoiled and the irradiation treated fillets, in either Scotchpack® or polyethylene were rated at the lower end of "good".

Spinelli et al.[182] reported that English sole fillets pretreated with tripolyphosphate prior to treatment with ionizing radiation at low-dose levels maintained their original color during refrigerated storage. The color of cooked, tripolyphosphate-and-irradiation-treated fillets was also better than that of the irradiated controls not receiving the phosphate treatment. Phosphate treatment did not change or improve the texture of the irradiated fillets and English sole fillets packaged in the presence of air eventually became rancid after subjection to radiation treatment regardless of whether or not they were pretreated with phosphate.

Liston and Matches[62] irradiated uncut English sole aboard ship at 50 krad and found that the storage life of this product at 0°C (32°F) was 22 days. The nonirradiated fish had a storage life of 14 days at this temperature. Fillets cut from the irradiated fish after storage on ice for 7 days then packaged and irradiated at a level of 100 krad had a storage life of more than 28 days at 0°C (32°F). Fillets cut from fresh nonirradiated fish had a storage life of 14 days at the temperature stated. Storage life times were determined by total aerobic bacterial plate counts. Miyauchi et al.[183] reported the storage life of English sole fillets held at 0.6°C (33°F) to be as indicated in the following: untreated, 4 to 6 days; treated with a dose of 100 krad of ionizing radiation, 14 to 21 days; treated with a dose of 200 krad of ionizing radiation, 21 to 28 days.

Spinelli et al.[184] treated English sole fillets with 200 krad of ionizing radiation and held them at 0.6°C (33°F) after packaging as indicated in the following: air packed in hermetically sealed cans; vacuum packed in hermetically sealed cans; packed in hermetically sealed cans in an atmosphere of carbon dioxide; packed in hermetically sealed cans in an atmosphere of nitrogen; and packed in hermetically sealed cans in an atmosphere of helium. The air packed samples became rancid in 7 days. The carbon dioxide- and nitrogen-packed samples became rancid in 10 days. The helium-packed samples were acceptable after 23 days and the vacuum-packed samples were best and judged to be acceptable after 35 days.

Liston et al.[185] irradiated uncut English sole with a single dose of 300 krad or double doses of 50 and 50 krad or 50 and 100 krad and held them on ice. Double doses of irradiation were applied 7 days apart. The shelf life of the fish given the double-dose treatment was equivalent or greater than that of fish given the single treatment. It was found that there was a minimal extension of the storage life of fish which was in the spoiling stage when treated with ionizing radiation, regardless of whether the single or double-dose technique was employed. A preirradiation iced storage time of 8 to 10 days was considered to be the limit for English sole if good results were to be obtained.

Miyauchi and Teeny[186] packaged 3- to 4-lb portions of English sole fillets in polyeth-

* Trade name for the chemically bound laminate of medium density polyethylene to polyethylene terephthalate (Mylar®)

ylene and placed them in styrofoam containers inside of double-walled, corrugated cartons. Quantities of 25 lb of product were placed in each carton. Some of the product packed in this manner had received no irradiation treatment, while the fillets in other cartons had been treated with ionizing radiation at levels of 100 or 200 krad. Some of each type of irradiated fillets and nonirradiated controls were held in Seattle, Wash., at 0.6°C (33°F) and some of the different types were shipped to Gloucester, Mass., where they were held at 0.6°C (33°F). Part of the product shipped to Gloucester was returned to Seattle and held at 0.6°C (33°F). In each case of shipment of product refrigerated trucks were used and the load temperature was held at −1°C (30°F). Shipment from west to east and the return trip required 14 days. It was found that the storage life of the nonirradiated samples was 5 to 13 days. The fillets treated with 100 krad of ionizing radiation had a storage life of 16 to 20 days, and those treated at a level of 200 krad had a storage life of 15 to 21 days.

Miyauchi[177] reported that the storage life of English sole fillets held at 0.6°C (33°F) is 4 to 6 days. After treating with ionizing radiation at a level of 100 krad, the storage life of the fillets at 0.6°C (33°F) was found to be 14 to 21 days. Fillets of this species treated with 200 krad of ionizing radiation had a storage life at the above temperature of 21 to 28 days. It was further found that packaging under vacuum in some material impervious to oxygen was a requisite for English sole fillets which were to be treated with ionizing radiation, otherwise they became rancid in 7 to 14 days of storage. Miyauchi also reported that English sole treated aboard ship with ionizing radiation at levels of 50 to 100 krad had better iced storage characteristics than did the nonirradiated fish. Prerigor irradiated fish had a longer iced storage life than those irradiated postrigor. Fillets from English sole treated with 50 krad of ionizing radiation and held iced for 14 days had an additional storage life at 0.6°C (33°F) 4 to 7 days longer than those cut from the nonirradiated fish which had been held on ice for the same period of time. It was also stated that English sole fillets treated with ionizing radiation in Seattle, Wash., and shipped to Gloucester, Mass., had a refrigerated storage life 10 days longer than that of the nonirradiated controls.

4. Gray Sole (*Glyptocephalus cynoglossus*)

Miyauchi et al.[179] reported that gray sole fillets were treated with ionizing radiation at levels of 100 or 200 krad in Gloucester, Mass., and shipped, along with the nonirradiated controls, in wet ice to Seattle, Wash. Upon arrival, the nonirradiated controls were spoiled, whereas the radiation treated fillets were in good condition and had an additional storage life on ice of 7 to 14 days. Ronsivalli et al.[173] packed gray sole fillets in commercial-type metal fish containers, irradiated the product at 50 or 100 krad and held them, along with controls, at different temperatures. The fillets were tested, after cooking by steaming, by a panel of expert tasters. At 0.6°C (33°F) the controls had a shelf life of 6 to 7 days while the irradiated fillets had a storage life of about 15 and 29 days, respectively. At 5.6°C (42°F) the control fillets had a storage life of 6 days, and the radiation-treated fillets had a storage life of 6 to 7 and 10 to 11 days, respectively. At 7.8°C (46°F) the control fillets had a storage life of 4 days; the irradiated products had storage life times of 6 and 6 to 7 days.

5. Atlantic Halibut (*Hippoglossus hippoglossus*)

Eukel and Huber[187] treated halibut steaks with 250 krad of ionizing radiation (electrons, 8 MeV) and determined the storage life of the treated product at 5.6°C (42°F) by measuring volatile reducing substances. It was stated that the irradiation treatment extended the storage life to about 20 days. Sieling[188] stated that halibut steaks treated with ionizing radiation at a level of 250 krad were normal in color and taste, but the odor was not normal. Rhodes[167] reported on the irradiation treatment of halibut steaks, vacuum packed in oxygen-impermeable plastic films. The product was irradiated at levels of 100, 300, 500, or 1000 krad and held at 0°C (32°F) or 4.4°C (40°F). A loss of the fresh flavor was noted in the steaks treated at 300 krad and a slight irradiation odor was present in product treated at 1000 krad. The taste of halibut irradiated at 500 krad was acceptable. The storage life of halibut steaks treated with 300 krad of ionizing radiation was longer than 30 days at 0°C (32°F) and more than 21 days at 4.4°C (40°F). Coleby and Shewan[164] stated that the maximum permissible dose for halibut steaks is 1000 to 2000 krep.

6. California Halibut (Paralichthys californicus)

Miyauchi et al.[189] packed halibut steaks in cans under vacuum and found that the optimum dose level was 200 krad. When held at 0.6°C (33°F) the storage life of the untreated control samples was 7 days, while that of the irradiation treated samples was 21 to 28 days. When a temperature of 5.6°C (42°F) was used, the control halibut had a storage life of about 4 days, while that of the irradiation treated halibut was 10 to 14 days. Miyauchi et al.[180] vacuum packed halibut steaks in cans, irradiated the product at various levels, and held it at 0.6°C (33°F). At a level of treatment of 100 krad the storage life was less than 14 to 21 days. When treated with 200 krad, the storage life was 14 to 56 days. Higher dose levels extended the storage life time, but because of induced off-flavors detected at the higher levels, 100 to 200 krad was judged to be the optimum irradiation dose level for Pacific halibut.

Dassow and Miyauchi[190] irradiated halibut steaks, vacuum packed in cans, at a level of 200 krad and held them, along with nonirradiated controls, at 0.6°C (33°F) or 5.6°C (42°F). The controls had a storage life of 4 to more than 9 days at the lower and 4 to 7 days at the higher temperature. The radiation-treated steaks had a storage life of 21 to 42 days at 0.6°C (33°F) and 14 to 28 days at 5.6°C (42°F). Miyauchi[177] stated that the recommended irradiation dose for halibut steaks is 200 krad, and that this treatment provided a shelf life at 0.6°C (33°F) of 14 to 35 days. It was also stated that it was necessary to eliminate air from the package in which halibut steaks were irradiated in order to minimize the development of rancidity during storage.

Steinberg[168] stated that halibut steaks, vacuum packed in cans, treated with ionizing radiation at a level of 200 krad and held at 0.6°C (33°F) had a storage life of 30 days, while the nonirradiated control samples handled under the same conditions had a storage life of 4 to 9 days. Slavin et al.[17] reported that halibut steaks treated with 200 krad of ionizing radiation had a shelf-life extension of 14 days beyond that of the nonirradiated control steaks. Miyauchi et al.[181,191] vacuum packed large pieces of halibut in cans or Scotchpak® or in air in "cryovac" bags, irradiated these samples at 250 krad, then stored them for 14 days at 0.6°C (33°F). These products were then tested for quality, cut into steaks, and repackaged in Scotchpak® under vacuum or in air on paperboard trays, overwrapped with cellophane. The large pieces held in cans or in Scotchpak® were of good quality after storage for 14 to 21 days. Product packed in "Cryovac®" was rancid after this period. Steaks from the canned or Scotchpak® packaged pieces had an additional storage life of 7 to 9 days as steaks, making the total storage life 21 to 23 days. Steaks packed in Scotchpak® had a storage life of up to 2 days longer than the cellophane wrapped steaks.

Large pieces of halibut were also vacuum packed in cans, treated with ionizing radiation at 300 krad, and held at 0.6°C (33°F) for 2 to 4 weeks to be compared with frozen halibut stored for the same period at −21°C (−5.8°F). These products were tested and compared after cooking by both steaming and deep-fat frying. The tasters found no appreciable difference between the radiation-treated and the frozen product. Spinelli et al.[182] found that pretreatment of halibut steaks with tripolyphosphate prior to irradiation treatment improved the color of the raw and cooked product but did not affect the texture and flavor. Tripolyphosphate-treated samples that were irradiated in the presence of air became rancid as did the irradiated samples without the phosphate treatment. Miyauchi et al.[183] found that the storage life of halibut steaks (untreated) was 4 to 18 days at 0.6°C (33°F) and less than 4 to 8 days at 5.6°C (42°F). The storage life of halibut steaks irradiated at 100 krad was 14 to 21 days at 0.6°C (33°F). When treated with ionizing radiation at a level of 200 krad, the storage life was 21 to 56 days at 0.6°C (33°F) and 14 to 21 days at 5.6°C (42°F).

7. Lemon Sole (Microstomus kitt)

Shewan and Liston[165] found that the best results with irradiation treatment of lemon sole were obtained when a dose level of 500 krad was applied to the fish in the frozen state or when a dose level of 250 krad was applied to the unfrozen fish. Coleby and Shewan[164] stated that the maximum permissible dose of ionizing radiation for lemon sole is 500 to 1000 krep.

8. Petrale Sole (Eopsetta jordani)

The U.S. Fish and Wildlife Service[192] reported that petrale sole fillets, vacuum packed in cans were treated with ionizing radiation at levels of 400 and 600 krad and held, along with nonirradiated control samples, at 0.6°C (33°F) or 5.6°C (42°F). At 0.6°C (33°F) the controls had a storage life of 7 days, the fillets treated with 400 krad had a storage life of 35 days, and those treated with 600 krad had a storage life of 42 days. At 5.6°C (42°F) the controls had a storage life of less than 5 days, while the irradiated fillets had a storage of 21 days. Vacuum packing was found to be a requisite for irradiated petrale sole fillets, otherwise the treated product became rancid during storage. It was stated that the storage life of radiation-treated petrale sole fillets was about twice as long when stored at 0.6°C (33°F) as when stored at 5.6°C (42°F).[193] Miyauchi et al.[189] packaged petrale sole fillets in air and under vacuum and treated the samples with ionizing radiation at levels of 200 or 300 krad. When stored at 0.6°C (33°F) or 5.6°C (42°F) the air-packed samples became rancid in 7 days. The vacuum-packed samples had a storage life of 28 to 49 days at 0.6°C (33°F) and 14 to 21 days at 5.6°C (42°F).[193] The nonirradiated vacuum-packed control fillets had a storage life of 4 to 10 days at 0.6°C (33°F) and 4 to 7 days at 5.6°C (42°F). Miyauchi et al.[189] packaged petrale sole fillets in Mylar®-polyethylene pouches under vacuum and irradiated samples at levels of 300, 400, or 500 krad of ionizing radiation. The treated fillets, along with nonirradiated controls, were held at 2.2°C (36°F). Control fillets had a storage life of less than 7 days while those treated with 300 krad of ionizing radiation had a storage life greater than 20 days and those treated at the higher irradiation levels had a storage life greater than 30 days.

Spinelli et al.[88] stated that the treatment of vacuum-packed petrale sole fillets with ionizing radiation at levels of 200 or 300 krad extended the storage life to 28 to 35 days at 0.6°C (33°F) and to 14 to 21 days at 5.6°C (42°F). They found that the treatment of petrale sole fillets with ionizing radiation at levels of 200 or 300 krad extended the storage life 3 to 5 times over that of the nonirradiated controls when these products were held at 0.6°C (33°F). It was also found that irradiation treatment at these levels caused minimal changes in the odor and flavor, but that such changes were noticeable only when irradiation doses greater than 300 krad were applied. Vacuum packaging was found to be necessary to prevent the development of rancidity in radiation-treated petrale sole fillets. Some autolysis occurred in product stored for long periods at 0.6°C (33°F). Vacuum packaging in Mylar®-polyethylene pouches was found to be as suitable as vacuum packing in cans in preventing the development of oxidative rancidity in petrale sole fillets during storage.

Dassow and Miyauchi[190] reported that untreated petrale sole fillets vacuum packed in cans had a storage life of 4 to 10 days at 0.6°C (33°F) and 4 to 7 days at 5.6°C

(42°F). Product packed in the same manner and treated with ionizing radiation at 200 or 300 krad had a storage life of 28 to 42 days at 0.6°C (33°F) and 14 to 17 days at 5.6°C (42°F). During storage there was some loss of the typical flavor in the irradiated fillets. Steinberg[168] packed petrale sole fillets in Mylar®-polyethylene pouches under vacuum and treated them with ionizing radiation at levels of 200 or 300 krad. Treated fillets, along with nonirradiated controls, were held at 0.6°C (33°F). The controls had a storage life of 4 to 10 days, the 200-krad-treated samples had a storage life of 21 days and the 300-krad-treated samples had a storage life of 42 days.

Miyauchi et al.[180] vacuum packed petrale sole fillets in cans, irradiated the product at levels of 100 or 150 krad and held them, along with nonirradiated controls, at 0.6°C (33°F). The control fillets had a storage life of 8 to 11 days, the 100-krad-treated samples had a storage life of 18 to 20 days and the 150-krad-treated product had a storage life of at least 21 days. It was concluded that treatment with ionizing radiation increased the storage life of petrale sole fillets at 0.6°C (33°F) by 2 to 3 times compared to that of the nonirradiated fillets. Slavin et al.[170] reported that the shelf life of petrale sole fillets treated with ionizing radiation at a level of 200 krad and held at 0.6°C (33°F) was 14 to 21 days.

Pelroy and Eklund[93] packed petrale sole fillets in cans or in Mylar®-polyethylene pouches under vacuum and treated them with ionizing radiation at levels of 100, 200, 300, or 400 krad then held them, along with nonirradiated controls, at 0.6°C (33°F). The controls spoiled in 7 days, the 100-krad-treated fillets spoiled in 14 days, the 200-krad-treated product spoiled in 28 days, the 300-krad-treated samples spoiled in 49 days, and those treated with 400 krad had a storage life of more than 59 days. The spoilage patterns for the controls and the 100-krad-treated fillets were such that both had ammoniacal and putrid odors. The spoilage pattern for the fillets irradiated at levels of 200 or 300 krad was different from that of the controls and was indicated by sour, stale, and acetic odors.

Slavin et al.[194] stated that vacuum packing in a material impervious to oxygen is a requisite for petrale sole fillets which are to be treated with ionizing radiation for purposes of extending the storage life at refrigerator temperatures above freezing. It was also stated that the onset of spoilage is less rapid for vacuum-packed fish treated with ionizing radiation and held under refrigeration than for air-packed fish treated in the same manner. Spinelli et al.[182] found that the pretreatment of petrale sole fillets with tripolyphosphate improved the color of the raw and cooked product after treatment with ionizing radiation and storage but did not affect the texture, odor, or flavor.

Miyauchi et al.[179] treated 12- or 25-lb quantities of petrale sole fillets packed in commercial metal fish cans or in water-resistant paperboard cartons, lined with oxygen-impermeable plastic, with ionizing radiation at levels of 100 or 200 krad. The treated product was shipped to different cities in the U.S. During shipment the temperature of the product varied between −2.2°C (28°F) and 5°C (41°F). The fillets were found to be in good condition upon arrival and had a total storage life of at least 14 days. Petrale sole fillets were also treated with ionizing radiation at levels of 100, 200, or 300 krad and held, along with nonirradiated controls, at 15.5°C (60°F), 10°C (50°F), 7.8°C (46°F), 5.6°C (42°F), 3.3°C (38°F), or 1.6°C (35°F). Little extension of storage life was obtained by irradiating, even at 300 krad, and holding the treated product at 15.5 or 10°C (60 or 50°F). When the holding temperature was 5.5°C (42°F) there was an appreciable extension of the storage life even with fillets treated at a radiation level of 100 krad. The greatest extension of storage life beyond that of the controls was obtained with irradiated fillets held at 1.6°C (35°F).

Miyauchi et al.[183] reported the shelf life of vacuum-packed petrale sole fillets to be 4 to 11 days at 0.6°C (33°F) and 3 to 7 days at 5.5°C (42°F). Fillets of this species treated with ionizing radiation at levels of 100 or 200 krad had a storage life of 20 to 42 days at 0.6°C (33°F) and 12 to 18 days at 5.6°C (42°F).

Ronsivalli et al.[175] reported on taste tests made by a consumer-type panel on frozen control and radiation-treated (200 krad) petrale sole fillets. No appreciable difference was indicated by organoleptic scores for the two products (See Table 3).

Miyauchi[177] reported the storage life of vacuum-packed untreated controls and irradiated petrale sole fillets (100 or 150 krad) held at 0.6°C (33°F) and 5.6°C (42°F). Storage life times were listed as follows: untreated controls 4 to 11 and 3 to 7 days, respectively; radiation-treated fillets 20 to 25 and 3 to 8 days, respectively. Vacuum packaging was found to be a requisite for these fillets treated with ionizing radiation, otherwise the product became rancid in 7 to 14 days.

9. *Yellowtail Flounder (Limanda ferruginea)*

Miyauchi et al.[179] reported that quantities of yellowtail flounder fillets, treated with ionizing radiation at levels of 100 to 200 krad, along with nonirradiated control fillets of the same species, were shipped and refrigerated by wet ice, from Gloucester, Mass., to Seattle, Wash., by commercial transportation. Upon arrival, the control samples were spoiled while the irradiated fillets were in good condition and had an additional storage life of 7 to 14 days.

F. Haddock *(Melanogrammus aeglefinus)*

Nickerson et al.[3] packaged haddock fillets in cellophane bags, treated the product with ionizing radiation at levels of 400, 500, 600 or 700 rep (3 MeV cathode rays), and held them at 2.2 to 4.4°C (36 to 40°F). The products were examined by aerobic TPCs (bacterial) and organoleptically at intervals of 14 days. Irradiation dose levels of 600 to 700 rep were required to prevent unacceptable increases in bacterial counts in a period of 42 days of storage. The results of organoleptic tests showed that the frozen control fillets (stored at −20°C (−4°F)) were always significantly different from the radiation-treated product, but the latter were judged to be acceptable (95% level of significance).

Sieling[188] reported that good quality was retained in haddock fillets when treated with ionizing radiation up to levels of 150 krad. Slavin and Ronsivalli[195] treated haddock fillets with 250 krad of ionizing radiation and stored them at 0.6°C (33°F). During storage the product was periodically deep-fat fried and tested by a consumer-type panel. After 45 days of storage, the panel judged the irradiated product to be acceptable. Fillets irradiated at 250 krad and stored for 14 days at 0.6 to 1.7°C (33 to 35°F) were judged to be "very good" by both an untrained and an expert panel (see Table 5).

Slavin et al.[196] packed haddock fillets in cans without vacuum and treated the product with ionizing radiation at levels of 50, 150, 250, or 350 krad. When held at 5.6°C (42°F) the 50-krad-treated product spoiled in 10 to 20 days while that treated at higher dose levels spoiled in 20 to 30 days. When held at 0.6°C (33°F), the 50-krad-treated product spoiled in 20 to 30 days while that treated at higher dose levels was not spoiled in 30 days. Fillets irradiated at 250 krad and held at 0.6°C (33°F) were judged by taste test to be similar to the frozen control (stored for the same period at −18°C (−0.4°F)) after 20 days of storage. The irradiated fillets held for 30 days at 0.6°C (33°F) were judged to be acceptable (rated as "like slightly"). The fresh nonirradiated control fillets spoiled in 12 days at 0.6°C (33°F). Vacuum packing did not increase the storage life or improve the taste panel scores of radiation-treated haddock fillets.

Nickerson[197] reported that the storage life of fresh haddock fillets is 12 days at 0 to 0.6°C (32 to 33°F) and 5 days at 6.1 to 7.2°C (43 to 45°F). Haddock fillets treated with ionizing radiation at a level of 200 krad were said to have a storage life of 18 days at 0 to 0.6°C (32 to 33°F) and 9 days at 6.1 to 7.2°C (43 to 45°F). Storage life times were determined by aerobic TPCs (bacterial).

Table 5
AVERAGE ORGANOLEPTIC SCORES ASSIGNED BY TRAINED AND UNTRAINED PANELS TO FOUR IRRADIATED SEAFOODS

Irradiated fishery product	Untrained panel	Trained panel
Clam chowder	4.1	3.9
Fried clams	4.3	3.9
Fried haddock	4.2	4.0
Finnan haddie	4.1	4.5

Note: Evaluated on a 5-point scale basis: 5 = excellent; 4 = very good; 3 = good; 2 = fair; 1 = poor.

From Slavin, J. W. and Ronsivalli L. J., U.S. Fish and Wildlife Technological Laboratory for U.S. Atomic Energy Commission Contract No. AT(49-11)-1889, Gloucester, Mass., 1963.

Slavin et al.[98] stated that the optimum dose level for treatment of haddock fillets with ionizing radiation is 250 krad. It was also stated that haddock responded better to treatment with ionizing radiation than did the fillets of pollock or ocean perch. They noted that taste panel testers indicated a greater preference for fish treated with ionizing radiation after deep-fat frying than after cooking by steaming. Irradiated fish products were said to have a much shorter shelf life at 5.6°C (42°F) than at 0.6°C (33°F). It was concluded that the fillets of fish which are to be treated with ionizing radiation for purposes of extending the storage life at refrigerator temperatures above freezing should be held on ice no longer than 7 to 9 days prior to filleting and irradiating the fillets; otherwise the shelf life of the irradiated product is shortened.

Power et al.[152] treated haddock fillets from 2-day-old fish with a chlortetracycline compound or with ionizing radiation at levels of 75, 125, or 250 krad and stored them on ice. Frozen control fillets stored at −26°C (−14.8°F) were judged to be of top quality over a 30-day storage period. The untreated fillets (iced fresh) were judged to be very poor and spoiled in 10 to 16 days. The chlortetracycline and 75-krad-treated fillets were very good after 16 days of storage and spoiled after 23 to 30 days of storage The 125- and 250-krad-treated fillets had lower initial taste test scores than the controls but maintained their quality for 23 days. After 30 days of storage there was a falling off of the quality of fillets treated with ionizing radiation at a level of 125 or 250 krad causing them to be judged as "borderline". It was concluded that the optimum dose of ionizing radiation for haddock fillets is 75 krad.

Slavin and Ronsivalli[166] air packed haddock fillets in No. 10 hermetically sealed cans, treated them with ionizing radiation at a level of 250 krad, and shipped them (refrigerated by wet ice) along with fresh controls (packaged and handled in the same manner) and along with frozen controls (kept frozen with dry ice) to Fort Lee, Va., for consumer-type taste tests. The taste tests indicated that there was no appreciable difference between frozen control and radiation-treated fillets stored for 15 or 29 days.

Ronsivalli and Slavin[198] stated that the optimum dose of ionizing radiation for haddock fillets air packed in cans is 250 krad and that for haddock fillets vacuum packed in cans is 150 krad. Coleby and Shewan[164] stated that the maximum permissible dose of ionizing radiation for haddock fillets is 600 krep. Ronsivalli et al.[85] confirmed the optimum radiation dose level for air or vacuum-packed haddock fillets reported by Ronsivalli and Slavin.[198] They reported that nonirradiated fillets cut from fresh haddock spoiled in 16 days on ice, while haddock fillets cut from 4-day-old fish and treated with 250 krad of ionizing radiation spoiled in 31 days on ice. Fillets cut from 11-day-old fish and irradiated at a level of 250 krad spoiled after 11 days of iced storage. It was concluded that the length of time during which fish are held on ice prior to filleting and treating the fillets with ionizing radiation is important in governing the refrigerated storage life of the irradiated product.

Slavin et al.[169] air-packed haddock fillets in cans, treated them with ionizing radiation at a level of 150 krad, and held this product at 0.6°C (33°F) to 1.6°C (35°F). After storage the fillets were cooked by steaming and compared by expert and nonexpert panelists to freshly frozen fish cooked in the same manner. In one of four tests the irradiated product (stored for 14 and 21 days) was found to be significantly different from the frozen control and was rated as "fair to good" and "less than fair", respectively. In the other three tests there was no significant difference between the frozen and irradiated fillets after storage.

Steinberg[168] reported that the storage life of haddock fillets air packed in cans was 10 to 14 days when held surrounded by melting ice. The storage life of haddock fillets packed in the same manner and treated with ionizing radiation at a level of 250 krad and held at melting ice temperature was found to be 30 days. Connors and Steinberg[199] packed haddock fillets in cans in air and under vacuum and treated them with ionizing radiation at levels of 150 krad (vacuum packed) and 250 krad (air packed). The irradiated product was stored at 0.6°C (33°F), taste tested periodically, and compared to frozen control samples. The taste-test values assigned to each product were analyzed statistically. At the start and after 10 days of storage the frozen and irradiated fillets were not significantly different and both were rated as "like very much". After 20 days of storage the two types of product were significantly different, the irradiated product being rated as "like moderately", the frozen product rated as "like very much". After storage for 30 days the vacuum-packed irradiated and the frozen fillets were both rated as "like moderately". It was stated that the vacuum-packed irradiated fillets and the frozen control fillets were considered to be highly acceptable throughout the 30-day test. No irradiation flavor was noted in the treated fillets, but flavor was given a lower score than that for appearance, odor, or texture.

Slavin et al.[170] reported the optimum dose level of ionizing radiation for haddock fillets to be 150 to 250 krad, and the storage life of fillets treated in this manner to be 21 to 28 days at 0.6°C (33°F). In a consumer-type taste test on haddock fillets irradiated at 250 krad, stored for 21 days at 0.6°C (33°F), and cooked by deep-fat frying, the end product was rated as "like very much" to "like moderately". It was stated that fillets from fish stored 7 to 9 days on ice are suitable for treatment with ionizing radiation but that the shelf life of irradiated fillets starts to decrease when cut from fish held longer than 4 days on ice and that this storage life decreases at an increasing rate with an increase of the age of the iced fish.

Miyauchi et al.[181] reported that haddock fillets cut from 4-day-old fish were packed in commercial metal fish tins, treated with ionizing radiation at a level of 200 krad, iced in barrels, and shipped from Gloucester, Mass., to Seattle, Wash. Nonirradiated control samples were shipped under the same conditions. The temperatures encountered during shipment were 0.6 to 2.8°C (33 to 37°F). Some of the control fillets were spoiled upon arrival and some were dark in color. Some of the irradiated fillets were pinkish in color, some had a strong irradiation odor and flavor, and some did not. When cooked by steaming the irradiated product was rated as "like moderately".

After 5 days of additional storage at 0.6°C (33°F) all of the control fillets were judged to be spoiled. The irradiated fillets were judged to be acceptable after 13 days of additional storage and borderline after 18 days of additional storage at 0.6°C.

Carver et al.[157] treated gutted haddock pre and postrigor with ionizing radiation aboard ship at levels of 50, 150, or 200 krad and stored them in ice. There was no difference in the storage life of the pre and postrigor irradiated fish, but the prerigor irradiated fish were judged to be of better quality organoleptically. The storage life of nonirradiated control fish was about 14 days and that of the irradiated fish 21 to 28 days when held in the iced condition. The irradiated fish held 10 days on ice was comparable in quality to nonirradiated fish held 2 days on ice.

Hembree and Burt[200] carried out troop acceptance tests on haddock fillets which had been treated with ionizing radiation at a level of 250 krad and held for 15 to 29 days at 0.6°C (33°F). The irradiated product was cooked by deep-fat frying and compared to frozen controls stored at −23°C (−10°F) cooked in the same manner. The irradiated and frozen control samples were rated as equal, organoleptically, after both storage periods. Totals of 316 and 272 people, respectively, participated in the tests.

Power et al.[152] stated that the treatment of haddock fillets with ionizing radiation is useful in extending the refrigerated storage life. They found that nonirradiated fillets held at 0°C (32°F) were unacceptable after 12 days and those treated with 75 krad of ionizing radiation were unacceptable after 25 days at this temperature. Fillets irradiated at 150 or 250 krad were of lower quality initially than the nonirradiated or 75-krad-treated fillets, but there was no significant decrease in the acceptability of the fillets treated at the higher dose levels during a storage period of 27 days on ice. The raw irradiated fillets developed an objectionable odor and appearance just prior to becoming objectionable to the taste panel as the cooked product.

Ronsivalli et al.[173] packed haddock fillets in polyethylene pouches and irradiated them at 50, 100, 200 krad. The irradiated fillets, along with fresh controls, were held at 0.6, 5.6, 7.8, or 10°C (33, 40, 46, or 50°F). When tested by an expert panel the storage life of the fresh controls was as follows: 15 days at 0.6°C (33°F); 11 days at 5.6°C (40°F); 7 days at 7.8°C (46°F); and 4 days at 10°C (50°F). The irradiated fillets were judged to have a storage life of 21 to 24 days at 0.6°C (33°F); 15 to 22 days at 5.6°C (40°F); 10 to 14 days at 7.8°C (46°F); and 7 to 14 days at 10°C (50°F), depending upon the level of ionizing radiation applied. Ampola and Ronsivalli[201] held commercially handled haddock on ice for 11 days. On each of the 11 days, fillets were cut from the fish, skinned, packed in cans, treated with ionizing radiation at a level of 250 krad, and held at 0.6 to 1.7°C (33 to 35°F). The irradiated fillets cut from fish held 4 days had a storage life of 29 days, while those cut from fish held 12 days had a storage life of 12 days. The nonirradiated commercially handled fish spoiled in 13 days. Commercially handled fish were also specially washed and iced (including placing ice in the belly cavity). On each of 14 days fillets were cut, skinned, packed in cans, and treated with ionizing radiation at a level of 250 krad. Irradiated fillets cut from specially handled fish held for 12 days had a storage life of 32 days at 0.6°C (33°F), while those cut from specially handled fish held for 14 days had a storage life at 0.6°C (33°F) of 14 days.

Metlitskii et al.[174] reported that the storage life on untreated haddock fillets was 5 to 7 days at 2°C (35.6°F), while haddock fillets treated with ionizing radiation at a level of 600 krad had a storage life of 30 to 40 days at this temperature. They also reported that the storage life of thawed haddock fillets held at 0°C (32°F) was significantly increased by treating with ionizing radiation at levels of 200 to 1500 krad.

Ronsivalli et al.[175] cut fillets from commercially handled haddock, packed them in commercial-type fish tins, and irradiated them at a level of 250 krad. These were shipped under commercial conditions (using trucks and trains), along with nonirradiated control samples, for distances of 2400 to 4800 km (approximately 1500 to 3000 mi). Under these conditions the samples treated with ionizing radiation were judged to have both a high level of quality and a shelf life that was 1 week longer than the similarly handled nonirradiated controls. Ampola and Ronsivalli[64] held haddock on ice for various periods and at 2- or 3-day intervals cut fillets from the fish, packed them in cans, and treated them with ionizing radiation at a level of 250 krad. The irradiated product was stored at 0.6 to 1.6°C (33 to 35°F), and samples were periodically cooked by steaming and compared with freshly frozen control fillets cooked in the same manner. Irradiated fillets from fish held 4 days on ice had a storage life of 13 to 23 days, those from fish held 7 days on ice had a storage life of 11 to 27 days, and those from fish held 11 days on ice had a storage life of 6 to 13 days. It was found that texture and appearance were of less significance in the overall quality score given to fillets treated with ionizing radiation than were odor and flavor. The taste panel was usually able to identify the irradiated and the nonirradiated samples. Fish cut from fillets of marginal quality were not improved by treatment with ionizing radiation.

Carver et al.[8] irradiated eviscerated haddock aboard ship with ionizing radiation at levels of 50 to 100 krad and reported that this treatment doubled the iced storage life. Irradiation of the fish, prerigor, appeared to be more suitable for extending the storage life than did irradiation in the postrigor state. Ronsivalli et al.[176] packed haddock fillets in polyethylene pouches, treated them with ionizing radiation at 50, 100, or 200 krad, and held them, along with nonirradiated control fillets, at various temperatures. The control samples had a storage life of 4 days at 10°C (50°F); 7 days at 7.8°C (46°F); 7 days at 6.7°C (44°F); and 14 days at 3.3°C (38°F). The 50-krad-irradiated samples were spoiled after 9 days at 10°C (50°F); 9 days at 7.8°C (46°F); 14 days at 6.7°C (44°F); and 19 days at 3.3°C (38°F). The 100-krad-treated samples were spoiled after 10 days at 10°C (50°F); 16 days at 7.8°C (46°F); 15 days at 6.7°C (44°F); and 24 days at 3.3°C (38°F). The 200-krad-treated fillets were spoiled after 15 days at 10°C (50°F); 17 days at 7.8°C (46°F); 18 days at 6.7°C (44°F); and 35 days at 3.3°C (38°F).

Miyauchi[177] reported that haddock fillets treated with ionizing radiation at levels of 100 or 200 krad were shipped, along with nonirradiated control fillets, from Gloucester, Mass., to Seattle, Wash. The irradiated fillets were in good condition upon arrival and had an additional storage life of 14 days at 0.6°C (33°F). The nonirradiated fillets were spoiled upon arrival. Learson et al.[56] air packed haddock fillets in cans at Gloucester, Mass., irradiated them at 100 or 200 krad, and shipped them along with nonirradiated control fillets, to Pascagoula, Miss., or to Lansing, Mich., for testing. Samples were also held at Gloucester. Storage temperatures of 15.5, 10, 7.8, 6.7, and 0.6°C (60, 50, 46, 44, and 33°F) were used. The nonirradiated samples were judged to have a storage life of less than 5 to 6 days at 15.5°C (60°F) to greater than 5 to 35 days at 0.6°C (33°F). The 100-krad-treated samples were judged to have a storage life of 5 to 12 days at 15.5°C (60°F) to 37 to 47 days at 0.6°C (33°F). The 200-krad-treated samples were judged to have a storage life of 6 to 15 days at 15.5°C (60°F) to 54 to 56 days at 0.6°C (33°F).

G. Herring, Sea Herring *(Clupea harengus harengus)*

Shewan and Liston[165] stated that sea herring stand up less well to treatment with ionizing radiation, particularly at high-dose levels, than do the lean fish. Inconsistent results were obtained with kippered herring (lightly salted and lightly smoked) possibly owing to the masking of irradiation flavors by the smoke. Rhodes[167] reported on the treatment of herring, vacuum packed in oxygen-impermeable wraps, with ionizing radiation. Dose levels of 100, 300, and 1000 krad were used. Irradiation treatment at these levels was found to have no effect on the appearance or odor of this species. Stansby and Kudo[202] air and vacuum packed herring in cans or in Mylar®-polyethylene pouches, treated them with ionizing radiation at a level of 300 krad, and stored them at 0.6°C (33°F). The vacuum-packed product was only slightly rancid after 14 to 42 days of storage, and rancidity was confined to the dark flesh portions. The air-packed product was slightly rancid after 14 days and quite rancid after 42 days. Coleby and Shewan[164] stated that the maximum permissible dose of ionizing radiation for kippered herring is 1000 krep.

Wittfagel[171] packed Bismarck herring (cured with salt) in plastic pouches and treated them with ionizing radiation at levels of 25 to 200 krad. The product stored at 4°C (39.2°F) was found to have a shelf life only 2 days longer than the nonirradiated controls. Bismarck herring treated with ionizing radiation at levels above 75 krad were found to have an off flavor. Schonborn et al.[203] cured herring fillets with acetic acid, salt, and spices, irradiated them with 160 krad of ionizing radiation, and held them, along with nonirradiated controls, at 15°C (59°F). It was found that the irradiated product had a storage life 33 to 99 days longer than the nonirradiated controls, provided that they were treated with ascorbic acid, sorbitol, and dry spices prior to irradiating.

H. Ling *(Molva molva)*

Coleby and Shewan[164] reported that the maximum permissible dose of ionizing radiation for ling is 500 to 1000 krep.

I. Mackerel

1. Atlantic Mackerel (Scomber scombrus)

Shewan and Liston[165] reported that mackerel held up less well to treatment with ionizing radiation than do lean fish. Slavin and Ronsivalli[166] treated mackerel fillets with ionizing radiation at various dose levels and concluded that the optimum level of treatment is 250 krad. Product packed in cans and treated with the optimum dose had a storage life of at least 30 days at 0.6°C (33°F). When held in this manner, cooked by broiling, and compared to frozen controls (stored at −18°C [−0.4°F] for 30 days), prepared in the same manner, the two products were not appreciably different. Coleby and Shewan[164] reported that the maximum permissible dose of ionizing radiation for mackerel is 1000 krep. Slavin et al.[170] reported that the optimum dose of ionizing radiation for mackerel is 250 krad and that this treatment provides a shelf life at 0.6°C (33°F) of 28 to 35 days.

2. Chub Mackerel (Rastrelliger neglectus)

Goresline[204] stated that in Thailand, chub mackerel are boiled in salt water and held at ambient temperatures (about 30°C [86°F]). At this temperature the storage life of the cooked product is only 2 days. When treated with ionizing radiation at levels of 100, 200, or 300 krad and held at ambient temperatures, this product was judged to be acceptable after 10, 15, and 17 days, respectively.

J. Ocean Perch

1. Ocean Perch (Sebastes marinus)

Lerke et al.[80] stated that there may be a problem with the development of rancidity in medium fat fish, such as ocean perch, when treated with ionizing radiation to extend the refrigerated storage life. It was also stated that concurrent treatment with chlortetracycline and ionizing radiation might further extend the storage life of this and other species. Slavin and Ronsivalli[195] reported that the optimum dose of ionizing radiation for ocean perch fillets packed in cans is 250 krad. The shelf life of product treated in this manner was greater than 30 days at 0.6°C (33°F) and about half that at 7.8°C (46°F). This was confirmed in another report (Slavin and Ronsivalli[166]) and it was further reported that the irradiated product was not appreciably different from the frozen controls held for the same period of time at −18°C (−0.4°F). Slavin et al.[98] found that ocean perch fillets treated with ionizing radiation at a level of 250 krad had a much greater acceptability when cooked by deep-fat frying than when cooked by steaming. Ronsivalli and Slavin[198] found that the optimum dose of ionizing radiation for ocean perch fillets packed in cans in air is 250 krad and that when packed in cans under vacuum the optimum irradition dose is 150 krad. Air-packed irradiated ocean perch fillets became slightly rancid after 30 days of storage at 0.6°C (33°F). The lower dose level required for vacuum-packed ocean perch fillets was not due to the fact that treatment at higher levels gave bad results, but because higher treatment levels were not required to extend the storage life when this product was packed under vacuum.

Coleby and Shewan[164] stated that the maximum permissible dose of ionizing radiation for ocean perch fillets is 800 krep. Steinberg[168] reported that ocean perch fillets, packed in cans without vacuum and treated with ionizing radiation at a level of 250 krad, had a storage life of 30 days at 0.6°C (33°F) and that the nonirradiated control fillets, otherwise handled in the same manner, had a storage life of 12 to 14 days at this temperature. Slavin et al.[170] stated that the optimum irradiation dose for ocean perch fillets is 150 to 250 krad and that the storage life at 0.6°C (33°F) of product treated in this manner is 28 days. Ampola et al.[90] treated air-packed ocean perch fillets in cans with ionizing radiation at 250 krad and tested them, along with frozen control fillets, after deep-fat frying. Testing was carried out by an expert panel after these products had been stored for 10, 20, and 30 days (the irradiated product was held at 1°C (34°F), the frozen one at −18°C (−0.4°F)). The results of taste tests, analyzed statistically, indicated that there was no significant difference between the fillets preserved by irradiation treatment and those which were frozen.

2. Pacific Ocean Perch (Sebastodes alutus)

Miyauchi et al.[181] packed 25-lb quantities of ocean perch fillets in polyethylene bags or vacuum packed them in Scotchpak®. These products were treated with ionizing radiation at a level of 200 krad and held for 7 or 14 days at 0.6°C (33°F). After each period the treated fillets were repacked (in 1-lb units) under vacuum in Scotchpak® or on paperboard trays overwrapped with cellophane. After holding the bulk packs for 7 days and repacking, all were of marketable quality after 5 days at 0.6°C (33°F), but the Scotchpak® product was of better quality. After holding the bulk packs for 14 days and repacking, the cellophane-wrapped product spoiled in 4 days, and the Scotchpak® samples spoiled in 7 days at 0.6°C (33°F). Pacific ocean perch fillets of different quality were also packaged and treated with ionizing radiation at levels of 100 or 200 krad. It was found that high quality fillets with high initial bacterial counts had a storage life of 8 and 19 days, respectively, at 0.6°C (33°F). High-quality fillets with low-initial bacterial counts had a storage life of 13 and 28 days, respectively, at this temperature. Low quality, high initial bacterial count fillets had storage life times of 7 and 17 days at 0.6°C (33°F) and low quality, low bacterial count fillets had storage life times of 8 and 18 days at this temperature.

Spinelli et al.[182] found that treatment of Pacific ocean perch fillets with tripolyphosphate prior to subjecting to ionizing radiation improved the color of the raw and cooked fillets but did not improve texture, odor, or flavor. The air-packed fillets treated with tripolyphosphate became rancid during storage as did those not treated with this compound. Miyauchi et al.[179] packed 12- or 25-lb quantities of Pacific ocean perch fillets in commercial-type fish tins or in water-resistant paperboard cartons lined with a plastic impermeable to oxygen and treated these products with ionizing radiation at levels of 100 or 200 krad. These fillets were shipped to different cities in the U.S. A temperature range of −2.2 to 5°C (28 to 41°F) was encountered by the product during shipment. The fillets were in good condition upon arrival at their destinations and had an additional refrigerator storage life of 14 or more days. Miyauchi et al.[183] reported that the storage life of untreated Pacific ocean perch fillets at 0.6°C (33°F) to be 6 to 7 days and that of fillets of this species treated with ionizing radiation at levels of 100 or 200 krad to be 25 to 28 days at this temperature. It was also stated that Pacific ocean perch fillets which are to be treated with ionizing radiation to extend the storage life time at refrigeration temperatures above freezing must be vacuum packed in an oxygen-impermeable material, otherwise they become rancid during storage.

K. Pollock *(Pollachius virens)*

Shewan[163] vacuum packed pollock fillets in plastic material impervious to oxygen and treated them with ionizing radiation at a level of 250 krad. It was found that product handled in this manner had a storage life 10 to 20 days longer at 1.5°C (34.7°F) than did the nonirradiated product packaged and held in the same manner. Carver and Steinberg[162] stated that the maximum acceptable dose of ionizing radiation and the dose of greatest acceptability for pollock fillets is about 230 krad.

Slavin and Ronsivalli[195] packed pollock fillets in cans without vacuum and treated them with ionizing radiation at various levels to determine the optimum dose and the storage life times of product treated at this level. The fillets were held at 0.6°C (33°F) or at 7.8°C (46°F). The product was tested together with frozen control fillets stored at −21°C (−5°F). The optimum dose level for pollock was stated as 150 krad. It was also stated that the nonirradiated control samples were always rated higher than that of the irradiated product by taste panels. There was a noticeable darkening of the product irradiated at 150 krad. Spoilage times for this product were greater than 30 days at 0.6°C (33°F) and less than 20 days at 7.8°C (46°F). Slavin and Ronsivalli[166] confirmed the optimum dose level of ionizing radiation for pollock previously reported and stated that taste panels found the cooked irradiated pollock fillets were not appreciably different from the cooked frozen control samples.

Slavin et al.[98] reported that haddock responded better to treatment with ionizing radiation than did pollock and that irradiated pollock had a much shorter shelf life at 7.8°C (46°F) than at 0.6°C (33°F). Irradiated fish was found to have a much greater acceptability when cooked by deep-fat frying than when cooked by steaming. It was stated that the shelf life of irradiated fish fillets is shortened when they are cut from fish held for longer than 7 to 9 days on ice. Coleby and Shewan[164] found the maximum permissible dose of ionizing radiation to be 500 to 700 krep for pollock. When treated at levels of 75 to 250 krep, the extension of the storage life of pollock fillets held at 0 to 1.5°C (32 to 35°F) beyond that of the controls was found to be 10 to 20 days. Ronsivalli et al.[85] stated that when packed in cans or in oxygen-impermeable plastic, with or without vacuum, the optimum dose of ionizing radiation for pollock fillets is 150 krad.

Steinberg[168] reported the storage life of pollock fillets packed in cans in air and treated with ionizing radiation at a level of 150 krad to be 30 days at the temperature of melting ice. Nonirradiated control fillets packed and held in the same manner had a storage life of 12 to 14 days. Slavin et al.[170] stated that the optimum dose of ionizing radiation for pollock fillets is 150 krad and that the storage life of product treated in this manner and held at 0.6°C (33°F) is 28 days. Metlitskii et al.[174] found that at 2°C (35.6°F) the storage life of pollock fillets (untreated controls) was 7 to 10 days and that when these fillets were treated with 600 krad of ionizing radiation the storage life at this temperature was 30 days. Ampola et al.[90] found that pollock fillets packed in cans, in air or under vacuum, and treated with ionizing radiation at a level of 150 krad did not differ in quality from frozen control fillets after holding the irradiated product for extended periods at 1°C (33.8°F). The frozen controls were stored at −18°C

(−0.4°F) for the same period of time as that of the irradiated product. The results of taste tests, analyzed statistically, showed that there was no significant difference between the frozen and the irradiated fillets.

L. Pomfret, Black Pomfret *(Stromateus niger);* Silver Pomfret *(Stromateus cinereus)*

Aiyar[161] reported that treatment of black and silver pomfret with ionizing radiation at dose levels of 100 to 200 krad increased the storage life of these fish at refrigerator temperatures above freezing three to fourfold.

M. Rockfish, Black Rockfish *(Sebastes melanops)*

Eukel and Huber[187] treated rockfish with ionizing radiation at a level of 250 krad (8 MeV electrons) and found that this treatment extended the storage life of the product to about 20 days at 5.6°C (42°F). Miyauchi[177] irradiated black rockfish at sea with 50 or 100 krad of ionizing radiation. The treated fish had better storage characteristics on ice than did the nonirradiated control fish. Prerigor irradiated fish had better iced storage stability than did postrigor irradiated fish.

N. Sablefish *(Anopoploma fimbria)*

Stansby and Kudo[202] vacuum and air packed sablefish portions in cans and in polyethylene-Mylar® film, treated them with 300 krad of ionizing radiation, and stored them, along with nonirradiated control portions, at 0.6°C (33°F). Frozen control portions were prepared and stored at −29°C (−20°F). The frozen controls were not significantly different from the freshly cut sablefish after 42 days of storage. The fresh nonirradiated controls spoiled after a period of 14 to 21 days. The air-packed irradiated sablefish was rancid after 7 days of storage, but the vacuum-packed irradiated fish portions were acceptable after 14 days of storage at 0.6°C (33°F). The average quality of the irradiated fish did not stand up during 42 days of storage at 0.6°C (33°F).

O. Argentine, Herring Smelt *(Argentina silus)*

Carver et al.[157] irradiated smelt, prerigor, aboard ship with ionizing radiation at levels of 50 or 100 krad and held them on melting ice. The storage life was increased from 16 (nonirradiated controls) to 21 days. When reirradiated with a dose of 100 krad, the storage life was increased to 20 days. It was stated that smelt are not affected organoleptically by treatment with ionizing radiation at levels of 50 or 100 krad.

P. Tuna (No Species Indicated); Big Eye Tuna *(Thunnus obesus)*

Coleby and Shewan[164] stated that the maximum permissible dose of ionizing radiation that can be applied to tuna is 1000 krep. Amano and Yamanaka[205] reported that during frozen storage tuna becomes discolored because of the oxidation of myoglobin to metmyoglobin. When oxidized tuna was treated with ionizing radiation at levels of 200 to 1000 krad, there was a 37 to 39% regeneration of the myoglobin in tuna containing 75% of the myoglobin pigment as metmyoglobin. This treatment was sufficient to change the color to bright red, judged organoleptically. Off-flavors were induced in tuna irradiated at levels above 200 krad. When air packed and stored for 168 hr under refrigeration after irradiation treatment, there was an almost complete reversion of the regenerated myoglobin to metmyoglobin, but when stored for 168 hr under vacuum after irradiation treatment, only about 15% of the regenerated myoglobin reverted to metmyoglobin.

Q. Whiting *(Merluccius bilinearis)*

Brooke and Steinberg[13] reported that when treated with ionizing radiation the best acceptability with whiting was obtained when this product was treated at levels of 450 krad or less. Carver and Steinberg[162] found that the maximum acceptable dose of ionizing radiation for whiting was approximately 115 krad. Coleby and Shewan[164] stated that the maximum permissible dose of ionizing radiation for whiting is 200 to 250 krep. Massa et al.[206] irradiated whiting with 300 krad of ionizing radiation and held the treated fish along with nonirradiated controls, at 2°C (35.6°F). The nonirradiated control fish were inedible after 11 days of storage. The irradiated fish kept well but had a slight irradiation odor.

R. Summary

It is evident that the treatment of a number of marine fish species with ionizing radiation could be used advantageously for purposes of extending the storage life at refrigerator temperatures above freezing. This treatment would provide for the availability of fresh fish of good quality in inland areas of the U.S. and certain other countries. There are also some fish species which cannot be shipped to inland areas of different countries because of spoilage under refrigeration during the period required for shipment. Treatment of such species with ionizing radiation could change this situation by extending the refrigerated storage life sufficiently to allow shipment, sale, and consumption of fish to and in remote areas. Tests have been carried out under commercial conditions which confirm this.

The treatment of fish products with ionizing radiation might also stabilize the marketing of marine fish species in that it would provide a more constant supply of the fresh product. Considering the fact that the U.S. and certain other countries have imposed a 200-mi fishing limit for foreign boats, a method of increasing the refrigerated storage life of fresh fish could, and probably will, become of great usefulness in increasing the consumption of marine species and thus prevent market gluts, hence economic losses.

The treatment of fish with ionizing radiation at optimum dose levels generally does not affect the texture or appearance of these products, but it may have some effect on the odor of the raw and cooked fish and on the flavor of the cooked fish. At optimum irradiation dose levels there may be a slight loss of the typical flavor of irradiated fresh fish, but at maximum and certainly at higher dose levels, unless irradiated in the frozen state under high vacuum, fish treated with ionizing radiation develop a "burnt", "irradiation" odor and flavor. Taste panel results (expert and consumer types), however, indicate that when suitable dose levels are applied there is little organoleptic difference between fresh fish treated with ionizing radiation and untreated fresh fish.

The literature indicates that there is a maximum and an optimum dose of ionizing radiation which can be applied to marine fish. The optimum dose level varies for different species between 100 and 250 krad. A table indicating the optimum dose level for fish and shellfish and the refrigerated storage life provided by this treatment has been included (see Table 6).

In addition to the evident efficacy of the radiation treatment of marine fish as fillets or steaks to extend the refrigerated storage life, the literature indicates that eviscerated but otherwise uncut or whole fish could be treated with ionizing radiation at levels of 50 to 100 krad aboard ship to provide a better quality and longer storage life for the iced product. However, owing to the weight and space required for shielding the radiation sources necessary for such treatment, there is little probability that this method of preservation will be applied in American ships in the near future. This method of preservation could be applied by certain foreign ships associated with fishing, which are larger than American fishing boats and have more available space. The literature indicates that the best results, in irradiating eviscerated fish aboard ship, are obtained when they are treated prerigor.

Table 6
OPTIMAL RADIATION DOSE LEVELS AND SHELF LIFE AT 33°F (0.6°C) FOR FISH AND SHELLFISH AEROBICALLY PACKED IN HERMETICALLY SEALED CANS

Seafood	Optimal irradiation dose (Mrad)	Shelf life (weeks)
Oysters (shucked)	0.20	3—4
Shrimp	0.15	4
Smoked chub	0.10	6
Yellow perch fillets	0.30	4
Petrale sole fillets	0.20	2—3
Pacific halibut steaks	0.20	2
King crab meat (cooked)	0.20	4—6
Dungeness crab meat (cooked)	0.20	3—6
English sole fillets	0.20—0.30	4—5
Soft-shell clam meats	0.45	4
Haddock fillets	0.15—0.25	3—4
Pollock fillets	0.15	4
Cod fillets	0.15	4—5
Ocean perch fillets	0.15—0.25	4
Mackerel fillets	0.25	4—5
Lobster meat (cooked)	0.15	4

From Slavin, J. W., Nickerson, J. T. R., Goldblith, S. A., Ronsivalli, L. J., Kaylor, J. D., Licciardello, J. J., *Isot. Radiat. Technol.*, 2, 365, 1966.

Regarding the extension of the refrigerated storage life of fillets or steaks cut from fish by treatment with ionizing radiation, it has been found that low-fat fish, such as cod, haddock, or pollock, should be prepackaged in an oxygen-impermeable container such as a hermetically sealed can or a plastic which prevents moisture loss and diffusion of oxygen. Fillets or steaks from fish higher in fat such as English sole, halibut, or petrale sole should be prepackaged under vacuum in the type of containers previously indicated, otherwise these products become rancid during refrigerated storage after treatment with ionizing radiation.

It may be advantageous to pretreat marine fish fillets or steaks with a solution of sodium tripolyphosphate prior to packaging and irradiating in order to prevent "drip" or a loss of fluids during refrigerated storage.

It is evident that the treatment of fillets or steaks with ionizing radiation under conditions in which the fish portions have been cut from fish which are alredy spoiled or in the spoiling stage will do nothing to improve the quality of such products. The best results will be obtained by the irradiation treatment of portions cut from very fresh fish.

The temperature at which fish fillets or steaks are held after packaging and treating with ionizing radiation is important in that it governs the storage life of the treated product. Such irradiation-treated materials should be held at temperatures as near the freezing point, without freezing, as is possible. The higher the temperature at which fish portions are held postirradiation, the shorter the storage life and little advantage is gained by irradiating fish products and holding them at high refrigeration temperatures. It has been shown that fish fillets treated with ionizing radiation and held at 0.6°C (33°F) have about twice the storage life as that of the same product held at 5.6°C (42°F).

Better results have been obtained with fillets or steaks from white-fleshed than those from dark-fleshed fish by treatment with ionizing radiation. It is also evident that the dark tissues of such species as sablefish may become rancid during refrigerated storage after preservation treatment of this kind.

X. FRESH-WATER AND ANADROMOUS FISH

A. Catfish, Channel Catfish *(Ictalurus punctatus)*

Sieling[188] reported that catfish treated with a dose of 500 krad of ionizing radiation had an abnormal odor, but the color and taste of the cooked product was quite acceptable. Emerson et al.[207] treated channel catfish with a dose of 100 to 600 krad of ionizing radiation in increments of 100 krad and also at 800 krad. The treated samples were breaded, deep-fat fried, taste tested by expert and consumer-type panels, and rated organoleptically. Fresh controls were also tested. The fresh controls were rated as "excellent", the fish irradiated at 100 krad was scored as "like slightly" to "like moderately", and that irradiated at 400 krad as "neither like nor dislike" to "like slightly". Cut catfish was held as a control at 0°C (32°F) along with the same type of fish which had been treated with ionizing radiation at levels of 200 or 300 krad. The storage life of these products, as determined by total plate bacterial counts, was 4, 20, and 27 days, respectively.

B. Chubs *(Coregonus* sp.)

Emerson et al.[208] treated smoked chubs with ionizing radiation at levels of 100 or 400 krad and held them, along with nonirradiated controls, at 2.2°C (36°F). The nonirradiated product was spoiled after 35 days of storage while the irradiated chubs were acceptable throughout a 42-day storage period. Emerson et al.[209] irradiated frozen −18°C (−0.4°F) raw chubs at 300, 780, 1000, and 1500 krad of ionizing radiation then defrosted, salted, and smoked them along with freshly defrosted controls. The smoked chubs were then rated organolepticaly. The controls were rated as "like very much", the product treated at 300 krad was rated as "like moderately" to "like very much", and the product treated at 780 krad was rated as "like slightly" to "like moderately". Chubs irradiated at the higher dose levels were judged to be unacceptable. Slavin et al.[17] stated that the optimum irradiation dose for smoked chubs is 100 krad and that the storage life of this product treated with ionizing radiation at this dose level is 42 days at 0.6°C (33°F).

C. Lake Herring *(Coregonus artedii)* Lesueur

Emerson et al.[210] treated lake herring fillets with ionizing radiation at a level of 300 krad and held them, along with nonirradiated controls, at 0.6°C (33°F). The controls were spoiled in 8 to 13 days. The irradiated fillets did not spoil in this period but were judged to be unacceptable after 8 days due to a loss of flavor.

D. Yellow Perch *(Perca Flavescens)*

Emerson et al.[209] treated perch fillets with ionizing radiation at levels of 200, 400, or 800 krad and had them tested organoleptically. The fillets treated with 800 krad had an irradiation off-flavor, but those treated at lower dose levels did not. Yellow perch fillets were also treated with ionizing radiation at levels of 300 to 1300 krad in various increments and tested organoleptically as the deep-fat fried or baked product. Irradiation flavor was more noticeable in the baked than in the deep-fat fried fillets, but the irradiation flavor was not objectionable in product irradiated at levels as high as 500 krad. Nonirradiated fillets spoiled in 7 to 14 days at 2.2°C (36°F), but the irradiated fillets were not spoiled after 3 weeks at this temperature. When nonirradiated fillets and fillets treated with ionizing radiation at levels of 300 or 600 krad were held at 5.6°C (42°F) the nonirradiated fillets were unacceptable after 5 to 8 days, the

300-krad-treated fillets were unacceptable after 19 to 22 days, and those treated with 600 krad were unacceptable after 42 days.

Emerson et al.[208] packed yellow perch fillets in "Cryovac®" bags (type S-304), placed them in screw-top glass jars, and treated them with ionizing radiation at levels of 200, 500, or 1000 krad. The treated samples, along with nonirradiated controls, were held at 1.1 to 2.2°C (34 to 36°F). The controls spoiled in 7 to 14 days whereas the irradiated fillets were acceptable after 21 days, and when taste tested there was no strong objection to the fillets irradiated at the highest dose level. In another test it was found that when cooked by deep-fat frying yellow perch fillets which had been treated with ionizing radiation at a level of 1000 krad were acceptable and when cooked by baking they were not objectionable after treatment at a level of 500 krad. Yellow perch fillets treated with 300 krad of ionizing radiation and stored at 0.6°C (33°F) were acceptable after holding for 44 days, and fillets treated with 600 krad were acceptable after holding for 51 days at this temperature. The nonirradiated controls spoiled in 13 days at 0.6°C (33°F).

Emerson et al.[211] treated yellow perch fillets with ionizing radiation at levels of 300 or 600 krad and held them at 0.6 or 5.6°C (33 or 42°F). Nonirradiated fillets were also held at these temperatures. Irradiation treatment increased the storage life four to fivefold. Storage of irradiated fillets at 5.6°C (42°F) greatly decreased the storage life time as compared to this product held at 0.6°C (33°F). A flavor difference was detected between the nonirradiated and samples irradiated at 600 krad. At 0.6°C (33°F) controls spoiled in 8 days, the 300-krad treated in 47 days, and the 600-krad treated in 55 days of storage. Slavin et al.[170] stated that the optimum dose of ionizing radiation is 300 krad for yellow perch fillets and that at 0.6°C (33°F) the storage life of these fillets irradiated at this level is 28 days.

Graikowski et al.[212] packaged yellow perch fillets in polyethylene bags, treated them with ionizing radiation at 100 krad, and stored them at 37°C (98.6°F). Samples were stored in air, in an atmosphere of 1.5% oxygen and 6% carbon dioxide or in an atmosphere of 1.5% oxygen and 12% carbon dioxide. The fillets stored in air spoiled in 3 days while those stored in a controlled atmosphere spoiled in 7 days. This experiment was repeated with the fillets being treated with ionizing radiation at levels of 100 or 300 krad and stored at 0.6°C (33°F). Nonirradiated controls were also prepared. Control fillets held in air spoiled in 10 days, and those held under controlled atmospheres spoiled in 13 days. Product irradiated at 100 krad and held in air was of borderline quality in 13 days and unacceptable after 24 to 28 days when held in a controlled atmosphere. The 300-krad-treated samples were of lower quality at the start than the controls or 100-krad-treated fillets but remained at this level of quality, in air or under a controlled atmosphere, throughout the 28-day storage period. It was concluded that there is a significant extension of the storage life of yellow perch fillets when they are stored in a controlled atmosphere and that the condition of the fresh fish at the start affects the storage life of yellow perch fillets treated with ionizing radiation.

Graikowski et al.[213] treated yellow perch fillets with ionizing radiation at levels of 100 or 200 krad and stored them, along with nonirradiated control fillets, at 10, 5.6, or 2.2°C (50, 42, 36°F). Spoilage times were determined by odor by expert and consumer-type panelists. At 10°C (50°F) the controls were judged to be spoiled in 3 to 12 days, the 100-krad treated in 12 to 15 days, and the 200-krad-treated in 8 to 19 days. At 5.6°C (42°F) the controls spoiled in 7 to 12 days, and the irradiation-treated samples spoiled in 19 to 26 and 20 to 38 days, respectively. At 2.2°C (36°F) the nonirradiated fillets were judged to be spoiled in 10 to 32 days and the radiation-treated samples were judged to be spoiled in 33 to 39 and 27 to 44 days, respectively. It was concluded that with yellow perch fillets treated at a level of 200 krad with ionizing radiation, bacterial growth did not produce the characteristic putrid odors commonly associated with the spoilage of the fresh (nonirradiated) product.

E. Salmon

1. Salmon (Species Not Identified)

Eukel and Huber[187] treated salmon steaks with ionizing radiation at a level of 150 krad (8 MeV) and determined the storage life by testing for volatile reducing substances. The storage life of the irradiated product was extended to about 20 days at 5.6°C (42°F). Sieling[188] stated that there was a distinct color loss in salmon steaks treated with ionizing radiation at levels of 200, 500, or 1000 krad. Coleby and Shewan[164] stated that the maximum permissible dose of ionizing radiation for salmon is less than 500 krep. Slavin et al.[194] stated that the bleaching of carotenoid pigments in salmon subjected to treatment with ionizing radiation may limit the acceptability of this method of processing for this group of fish. Metlitskii et al.[174] reported that when stored at 2.2°C (36°F) the storage life of salmon is increased from 3 to 4 to 20 days by treatment with ionizing radiation at levels of 150 krad.

2. King Salmon (Oncorhynchus tshawytscha); Chum Salmon (O. keta); Pink Salmon (O. gorbuscha); Silver Salmon (O. kisutch); and Sockeye Salmon (O. nerka)

Stansby and Kudo[202] packed salmon steaks in cans or in Mylar®-polyethylene pouches under vacuum or in air, treated them with ionizing radiation at a level of 300 krad, and held them at 0.6°C (33°F). After irradiation treatment, the dark tissues of king salmon were slightly rancid whether or not they were packed in air or under vacuum. The chum salmon was rancid after irradiation treatment and storage for 7 days when packed under vacuum. The pink salmon was rancid after irradiation treatment in air and after irradiation treatment and storage for 7 days when packed under vacuum. The silver and sockeye salmon were rancid both when packed in air and under vacuum immediately after irradiation treatment.

3. Fourfinger, Threadfin (Eleutheronesma tetradactylum)

Aiyar[161] stated that the storage life of this species is increased three to fourfold by treatment with ionizing radiation at levels of 100 to 250 krad.

F. Trout

1. Trout (No Species Indicated)

Jorgensen and Hanson[89] packed gutted trout under vacuum in cans, treated them with ionizing radiation at levels of 50, 100, or 200 krad, and stored them at 0°C (32°F). The nonirradiated control fish were spoiled in 21 days. The irradiated fish were of first class quality for only 21 days but were still edible after storage for 28 days. Hansen[214] stated that vacuum packing is valuable in preventing the development of rancidity in gutted trout which are to be treated with ionizing radiation to extend the refrigerated storage life. A short dip in an 0.2% solution of ascorbic acid was also found to be helpful in this respect.

2. Rainbow Trout (Salmo gairdneri)

Stansby and Kudo[202] air and vacuum packed rainbow trout in cans and in Mylar®-polyethylene pouches, irradiated the fish at 300 krad, and stored them at 0.6°C (33°F). The air-packed samples were rancid after 7 days of storage, but the vacuum-packed fish were only slightly rancid after storage for 42 days.

3. Lake Trout (Salvelinus namaycush)

Graikowski et al.[212] treated lake trout with ionizing radiation at levels of 100, 300, 400, 500, 600, 700, 800, and 1000 krad. There was no significant bleaching of color in trout which originally were deep red in color when these were treated at levels as high as 700 krad. Beyond this dose level there was significant to complete bleaching of color. In trout which originally were light red or pink in color, there was no significant bleaching even when treated with ionizing radiation at a level of 1000 krad. Lake trout were treated with 200 or 300 krad of ionizing radiation and held, along with nonirradiated control samples, at 0.6°C (33°F). The controls were judged to be putrid after 8 days. The irradiated samples were judged to be unacceptable after 26 days of storage. No typical putrid odors were observed in the samples treated with ionizing radiation. The optimum irradiation dose for lake trout was judged to be 300 krad.

G. Whitefish *(Coregonus clupeaformis)*

Ostovar et al.[215] treated eviscerated whitefish with ionizing radiation at 75, 150, or 300 krad and held them in aluminum boxes at 0°C (32°F). Fresh controls were held in the same manner and frozen controls were prepared and held at −30°C (−22°F). Samples were tested by odor in the raw state and taste tested after cooking by baking. After irradiation treatment the samples were judged to be acceptable, except for the fish treated at 300 krad which had a "burnt" odor and which changed in color and texture after 1 day of storage. The fresh control spoiled in 16 days and after this period the 75-krad-treated samples had spoilage odors. The frozen control was slightly discolored but, with one exception, was acceptable throughout the experiment. All irradiated samples received lower initial scores than did the controls. The "burnt" odor in the samples treated with 300 krad of ionizing radiation was not noted after 8 days of storage. A dose level of 150 krad was judged to be most suitable for the treatment of whitefish with ionizing radiation. This treatment extended the acceptable storage life of the product to 29 days at 0°C (32°F).

Over the course of the experiment there was a decline in the composite grades of the irradiated samples. This decline was less marked than that which occurred in the fresh controls but was more noticeable than that which occurred in the composite grade of the frozen control fish. It was also stated that the storage life of whitefish is extended considerably by treatment with low-dose levels of ionizing radiation. Product treated at 150 krad was acceptable after 29 days on ice and that treated with 75 krad was acceptable after 22 days on ice. The 300-krad-treated fish received lower sensory scores than those which were treated at lower dose levels.

Graikowski et al.[212] treated whitefish with ionizing radiation at various levels below 900 krad and compared them with nonirradiated fresh or frozen controls. Products treated at 300 krad or less were given organoleptic scores comparable with those given to the fresh controls. In samples held at 0.6°C (33°F) the nonirradiated controls were judged, in the raw state, to be putrid after 12 days and unacceptable as the cooked fish in 15 days. The fish treated at a level of 200 krad had no off-odor raw or cooked, but were judged to be unacceptable after the 15th day of storage. The fish treated with ionizing radiation at a level of 300 krad were judged to be unacceptable in the raw state after the 29th day of storage but were acceptable organoleptically (cooked) after 26 days of storage. Putrid odors were not observed in the raw whitefish which had been treated with 200 krad or at higher dose levels of ionizing radiation. The organoleptic ratings given the 200- or 300-krad-treated whitefish were comparable to those assigned to the frozen control samples.

H. Summary

The treatment of fresh-water or anadromous fish with ionizing radiation to extend their storage life at refrigerator temperatures above freezing may be less advantageous than for salt-water species because of the limited availability of the various species. If sufficient quantities of some fresh-water fish can be obtained to warrant a wider distribution, there may be good reasons to apply radiation preservation to these types;

yellow perch, whitefish, and lake trout would be included in this group. Catfish, chubs (as smoked), and trout are already widely distributed and radiation preservation might be helpful in providing a product of better quality to the consumer as far as these species are concerned. According to results reported in the literature, there would appear to be no advantage in the treatment of lake herring with ionizing radiation because processing in this manner causes an extreme loss of the typical flavor of this fish. It has also been shown that salmon are not suitable for radiation preservation treatment owing to the development of rancidity, even in vacuum-packed samples, and because of a bleaching of the carotenoid pigments when this treatment is applied.

Radiation preservation of some fresh-water or anadromous fish, when applied at optimum (low-dose) levels, does not affect the color or texture (catfish, chubs, yellow perch, trout, lake trout, or whitefish), but it does affect the color of salmon causing it to bleach. There may be a slight lessening of the typical flavor of all species of the type indicated when treated with ionizing radiation at optimum dose levels and this may be excessive in lake herring. Salmon becomes rancid in refrigerated storage when treated with ionizing radiation at optimum (low-dose) levels.

In treating fresh-water fish with ionizing radiation to extend the refrigerated storage life, the most suitable level of treatment has been found to be in the range of 100 to 300 krad.

Fresh-water fish which are to be preserved with ionizing radiation should be prepackaged in an oxygen and moisture-impermeable material and those species, such as trout, which may become rancid during refrigerated storage, should be prepackaged under vacuum.

As is the case with salt-water fish, fresh-water or anadromous fish or fillets or steaks cut from these fish should be strictly fresh when treated with ionizing radiation since processing in this manner will do nothing to improve the quality of a spoiled or spoiling fish, and the refrigerated storage life of product of poor quality will be significantly shorter than that of fish of good quality, after radiation preservation treatment.

The refrigeration temperature above freezing at which fresh-water fish are held after treatment with ionizing radiation to extend storage life is of great importance in that spoilage occurs much faster in product held in the higher temperature ranges. It has been shown that fresh-water fish treated with ionizing radiation at low-dose levels spoils about twice as fast when held at 5.6°C (42°F) as when held at 0.6°C (33°F).

XI. MOLLUSCS, BIVALVE MOLLUSCS

A. Clams

1. Soft-Shelled Clams (Mya arenaria)

Nickerson[197] reported that the storage life of untreated soft-shelled clam meats is 6 days at 0 to 0.6°C (32 to 33°F) and 4 days at 6.1 to 7.2°C (43 to 45°F). Product of this kind treated with 200 krad of ionizing radiation had a storage life of 17 days at 0 to 0.6°C (32 to 33°F) and 13 days at 6.1 to 7.2°C (43 to 45°F). When treated with 800 krad of ionizing radiation, clam meats had a storage life of 85 days at 0.6°C (33°F) and 60 days at 6.1 to 7.2°C (43 to 45°F). When taste tested by a nonexpert panel after deep-fat frying, the results, analyzed statistically, indicated that there were no significant differences in the grades assigned to fresh control samples (not stored) and product treated with 800 krad of ionizing radiation and stored for 85 days at 0°C (32°F) or stored for 40 days at 6.1 to 7.2°C (43 to 45°F). Also unstored control samples and clam meats treated with ionizing radiation at a level of 450 krad and stored at 14 days at 0°C (32°F) or 12 days at 6.1 to 7.2°C (43 to 45°F) were not significantly different when cooked by deep-fat frying.

Slavin and Ronsivalli[195] found that the optimum dose level of ionizing radiation for clam meats is 450 krad. When treated at this level, stored for 15 days at 0.6 to 1.7°C (33 to 35°F), and prepared as clam chowder the product was rated as "good" to "very good" by expert and untrained panelists. When treated in this manner and prepared as fried clams, both types of panelists rated the product as "very good" (see Table 5). Slavin et al.[196] packed soft-shell clam meats in cans without vacuum and treated them with ionizing radiation at various dose levels below 600 krad. When treated at a level of 350 krad, the product was not spoiled after 30 days at 0.6°C (33°F). When held at 5.6°C (42°F), these clam meats were not spoiled after 20 days but were spoiled after 30 days. Clam meats treated with 450 krad of ionizing radiation and stored for 20 days at 0.6°C (33°F) were judged to be of equal quality with fresh unstored control samples. After storage for 30 days at this temperature, they were rated as slightly less acceptable than strictly fresh control samples. Slavin et al.[98] reported that the optimum dose of ionizing radiation for soft-shelled clam meats packed in cans without vacuum is 450 krad. They also reported that clams treated in this manner have a much shorter shelf life at 5.6°C (42°F) than at 0.6°C (33°F).

Ronsivalli et al.[85] stated that the optimum dose of ionizing radiation for soft-shelled clam meats packed in oxygen-impervious containers in air is 450 krad and that when packed in oxygen-impervious containers under vacuum the optimum irradiation dose is 350 krad. Clam meats irradiated with 450 krad of ionizing radiation, held for 15 to 30 days at 0.6°C (33°F), and deep-fat fried were rated as "very good" by consumer-type and expert panelists. Slavin et al.[169] found that sulfides increased in the head space gas of nonirradiated clam meats packed in cans under vacuum and held at 0.6°C (33°F). In clam meats otherwise handled in the same manner but treated with ionizing radiation at a level of 350 krad, there was no increase in sulfides in the head space gas during storage.

Steinberg[168] stated that nonirradiated soft-shelled clams packed in cans in air have a storage life of 5 to 8 days at 0.6°C (33°F). When packed in the same manner and treated with ionizing radiation at a level of 450 krad, the storage life at 0.6°C (33°F) was said to be 30 days. Slavin et al.[170] listed the optimum dose of ionizing radiation for soft-shelled clam meats as 450 krad when packed in cans in air and the storage life at 0.6°C (33°F) as 28 days. Gadbois et al.[216] air and vacuum packed soft-shelled clam meats in cans and treated them with ionizing radiation at levels of 450 and 350 krad, respectively. Both types were edible after 21 days at 0.6 to 1.7°C (33 to 35°F) when deep-fat fried. After holding for 31 days at this range of temperatures, they were considered to be inedible. When air packed in cans and irradiated or heated, carbonyls increased immediately in the head space gas of canned clam meats. There was a gradual increase in carbonyls in the head space gas of nonirradiated, raw clam meats until the 20th day of storage at 0.6 to 1.7°C (33 to 35°F) after which the concentration of carbonyls decreased.

Ronsivalli et al.[217] found that dimethyl sulfide is the chief compound present in the volatile substances obtained from soft-shelled clam meats which are stored at refrigerator temperatures above freezing, cooked, treated with ionizing radiation at low-dose levels, or treated with ionizing radiation at high-dose levels. Mendelsohn and Brooke[218] packed soft-shelled clam meats in cans in air and under vacuum and irradiated them at 450 and 350 krad, respectively. The head space gas was tested for chemical components during storage at 0.6°C (33°F). It was found that dimethyl sulfide was the most dominant head space gas component. Numerous other chemical compounds were present in the head space gas, but it was concluded that dimethyl sulfide is responsible for the typical odor of soft-shelled clam meats.

2. Surf Clams (*Spisula solidissima*)

Carver et al.[157] packed surf clam meats at sea in cans in air and treated them with ionizing radiation at levels of 100, 200, or 450 krad then held them at 0.6°C (33°F) along with nonirradiated control samples. The controls had a storage life of 10 days, the 200-krad-treated samples had a storage life of 39 days, and the 450-krad treated clam meats had a storage life of 53 days. During storage the 450-krad-treated samples were given an organoleptic score only slightly lower than freshly-frozen, nonirradiated, surf clam meats, when these products were cooked and tested. Carver et al.[8] stated that the refrigerated storage life of surf clam meats is doubled or tripled when treated with ionizing radiation at low-dose levels.

B. Oysters

1. Gulf Oysters (Crassostrea virginica)

Novak et al.[219] irradiated oyster meats in cans in air at 200 krad. The treated product, along with nonirradiated controls in cans, were stored in ice and tested after 0, 7, 14, 21, and 23 days of storage. Most samples were tested at Louisiana State University and some were shipped to Lafayette, La., Chicago, Ill., Indianapolis, Ind., and Memphis, Tenn., for testing by experienced tasters. Most of the nonirradiated control samples were rated as spoiled in 21 days of storage and some of them were rated as spoiled after 7 days of storage. The irradiated oyster meats were judged to be acceptable throughout the 23-day storage period. At the start the irradiated and nonirradiated oyster meats were given the same organoleptic score, but after 7 days of storage and thereafter the samples treated with ionizing radiation were given significantly higher scores than the nonirradiated control samples. It was found that a yellowing occurs in shucked oysters treated with ionizing radiation at levels higher than 200 krad. Also it was stated that the conditions under which oysters are shucked, packed, and stored and the length of time that they are held prior to treatment with ionizing radiation are all important in determining the effectiveness of irradiation treatment to improve the storage qualities of this product at refrigerator temperatures above freezing. It was stated that the low-dose level treatment of oyster meats with ionizing radiation to extend the refrigerated storage life is feasible and economical.

Novak and Liuzzo[220] reported that organoleptic tests on oyster meats made in inland cities of the U.S. showed that there was a preference for meats treated with ionizing radiation over those which had not been irradiated. The irradiated oyster meats were preferred when eaten in both the raw state and when cooked by frying. The organoleptic ratings given to oyster meats as the raw or cooked product were not greatly different from each other. Novak and Liuzzo[221] reported that hundreds of taste-test panelists, both lay and professional types, approved the quality of oysters which had been treated with ionizing radiation and held in the iced state for 25 and 29 days. Novak and Liuzzo[222] found that treatment of oyster meats with 200 krad of ionizing radiation did not affect the proximate composition, but the ash and glycogen contents of the irradiated or nonirradiated product decreased during iced storage. It was stated that the successful preservation of oyster meats by treatment with ionizing radiation requires that this product is of good quality when treated.

2. Florida Oysters (Crassostrea virginica)

Gardner and Watts[223] treated oyster meats with 630, 830, or 3500 rep of ionizing radiation and found that these levels of treatment were sufficient to prevent spoilage when the treated product was held at room temperature. When treated at levels of 830 rep or higher, the oyster meats took on a "grassy" odor, but this disappeared upon cooking. The odor of the irradiated oyster meats was found to be stronger than that of the untreated product, but this odor was not foreign or undesirable. It was stated that the preservation of oyster meats with ionizing radiation does not hold great promise since enzyme action takes place in this product even when held at temperatures as low as 5°C (41°F).

3. Pacific Oysters (Crassostrea gigas)

Coleby and Shewan[164] stated that the maximum permissable dose of ionizing radiation for Pacific oyster meats is 500 rep. Miyauchi et al.[179] vacuum packed oyster meats in cans and glass jars and treated them with ionizing radiation. The product irradiated at 100 krad had a 10-day longer storage life than the nonirradiated control samples at 0.6°C (33°F), but the storage life was the same for both types at 5.6°C (42°F). The product irradiated at 200 krad had a 10-day longer storage life both at 0.6 and 5.6°C (33 and 42°F) than did the nonirradiated controls. Schultz and Lee[224] stated that when it is desirable to extend the refrigerated storage life of this product, treatment with ionizing radiation might be helpful to the shucked oyster industry.

4. Oysters (No Species Stated)

Steinberg[168] stated that the shelf life of oyster meats, air packed in cans, is 15 days at 0°C (32°F), while that of oyster meats packed and held in the same manner but treated with ionizing radiation at a level of 200 krad is 21 days. Slavin et al.[170] stated that the optimum dose of ionizing radiation for oyster meats is 200 krad and that the storage life of this product at 0.6°C (33°F) is 21 to 28 days. Metlitskii et al.[174] stated that the storage life of shucked oysters at 2°C (35.6°F) is 7 to 9 days and that oyster meats treated with 500 krad of ionizing radiation and held at 2°C (35.6°F) had a storage life of 60 days.

C. Scallops

1. Sea Scallops (Placopecten magellanicus)

Sieling[188] stated that sea scallops lose little in quality when treated with dose levels of ionizing radiation up to 750 krad. Power et al.[225] packed sea scallop meats in plastic containers in air and treated them with ionizing radiation at levels of 75, 150, 300, 400, or 800 krad. The irradiated products, along with nonirradiated control samples, were stored on ice. The optimum dose level was found to be 75 krad which extended the storage life from 13 to 17 days (controls) to 28 to 43 days. Frozen control scallop meats stored at −26°C (−14.8°F) were judged to be much better in quality than the fresh (nonirradiated) or the irradiated products.

2. Alaskan Scallops (Pecten caurinus)

Schultz and Lee[224] stated that treatment with ionizing radiation might be helpful in extending the refrigerated storage life of scallop meats. They further stated that scallop meats treated with ionizing radiation are of acceptable quality.

D. Summary

There is good evidence that soft-shelled clam and sea clam meats can be treated with ionizing radiation at levels of 350 to 450 krad to provide an extension of storage life of several weeks at refrigerator temperatures above freezing. There is, however, some question as to whether such a storage life extension is a requirement or necessary for sea clams and there is also the question of whether or not sufficient quantities of soft-shelled clam meats are available to require an extension of storage life which would enable the distribution of this product to inland areas. If and when some clam producing areas, such as the mouth of the Merrimack River in Massachusetts, are depurated, greater quantities of soft-shelled clams will become available, and this would ensure the suitability of treatment with ionizing radiation for this product. Also, the irradiation treatment of clam meats at the levels stated would probably eliminate certain types of food-borne disease-causing bacteria which seem to be present sometimes in clams even though they may be taken from approved areas. The treatment of oyster meats with ionizing radiation at a level of 200 krad has been shown to be suitable for extending the refrigerated storage life of this product. Sufficient quantities of oysters are available to allow for the distribution of the meats to inland areas well away from the coast. When treated with ionizing radiation, oyster meats could be prepared for consumption as the fried or stewed product, or consumed in the raw state. While there is evidence that scallop meats could be treated with ionizing radiation at levels of 100 krad or higher to extend the refrigerated storage life, this product can be frozen with good results and irradiation treatment, unless less costly than freezing and storage, may not be desirable.

At dose levels of ionizing radiation considered to be optimum, there is little, if any, effect on the texture or appearance of mollusc meats. At higher dose levels than those which have been recommended, there may be a development of "irradiation flavor" in these products. Also oyster meats treated with ionizing radiation at levels higher than 200 krad have been found, in some instances, to become discolored. When treated at very high-dose levels, oyster meats have been found to take on a "grassy" odor. Scallop meats treated with ionizing radiation at levels higher than 100 krad have been found, in some instances, to undergo a loss of their typical flavor.

When mollusc meats are treated with ionizing radiation to extend their refrigerated storage life, they should be prepackaged in materials impervious to moisture and oxygen. The metal cans presently used to hold clam and oyster meats in commercial practice might be suitable for the packaging of these products for irradiation treatment. It has been found that the packaging of these products under vacuum is not a requisite.

As is the case with fish, the treatment of mollusc meats with ionizing radiation does not improve their quality; it does help to retain the quality as packed. For this reason only mollusc meats of good or very good quality should be subjected to this type of preservation.

After mollusc meats have been treated with ionizing radiation, they should be held at temperatures as near 0°C (32°F) as is possible without freezing. When held at relatively high refrigerator temperatures (4.4°C and above) the storage life of radiation-treated shucked clams or oysters is greatly decreased.

XII. CRUSTACEANS

A. Crabs

1. Dungeness Crabs (Cancer magister)

Lerke et al.[80] stated that cooked dungeness crab meat is markedly preserved by treatment with ionizing radiation (5 MeV beta rays) and that the refrigerated storage life of product treated in this manner is dependent upon the dose level applied. Scholz et al.[226] treated cooked dungeness crab meat with ionizing radiation at a level of 250 krad and found that it had a storage life of 28 days at 3.3°C (38°F). At Seattle, Wash.,[192] cooked dungeness crab meat was air and vacuum packed in cans, frozen, and shipped to Davis, Calif., to be treated with ionizing radiation at levels of 200, 300, 500, or 650 krad, then shipped back to Seattle packed in wet ice. Control samples which were not irradiated were included. All of the irradiated samples had much longer storage life times than the nonirradiated controls at 0.6, 3.9, or 7.8°C (33, 39, or 46°F). Judged organoleptically, all of the irradiated products were found to be different from the nonirradiated controls but were rated as acceptable.

Miyauchi et al.[189] packed cooked dungeness crab meat in cans, in air and under vacuum, and treated it with ionizing radiation at levels of 200, 350, 500, or 650 krad. The treated product, along with nonirradiated control samples, was held at 0, 3.9, or 7.8°C (32, 39, or 46°F). Control samples spoiled in 13 to 22 days at 0°C (32°F), 6 to 8 days at 3.9°C (39°F), and in less than 4 days at 7.8°C (46°F). Product treated at 200 krad or higher had storage life times of 20 to 50 days at 3.9°C (39°F). The optimum dose level was considered to be 200 to 350 krad. When tested by an expert panel, statistical analyses of results indicated that the irradiated product was always significantly different from the fresh control samples but was of acceptable quality in all tests. The vacuum-packed crab meat was not found to be different from that packed in air after treatment with ionizing radiation.

Miyauchi et al.[189] air and vacuum packed cooked dungeness crab meat in cans, treated these products with ionizing radiation at various levels, and held them, along with nonirradiated control samples, at 0.6 or 5.6°C (33 or 42°F). An irradiation dose level of 200 krad was judged to be optimum. Vacuum- and air-packed samples were not different after irradiation treatment. The nonirradiated control samples had a storage life of 6 days at 0.6°C (33°F) and less than 4 days at 5.6°C (42°F). Product irradiated at a level of 200 krad had a storage life of 21 to 42 days at 0.6°C (33°F) and 7 to 14 days at 5.6°C (42°F).

Miyauchi et al.[191] air and vacuum packed cooked dungeness crab meat in cans and treated it with ionizing radiation at levels of 100 or 200 krad. The irradiated crab meat, along with nonirradiated control samples, was held at 0.6°C (33°F). The irradiated samples were found to have storage life times up to five times greater than those of the nonirradiated control samples. Air and vacuum packaging were equally effective as far as the shelf-life extension of the irradiated product was concerned. Three different recipes were prepared from dungeness crab meat frozen and stored for 14 days at −18°C (−0.4°F) and from dungeness crab meat irradiated at a level of 200 krad and stored for the same period at 0.6°C (33°F). The recipes made from the irradiated crab meat were judged to be of the same quality as those prepared with the frozen product.

Dassow and Miyauchi[190] stated that cooked dungeness crab meat, air packed in cans and treated with ionizing radiation at a level of 200 krad, had storage life times of 28 to 42 days at 0.6°C (33°F) and 14 to 21 days at 5.6°C (42°F). The nonirradiated control samples had storage life times of 6 to 11 days at 0.6°C (33°F) and 2 to 5 days at 5.6°C (42°F). Miyauchi et al.[180] vacuum packed cooked dungeess crab meat in polyester-polyethylene pouches and in cans and irradiated these products at levels of 100 or 200 krad. The treated samples, along with nonirradiated controls, were held at 0.6 or 5.6°C (33 or 42°F). The samples irradiated at a level of 200 krad had a significantly longer storage life at either temperature than did those treated with 100 krad of ionizing radiation, and samples treated at either level with ionizing radiation had significantly longer storage life times at either temperature than did the nonirradiated controls. Products packed in cans or in plastic pouches were not noticeably different. Miyauchi et al.[181] air packed cooked dungeness crab meat in cans, treated the product with ionizing radiation at a level of 200 krad, and shipped it by air, under refrigeration, from Seattle, Wash. to Gloucester, Mass. In Gloucester the crab meat was held for 14 days at 0.6°C (33°F) then tested organoleptically by a consumer-type panel. The panelists rated the product as "like moderately".

Miyauchi et al.[227] packed cooked dungeness crab meat in No. 10 cans and in polyester-polyethylene pouches in air and under vacuum, treated these products with 100, 200, 300, 400, or 500 krad of ionizing radiation, and stored them, along with nonirradiated control samples, at 0.6°C (33°F) or 5.6°C (42°F). There was no difference in the appearance or texture of the fresh product and that of the irradiated at any level. An "irradiation flavor" was barely perceptible in product treated at a level of 100 krad, and there was a slight "irradiation flavor" in product treated with 200 krad of ionizing radiation. "Irradiation flavor" was stronger in product treated at a level of 300 krad and higher. There was a slight loss of the natural odor and flavor in the crab meat treated with ionizing radiation, and as in the fresh, nonirradiated product, loss of flavor continued during refrigerated storage. Irradiation treatment of dungeness crab meat at levels of 100 or 200 krad produced a "fresh-like" product and increased the refrigerated storage life two to five times over that of the nonirradiated control samples.

Eklund et al.[118] air and vacuum packed cooked dungeness crab meat in cans, treated it with ionizing radiation at levels of 200 or 400 krad, and stored these products, along with nonirradiated control samples, at 0.6 and 5.6°C (33 and 42°F). Bacterial growth soon spoiled the control samples. Yeasts increased rapidly in the air packed irradiated crab meat but not in the vacuum-packed irradiated product. Miyauchi et al.[179] treated cooked, whole (in the shell) dungeness crabs with ionizing radiation at a level of 200 krad and held them, along with nonirradiated control samples, at 0.6°C (33°F). The controls had a storage life time of 10 days, while that of the irradiated samples was 14 days. When the crabs were eviscerated prior to irradiation treatment, the storage life of the samples so treated was 24 days.

Miyauchi et al.[183] reported that the storage life of fresh cooked dungeness crab meat is 6 to 14 days at 0.6°C (33°F) and 2 to 9 days at 5.6°C (42°F). Cooked dungeness crab meat treated with ionizing radiation at a level of 100 krad had a storage life of 14 to 35 days at 0.6°C (33°F) and 14 to 21 days at 5.6°C (42°F). When treated with 200 krad of ionizing radiation the crab meat had a storage life of 21 to 56 days at 0.6°C (33°F) and 21 days at 5.6°C (42°F).

2. King Crab (Paralithodes camtschatica)

At Seattle[192] cooked king crab meat was packed in cans and treated with ionizing radiation at levels of 200, 400, or 600 krad. These products were held, along with nonirradiated control samples, at 0.6°C (33°F) or 5.6°C (42°F). The 200-krad-treated samples were the least different from the nonirradiated controls. At higher irradiation levels than 200 krad, the crab meat took on an "irradiation flavor" which decreased in intensity with increasing storage time. At 0.6°C (33°F) the nonirradiated control samples had a storage life of 7 days, while product treated with 200 krad of ionizing radiation was acceptable after 35 days. At 5.6°C (42°F) the fresh controls spoiled after 5 days, while the 200-krad-irradiated product spoiled after 14 days.

Miyauchi et al.[189] stated that the optimum dose of ionizing radiation for cooked king crab meat is 200 krad and that product irradiated at this level has good sensory characteristics. Prepackaging under vacuum in material impervious to oxygen and moisture was found to be superior to air packing for the radiation-treated crab meat, extending the storage life at 0.6°C (33°F) for 7 days longer than that which was air packed. The storage life of the nonirradiated control samples was found to be 7 days at 0.6°C (33°F). Product irradiated at 200 krad had a storage life of 28 to 42 days at this temperature.

Miyauchi et al.[189] stated that cooked king crab meat had a noticeable amount of "irradiation" odor and flavor. This was especially true when treated with ionizing radiation at levels higher than 200 krad. This off-flavor decreased, however, during refrigerated storage of the treated product.

Miyauchi et al.[99] packed cooked king crab meat in cans under vacuum and treated this product with ionizing radiation at levels of 100, 200, 300, 400, or 600 krad. The treated products, along with nonirradiated control samples, were held at 0.6 or 5.6°C (33 or 42°F) and tested periodically by expert taste panelists. It was concluded that an irradiation dose of 200 krad was that which resulted in a product of high quality and provided a significant extension of the refrigerated storage life. The odor and flavor of the vacuum-packed irradiated samples were better than that of the air-packed irradiated crab meat, and the vacuum-packed product also had a longer refrigerated storage life. After long storage times, compared to that of the nonirradiated control samples, crab meat irradiated with 100 or 200 krad of ionizing radiation developed bacterial spoilage odors and flavors.

Spinelli et al.[150] found that in cooked, vacuum-packed king crab meat treated with ionizing radiation at levels of 200 or 400 krad and held at 0.6 or 5.6°C (33 or 42°F), bacterial plate counts near 1 × 10^8/g occurred before sensory spoilage changes developed. The bacterial counts of nonirradiated samples were only slightly higher than 1 × 10^6/g when such changes were observed. The sensory characteristics of crab meat were found to correlate well with trimethylamine and total volatile base values for this product until these exceeded 0.9 and 12.0 mg%, respectively.

Eklund et al.[118] stated that yeasts increased rapidly in crab meat packed in air in cans and treated with ionizing radiation then held at 0.6 or 5.6°C (33 or 42°F). Vacuum packing inhibited the growth of yeasts in the irradiated product. Bacterial growth was the cause of spoilage of nonirradiated cooked king crab meat held under refrigeration. Miyauchi et al.[183] reported that cooked king crab meat packaged in cans under vacuum and treated with ionizing radiation at a level of 100 krad had a storage life of 21 days at 0.6°C (33°F) and 7 to 14 days at 5.6°C (42°F). The same product irradiated at a level of 200 krad had a storage life of 28 to 42 days at 0.6°C (33°F) and 14 days at 5.6°C (42°F). The nonirradiated control samples had storage life times of 5 to 14 days at 0.6°C (33°F) and 3 to 7 days at 5.6°C (42°F).

3. Crabs (No Species Indicated)

Shewan[163] stated that cooked crab meat treated with ionizing radiation at levels of 400 to 850 krad had a storage life of 40 days at 0°C (32°F). Eukel and Huber[187] found that cooked crab meat treated with ionizing radiation at a level of 250 krad (8 MeV electrons) had a storage life of up to 20 days at 5.6°C (42°F). Rhodes[167] stated that the appearance of cooked crab meat was not affected by dose levels of ionizing radiation at levels up to 300 krad. There was no change in the odor of this product irradiated at 100 krad, but there was a slight loss of the typical odor of cooked crab meat treated with ionizing radiation at a level of 300 krad.

B. Lobsters

1. American Lobster (Homarus americanus)

Goldblith and Nickerson[74] packed low bacterial count cooked lobster meat in cans without vacuum and treated it with ionizing radiation at a level of 125 krad while holding it at room temperature. The storage life of the treated product at 0.6°C (33°F) was effectively increased. The same results were obtained with high bacterial count cooked lobster meat by irradiation treatment at the same dose level while holding the product at 60°C (140°F). There was little effect on the flavor of the lobster meat treated in this manner.

Power et al.[69] air packed cooked lobster meat in plastic bags and treated it with ionizing radiation at 75, 150, or 250 krad. The treated lobster meat, along with nonirradiated control samples, was held on ice. Frozen control samples, held at −26°C (−15°F), were also tested. There was no oxidation in the lipids of the lobster meat during the 42 days of storage used for the test. During storage the frozen control samples deteriorated slightly in quality. The lobster meat treated with 150 or 250 krad of ionizing radiation lost some of the typical flavor at the start. The 150-krad-treated product was unacceptable after 35 days, and the 250-krad-treated product was unacceptable after 35 to 42 days on ice. The lobster meat irradiated at a level of 75 krad did not differ from the fresh control at the start, but the organoleptic quality gradually decreased during storage and after 28 days on ice it was judged to be slightly stale. The iced fresh control samples had a storage life of 14 to 21 days and after 21 days had a rubbery texture and were considered to be inedible. No differences were observed in the quality of claw and tail meat after irradiation treatment.

Dagbjartsson and Solberg[228] packed cooked lobster meat under vacuum in Mylar®-saran-polyethylene bags, treated it with ionizing radiation at 75 or 100 krad, and held the samples and nonirradiated controls at 0.6 to 2.2°C (33 to 36°F). Frozen control samples were held a −27°C (−16.6°F). Storage life times were determined by aerobic total plate counts (bacterial) and texture was measured by testing with a Warner-Bratzler shear device and organoleptically by chew count. The storage life times were found to be 10 days for the fresh control, 17 days for product irradiated at a level of 75 krad, and 30 days for product irradiated at a level of 100 krad. There was no difference between the texture of the fresh and the 75-krad-treated product at the start, but the change in the texture of the 100-krad-treated product was equivalent to that which occurred in the frozen product during 10 to 12 months at −27°C (−16.6°F). Most of the taste testers could detect differences between the odor and flavor of the fresh lobster meat and those of the meat treated with ionizing radiation.

2. *European Lobster (Homarus gammarus)*

Rhodes[167] reported that there was no difference in the appearance of fresh cooked and cooked irradiated lobster meat when the dose of ionizing radiation was no greater than 300 krad. There was a slight yellowing in cooked lobster meat treated with 1000 kilorad of ionizing radiation. The odor of cooked lobster meat treated with 100 krad of ionizing radiation was comparable to that of the fresh nonirradiated product, but there was a slight loss of the typical odor in product treated at a level of 300 krad. Product treated with ionizing radiation at a level of 1000 krad had an "irradiation" odor.

3. *Norwegian Lobster (Nephrops norvegicus)*

Goresline[204] stated that the storage life of lobster tails, preheated to prevent melanosis and treated with ionizing radiation, when held at refrigerator temperatures above freezing, was eight times longer than that of the nonirradiated product.

C. Shrimp

1. *White Shrimp (Penaeus setiferus)*

Scholz et al.[226] irradiated unshelled white shrimp with ionizing radiation at levels of 500 or 750 krad and held them at 3.3°C (38°F). This product remained in good condition for a period of 126 days. At these levels of treatment an "irradiation" flavor was acquired by the shrimp, but this off-flavor declined during storage.

Awad et al.[81] treated defrosted, unshelled white shrimp with a 5 ppm solution of chlortetracycline or with chlortetracycline and ionizing radiation (500 or 750 krad) or with ionizing radiation alone at the same levels. The shrimp were air packed in cans. Untreated control samples were also prepared and all samples were held at 3.3°C (38°F). Frozen control samples were also held at −18°C (−0.4°F). There were significant increases in the trimethylamine, volatile base, and indole contents of the defrosted, nonirradiated control samples in 7 and 14 days of storage. No increases in these compounds occurred in the shrimp treated with ionizing radiation during 70 days of storage. Shrimp treated with 500 krad of ionizing radiation alone was judged to be inferior to the frozen control samples after 28 to 35 days of storage. Product irradiated at a level of 500 krad after treatment with chlortetracycline and all product treated with 750 krad of ionizing radiation were judged to be inferior to the frozen control samples after 56 to 70 days of storage. It was concluded that treatment with ionizing radiation provided a four to fivefold extension of the refrigerated storage life of defrosted shrimp.

Dassow and Miyauchi[190] packaged unshelled white shrimp in cardboard cartons with a Mylar® overwrap and treated them with ionizing radiation at a level of 150 krad. These were held, along with nonirradiated control samples, at 0.6 to 1.1°C (33 to 34°F). The control samples had a storage life of 14 days while the storage life of the irradiated product was 24 to 35 days.

Kaylor et al.[229] packed unshelled white shrimp in air in cans, shipped them by air from Florida to Gloucester, Mass., repacked them in Mylar® bags, and treated them with ionizing radiation at levels of 50, 100, 200, or 300 krad. Untreated controls were also repacked. Some of each type of product were held in Gloucester at 0.6 to 1.1°C (33 to 34°F), and some were shipped back to Florida, iced in insulated containers. Product reshipped to Florida was repacked and shipped to Philadelphia, Pa., by truck and again repacked and shipped to Gloucester by truck where it was held at 0.6 to 1.7°C (33 to 35°F). Product originally held in Gloucester had the following storage life times: nonirradiated, 12 to 16 days; product treated with ionizing radiation, acceptable after 27 days. The products reshipped to Florida and back to Gloucester had the following storage life times: untreated control samples, 12 to 16 days; 50-krad treated, 23 days; 100-krad-treated samples, 20 to 23 days; 200-krad-treated samples, 25 days; 250-krad-treated samples 25 days.

In another test, 1- to 5-day-old unshelled white shrimp were shipped by air, iced in insulated containers, from Florida to Gloucester, Mass. They were then packed in Mylar® bags and treated with ionizing radiation at 25, 50, 100, or 200 krad or not irradiated. Samples of each type were then shipped by air, iced in insulated containers, to New Orleans, La., from which they were brought to Louisiana State University at Baton Rouge, La., where they were held at 0.6 to 1.7°C (33 to 35°F). The various products were judged to be acceptable after the following periods: not treated, 15 days; 25-krad treated, 22 days; 50-krad treated, more than 24 days; 100- or 200-krad treated, more than 24 days. Learson et al.[56] reported the following storage life times for unshelled white shrimp: untreated, 5 to 7 days at 3.3°C (38°F); 7 to 8 days at 5.6°C (42°F); less than 5 days at 7.8°C (46°F); and less than 5 days at 12.8°C (55°F); treated with 100 krad of ionizing radiation, more than 9 days at 3.3°C (38°F); 14 days at 5.6°C (42°F); 7 days at 7.8°C (46°F); and 5 days at 12.8°C (55°F); treated with 200 krad of ionizing radiation, 16 days at 3.3°C (38°F); 16 days at 5.6°C (42°F); 9 days at 7.8°C (46°F); and less than 5 days at 12.8°C (55°F).

Roy and Kaylor[82] treated headless shelled white shrimp with turmeric and salt solution and subjected this product to ionizing radiation at levels of 100, 200, or 300 krad. Samples were also irradiated without pretreatment with turmeric and salt solution. At 0.6°C (33°F) it was found that shrimp treated with ionizing radiation had storage life times that were 14 to 28 days longer than the nonirradiated controls. The spice treatment masked any off-odor and off-flavor due to treatment with ionizing radiation and also inhibited melanosis or the development of "black spot".

2. Brown Shrimp (*Penaeus aztectus*)

Kaylor et al.[229] shipped unshelled brown shrimp from Florida to Gloucester, Mass., repackaged and treated them with ionizing radiation, returned them to Florida with nonirradiated controls, and shipped them back to Gloucester as reported in the experiment on white shrimp, excepting that the second shipment from Florida was via New York City. Brown unshelled shrimp were also treated and shipped to Louisiana State University as reported for white shrimp. Among the shrimp held in Gloucester at 0.6 to 1.7°C (33 to 35°F), the nonirradiated samples were unacceptable in 5 days; the 50-krad-treated samples were unacceptable after 16 days; the 100-krad-treated samples were acceptable after 30 days; and the 200-krad-treated samples were acceptable after 30 days. In product shipped to and from Florida the nonirradiated controls were inedible after 7 days, the 100-krad-treated samples were acceptable after 23 days; the 200-krad-treated samples were acceptable after 21 days; and the 300-krad-treated samples were acceptable after 15 days. Among the products tested at Louisiana State University held at 0.6 to 1.7°C (33 to 35°F) the nonirradiated controls were acceptable after 19 days and inedible after 23 days (total storage times including shipment times); the 50-

krad-treated samples were acceptable after 23 days and inedible after 26 days; the 100-krad-treated samples were acceptable after 23 days and inedible after more than 26 days. The 200-krad-treated product was not given final tests due to a lack of samples.

Learson et al.[56] found that the storage life times for nonirradiated and irradiated unshelled brown shrimp was the same as for white unshelled shrimp treated in the same manner when held at the same temperatures. Ronsivalli et al.[176] reported the following storage life times for brown unshelled shrimp — untreated; 3 days at 12.8°C (55°F), 6 days at 8.9°C (48°F), 13 days at 5.6°C (42°F), and 15 days at 3.3°C (38°F); treated with 50 krad of ionizing radiation: 9 days at 12.8°C (55°F), 8 days at 8.9°C (48°F), 15 days at 5.6°C (42°F), and 27 days at 3.3°C (38°F); treated with 100 krad of ionizing radiation: 8 days at 12.8°C (55°F), 10 days at 8.9°C (48°F), 24 days at 5.6°C (42°F), and 35 days at 3.3°C (38°F); treated with 200 krad of ionizing radiation: 10 days at 12.8°C (55°F), 10 days at 8.9°C (48°F), 27 days at 5.6°C (42°F), and 30 days at 3.3°C (38°F).

3. *European Brown Shrimp (Crangon vulgaris)*

Lievens[230] packed brown shrimp in polyethylene and irradiated the product at a level of 100 krad. It was found that the nonirradiated control samples had a storage life of 6 to 9 days while the irradiated product had a storage life of 23 ± 2 days when held at refrigerator temperatures above freezing.

4. *Pink Shrimp (Penaeus duorarum)*

Ronsivalli et al.[176] packaged unshelled pink shrimp in polyethylene pouches, treated them with ionizing radiation at various levels and held them, along with nonirradiated control samples, at various temperatures. Storage life times were determined by odor and reported to be as follows— nonirradiated: 2 days at 12.8°C (55°F), 5 days at 8.9°C (48°F), 10 days at either 5.6°C (42°F) or 2.8°C (37°F), and 17 days at 0.6°C (33°F); treated with 50 krad of ionizing radiation: 4 days at 12.8°C (55°F), 6 days at 8.9°C (48°F), 12 days at 5.6°C (42°F), 20 days at 2.8°C (37°F), and 27 days at 0.6°C (33°F); treated with 100 krad of ionizing radiation: 6 days at 12.8°C (55°F), 10 days at 8.9°C (48°F), 12 days at 5.6°C (42°F), 25 days at 2.8°C (37°F), and 28 days at 0.6°C (33°F); treated with 200 krad of ionizing radiation: 7 days at 12.8°C (55°F), 13 days at 8.9°C (48°F), 26 days at 5.6°C (42°F), 33 days at 2.8°C (37°F), and 44 days at 0.6°C (33°F).

5. *Shrimp (Metapenaeus affinis), (Penaeus indicus), and (Parapeneopsis stylifera)*

Kumta et al.[70] packed shrimp in cans in polyethylene pouches, some in brine (1 to 3%) and some cooked, and treated them with ionizing radiation at levels of 100 to 1000 krad. These products were stored, along with nonirradiated control samples at 10 to 12°C (50 to 54°F) or at 2 to 4°C (36 to 39°F). The odor of the 150-krad-irradiated samples, raw or cooked, was normal. The shrimp treated with 250 krad of ionizing radiation had a slight rancid odor and this was distinct in shrimp treated with 500, 750, or 1000 krad. The fresh nonirradiated shrimp spoiled in 4 days at 10 to 12°C (50 to 54°F). Shrimp treated with 150 or 250 krad of ionizing radiation spoiled in 10 or 21 days, respectively, at 10 to 12°C (50 to 54°F). The cooked nonirradiated shrimp spoiled in 30 days at 10 to 12°C (50 to 54°F). The cooked irradiated shrimp (150 krad) spoiled in 60 days at 10 to 12°C (50 to 54°F) and 130 days at 2 to 4°C (36 to 39°F).

6. *Shrimp (No Species Indicated)*

Eukel and Huber[187] stated that the storage life of shrimp treated with ionizing radiation at a level of 250 krad (8 MeV electrons) was about 20 days at 5.6°C (42°F). Sieling[188] packed raw shrimp in cans without vacuum and treated them with ionizing radiation at levels of 100, 250, and 500 krad. He reported that there was no loss of quality at the lower and a slight loss of quality at the higher dose level. Nonirradiated samples spoiled in 16 days at 1.1 to 2.2°C (34 to 36°F). The irradiated samples were acceptable after 72, 96, and 100 days at the temperatures stated, depending upon the irradiation dose applied.

Novak and Luizzo[231] stated that, when handled under the best conditions, the storage life of fresh unshelled shrimp under refrigeration is 14 to 16 days. They also stated that treatment of fresh shrimp with ionizing radiation at low-dose levels should extend the storage life of this product, handled under the same conditions, for periods up to 30 days. Novak et al.[232] treated unshelled raw shrimp with ionizing radiation at a level of 100 krad and froze them. Frozen nonirradiated control samples were also prepared. After 3 months of frozen storage the samples which had been irradiated had less "black spot" and a better general appearance than the nonirradiated shrimp. Novak[233] reported that fresh, raw unshelled shrimp, treated with ionizing radiation at a level of 100 to 150 krad, had a 10-day longer storage life at refrigerator temperatures above freezing than did the nonirradiated control samples.

Rhodes[167] stated that raw prawn, treated with 300 krad of ionizing radiation, had a slight "irradiation" odor but was normal in appearance. The "irradiation" odor was definite in prawn treated with 1000 krad. Coleby and Shewan[164] stated that the maximum permissible dose of ionizing radiation for raw shrimp is 1000 krep.

Steinberg[168] packed raw unshelled shrimp in Mylar® bags and treated them with 150 krad of ionizing radiation. The treated samples, along with nonirradiated control shrimp, were held at 0.6°C (33°F). The controls had a shelf life of 14 days while the irradiated shrimp had a shelf life of 25 days. Slavin et al.[17] stated that the optimum dose of ionizing radiation for raw unshelled shrimp is 150 krad and that this treatment extends the storage life to 28 days at 0.6°C (33°F). Metlitskii et al.[174] reported that cooked shrimp had a storage life of 4 to 16 days at 2°C (36°F) but that this product treated with 250 to 450 krad of ionizing radiation had a storage life of 19 to 110 days at this temperature.

Novak et al.[67] treated headed, unshelled raw shrimp with ionizing radiation, aboard ship, at levels of 100, 200 or 300 krad. The shelf life of this product, on ice, was 10 to 20 days. Fresh unshelled shrimp were also treated with 200 krad of ionizing radiation on shore and shipped to Fort Lee, Va., along with nonirradiated control samples, for consumer-type taste testing. The irradiated shrimp were judged to be slightly better than the nonirradiated samples. See Table 7 for a listing of approved irradiated marine and fresh-water foods.

D. Summary

It has been found that cooked king and dungeness crab meats can be treated with ionizing radiation at a level of about 200 krad to provide an appreciable extension of their storage life times at refrigerator temperatures above freezing. Since king crabs are not caught in sufficient quantities to supply the consumer demand for the canned or frozen meat of this species, there appears to be no object in applying radiation preservation treatment to this product unless this method of processing is less costly than canning or freezing. Dungeness crab meat, which cannot be canned or frozen with good results, could be treated with ionizing radiation to provide an excellent refrigerated product which could be distributed to both coastal and inland areas.

There is indication that cooked lobster meat could be treated with ionizing radiation at a level of 75 to 100 krad to provide a suitable extension of the refrigerated storage life of this product. Although refrigeration may suffice for preservation of lobster meat for the domestic market in the U.S., there has been commercial interest in the use of ionizing radiation to accommodate the European market by sea shipment rather than by the more expensive air freight.

Table 7
IRRADIATED MARINE AND FRESH-WATER FOODS WHICH HAVE RECEIVED APPROVAL FOR CONSUMPTION BY HUMANS

Clearance for:	Type of clearance	Date of approval	Ref.
Cod and haddock fillet radurization (Canada)	Test marketing	October 1973	*Food Irradiat. Newsl.*, 4(3), 16, 1980.
Shrimp radurization (Netherlands)	Test marketing	June 1976	*Food Irradiat. Newsl.*, 4(3), 16, 1980.
Frog legs radicidation (Netherlands)	Temporary acceptance	September 1978	*Food Irradiat. Newsl.*, 4(3), 16, 1980.
Haddock fillet radurization (Netherlands)	Test marketing	September 1976	*Food Irradiat. Newsl.*, 4(3), 16, 1980.
Cod and redfish radurization (World Health Org. Joint FAO/IAEA/WHO Expert Committee, 1976)	Provisional acceptance	September 1976	Wholesomeness of Irradiated Food, Report of a joint FAO/IAEA/WHO expert committee, WHO Technical Report Series 604, World Health Organization, Geneva, 1977.

Raw unshelled shrimp can be treated with ionizing radiation at levels of 100 to 200 krad to provide a considerable extension of the storage life at refrigerator temperatures above freezing. Such treatment would allow shipment of fresh shrimp in the unfrozen state from coastal areas to different inland areas within the U.S. and other countries. This would do away with the costs of freezing, storage, and defrosting, resulting in saving energy. The development of "black spot" in shrimp treated with ionizing radiation, while held under refrigeration, could be a deterring factor, but this particular type of deterioration can be taken care of by other means. Since shrimp tend to deteriorate aboard ship, the application of ionizing radiation to this product should preferably be carried out on shipboard. However, the size and capacity of U.S. shrimp catching boats are not such as to make shipboard irradiation practicable.

The treatment of cooked or raw crustacean products with ionizing radiation at suitable dose levels causes a slight loss of the typical flavor but no change in the appearance or texture and no objectionable "irradiation" flavor. When levels of ionizing radiation higher than those which have been found to be optimum are applied to these crustacean products, they may take on an "irradiation" or "burnt" odor and flavor.

It is evident that crustacean meats should be of the best quality when treated with ionizing radiation to extend the refrigerated storage life. This type of preservation treatment, like others, will not improve the quality of a poor or spoiling food, and the storage life extension of irradiated crustacean products may be shortened when they are in stages of incipient spoilage when treated.

The literature indicates that cooked crab and lobster meat should be prepackaged in a material that is impervious to moisture and oxygen prior to treatment with ionizing radiation and that king crab meat should be packaged under vacuum prior to treatment in this manner. While the prepackaging of raw shrimp prior to irradiation treatment has been noted as requiring material only impervious to moisture, it is probable that prepackaging in material also impervious to oxygen would provide for an extended storage life at refrigerator temperatures above freezing.

As in the case of fish and molluscs, crustacean products, after treatment with ionizing radiation, should be held at temperatures near the freezing point to obtain the greatest extension of their storage life time. When such foods are treated with ionizing radiation and held at temperatures of 4.4°C (40°F) or above, they soon spoil and the storage life extension attained is minimal.

REFERENCES

1. **Proctor, B. E., Van de Graaff, R. J., and Fram, H.,** *Reports on Quartermaster Contract Projects, July 1, 1942 — June 30, 1943,* Food Technology Laboratories, Massachusetts Institute of Technology (MIT) Cambridge, 1943, 217.
2. **Nickerson, J. T. R., Goldblith, S. A., and Proctor, B. E.,** A comparison of chemical changes in mackerel tissues treated by ionizing radiation, *Food Technol. (Chicago),* 4, 84, 1950.
3. **Nickerson, J. T. R., Lockhart, E. E., Proctor, B. E., and Licciardello, J. J.,** Ionizing radiations for the control of fish spoilage, *Food Technol. (Chicago),* 8(1), 32, 1954.
4. **Goldblith, S. A.,** Historical development of food irradiation, in *Food Irradiat.,* Proc. Int. Symp. Food Irradiat., Karlsruhe, International Atomic Energy Agency (IAEA), Vienna, 1966.
5. **Proctor, B. E., Goldblith, S. A., Nickerson, J. T. R., and Farkas, D. S.,** Evaluation of the Technical Economic and Practical Feasibility of Radiation Preservation of Fish, Performed under Task VI and Contract Number AT(30-1)-2329 with the U.S. Atomic Energy Commission, MIT, Cambridge, 1960.
6. **Nickerson, J. T. R., Goldblith, S. A., Miller, S. A., Licciardello, J. J., and Karel, M.,** Outline of Projects to Determine the Feasibility of Radiation Preservation of Marine Products, Performed under Task X and Contract Number AT(30-1)-2329 with the U.S. Atomic Energy Commission, MIT, Cambridge, 1961.
7. **Dietz, G. R.,** Facilities supporting AEC food irradiation research, in *Radiation Preservation of Foods,* Publ. 1273 National Research Council, National Academy of Sciences, Washington, D.C., 1965.
8. **Carver, J. H., Connors, T. J., and Slavin, J. W.,** Irradiation of fish at sea, in *Freezing and Irradiation of Fish,* Kreuzer, R., Ed., Fishing News (Books), London, 1969.
9. **Goresline, H. E., Ingram, M., Macuch, P., Mocquot, G., Mossel, D. A. A., Niven, C. F., and Thatcher, F. S.,** Tentative classification of food irradiation processes with microbiological objectives, *Nature (London),* 204(4955), 237, 1964.
10. Federal Register 32(4), Part 121, Section 121.3002 through 121.3007, January 7, 1967; Federal Register 42(50), Part 179, Subpart B, Section 179.22, March 15, 1977.
11. Federal Register 46(59), 18992, March 27, 1981.
12. **Silverman, G. J. and Sinskey, T. J.,** The destruction of microorganisms by ionizing radiation, in *Disinfection, Sterilization and Preservation,* Lawrence, C. A. and Block, S. S., Eds., Lea & Febiger, Philadelphia, 1968, chap. 44.
13. **Brooke, R. O. and Steinberg, M. A.,** Preservation of fresh unfrozen fishery products by low-level radiation. I. Introduction, *Food Technol. (Chicago), 18, 1056, 1964.*
14. **Yankelovich, D.,** Cost Benefit Study of Selected Products in Atomic Energy Commission's Low-Dose Food Irradiation Program, Report to U.S. Atomic Energy Commission, New York, December 16, 1966.
15. **McQuade, L. C. and Borum, R. L.,** Report by Business and Defense Services Administration, U.S. Department of Commerce, TID-24058, Washington, D.C., March 1968, 26.
16. National Academy of Sciences, *Radiation Preservation of Foods,* Publ. 1273, National Research Council, National Academy of Sciences, U.S. Government Printing Office, Washington, D.C., 1965.
17. **Slavin, J. W., Ronsivalli, L. J., and Kaylor, J. D.,** Radiopasteurization of fishery products, *Act. Rep.,* 17, 120, 1966.
18. **Ronsivalli, L. J., Gorga, C., Kaylor, J. D., and Carver, J. H.,** Guaranteed Quality Report, National Marine Fisheries Service, Gloucester, Mass., May 15, 1977.
19. **Goldblith, S. A.,** Sane perspectives regarding radiation effects, *Food Technol. (Chicago),* 24(3), 47, 1970.
20. **Meyer, R. A.,** Induced radioactivity in food and electron sterilization, *Health Phys.,* 12, 1027, 1966.
21. **Glass, R. A. and Smith, H. D.,** Radioactivities Produced in Foods by High Energy Electrons, Rep. No. 10, Stanford Research Institute to U.S. Army, 1960, S572.
22. **Richardson, L.,** Progress Report by the U.S. Army Surgeon General, Contract No. DA 49-007-MD-582, 10/10/55 to 1/1/56, 1956.
23. **Mead, J. F.,** The Effect of Ionizing Radiation upon the Vitamins in Radiation Sterilization. Review of the Literature in Selected Fields, Compiled by Radiation Sterilization Project, U.S. Army Quartermaster Corps, Chicago, February 1955.
24. **Poling, C. E., Warner, W. D., Humberg, F. R., Reber, E. F., Urbain, W. M., and Rice, E.,** Growth, reproduction, survival, and histopathology of rats fed beef irradiated with electrons, *Food Res.,* 20, 193, 1955.
25. **Proctor, B. E. and Goldblith, S. A.,** Effect of soft X-rays on vitamins (niacin, riboflavin, and ascorbic acid), *Nucleonics,* 4(3), 56, 1949.
26. **Elliot, A. M., Brownell, L. E., and Gross, J. A.,** Use of tetrahymena to evaluate the effects of γ-radiation on essential nutrilites, *J. Protozool.,* 1, 193, 1954.

27. **Lemmon, R. M., Mazzetti, F., and Parsons, M. A.**, Report by University of California Radiation Laboratory, No. UCRL 3068 to U.S. Atomic Energy Commission, Contract No. W-7405-eng-48 July 1955.
28. **Hummel, R.**, Effect of X-ray irradiation on cholesterol and its fatty acid esters in vitro and in vivo, *Strahlentherapie*, 36, 533, 1930.
29. **Coloves, G. C. and Woods, O. R.**, Symp. Electron Beam Radiation, Milwaukee, Wis., March 21—22, 1955.
30. **Metta, V. C. and Johnson, B. C.**, Progress Report, U.S. Army Surgeon General, Contract No. DA 49-007-MD-544, 7/12 to 10/3, 1955.
31. **Sheffner, A. L., Adachi, R., and Wills, L.**, Progress Report, U.S. Army Quartermaster Corps, File No. 090-15E, 2/1 to 5/1, 1956.
32. **King, C. G. and Becker, R. R.**, Progress Report, U.S. Army Surgeon General, Contract No. DA 49-007-MD-550, 10/15/54 to 1/15/55, 1955.
33. **Salmon, W. D., Sauberlich, H. E., Hove, E. L., Alexander, H. D., and Day, E. J.**, Progress Report, U.S. Army Surgeon General, Contract No. DA 49-007-MD-543, 10/15/55 to 1/15/56, 1956.
34. **Proctor, B. E., Nickerson, J. T. R., and Goldblith, S. A.**, Progress Report, U.S. Army Quartermaster Corps, Contract No. DA 19-129-QM-521, 9/1/55 to 12/31/55, 1955.
35. **Brooke, R. O., Ravesi, E. M., Gadbois, D. F., and Steinberg, M. A.**, Preservation of fresh unfrozen fishery products by low-level radiation. III. The effects of radiation pasteurization on amino acids and vitamins in clams, *Food Technol. (Chicago)*, 18, 1060, 1964.
36. **Brooke, R. O., Ravesi, E. M., Gadbois, D. F., and Steinberg, M. A.**, Preservation of fresh unfrozen fishery products by low-level radiation. V. The effects of radiation pasteurization on amino acids and vitamins in haddock fillets, *Food Technol. (Chicago)*, 20(11), 99, 1966.
37. **Porter, G. and Festig, M.**, A comparison between irradiated and autoclaved diets for breeding mice with observations on palatability, *Lab. Anim.*, 4, 203, 1970.
38. **Kraybill, H. F.**, Are irradiated foods harmful?, *Nucleonics*, 18, 112, 1960.
39. **Hickman, J. R.**, United Kingdom food irradiation programme — wholesomeness aspects, in *Food Irradiation*, Proc. Int. Symp., Karlsruhe, IAEA, Vienna, 1966, 101.
40. **Keplinger, M. L., Kinoshita, F. K., Kennedy, G. L., Marias, A. J., and Oscarson, E.**, International Project in the Field of Food Irradiation, Rep. No. IFIP-R38, Karlsruhe, 1976.
41. **Keplinger, M. L., Kennedy, G. L., and Smith, S.**, International Project in the Field of Food Irradiation, Rep. No. IFIP-R27, Karlsruhe, 1974.
42. **Keplinger, M. L., Kennedy, G. L., and Smith, S.**, International Project in the Field of Food Irradiation, Rep.No. IFIP-R23, Karlsruhe, 1974.
42a. **Oscarson, E. and Marias, A. J.**, Final Rep. No. IFIP-R38, Industrial Bio-Test Labs, Northbrook, Ill., May 1976.
43. **Alexander, H. D. and Salmon, W. D.**, Long-term Rat and Dog Feeding Tests on Irradiated Sweet Potatoes and Codfish, Rep. from Alabama Polytechnic Institute to U.S. Army, Contract No. DA-49-007-MD-543, Auburn, September 1959.
43a. **West, T. E. and Burtner, B. R.**, Final Rep. No. IFIP-R41, Industrial Bio-Test Labs., Northbrook, Ill.,, June 1976.
44. **Paynter, O. E.**, Long-Term Feeding and Reproduction Studies on Irradiated Corn and Tuna Fish, Final rep. by Hazelton Labs, Falls Church, Va., Contract No. DA-49-007-MD-788 with U.S. Army, September 1959.
45. **Ross, M. A. and Hood, E.**, Radiation and Sterilization of Foods, Rep. by Hazelton Labs, Falls Church, Va., Contract with Armed Forces Institute of Pathology, Washington, D. C., Project No. 6x60-01-001-02, September 1963.
46. **Maurer, F. D., Ross, M. A., Hood, E., and Tucker, W. E.**, Radiation and Sterilization of Foods, Rep. by Virginia Polytechnic Institute, Blacksburg, Va., Contract with Armed Forces Institute of Pathology, Washington, D. C., Project No. 6x60-11-001, September 1961.
47. **Phillips, A. W., Newcomb, H. R., and Shanklin, D.**, Long-Term Rat-Feeding Studies: Irradiated Shrimp and Oranges Diet, Final rep., Biological Research Labs, Syracuse University Research Institute, New York, Contract No. DA-49-007-MD-791 with U.S. Army, June 1961.
48. **Taub, I. A., Angelini, P., and Merritt, C., Jr.**, Irradiated food; validity of extrapolating wholesomeness data, *J. Food Sci.*, 41, 942, 1976.
49. **Ronsivalli, L. J., King, F. J., Ampola, V. G., and Holston, J. A.**, Study of irradiated-pasteurized fishery products, *Isot. Radiat. Technol.*, 8(3), 321, 1971.
50. **Licciardello, J. J., Ronsivalli, L. J., and Slavin, J. W.**, The effect of oxygen tension on the spoilage microflora of irradiated and nonirradiated haddock *(Melanogrammus aeglefinus)* fillets, *J. Appl. Bacteriol.*, 30, 239, 1967.
51. **Lerke, P., Adams, R., and Farber, L.**, Bacteriology of spoilage of fish muscle. III. Characterization of spoilers, *Appl. Microbiol.*, 13, 625, 1965.

52. **Thornley, M. J.**, Radiation resistance among bacteria, *J. Appl. Bacteriol.* 26, 334, 1963.
53. **Schmidt, C. F., Lechowich, R. V., and Folinazzo, J. F.**, Growth and toxin production by Type E *Clostridium botulinum* below 40°F, *J. Food Sci.*, 26, 626, 1961.
54. **Schmidt, C. F., Nank, W. K., and Lechowich, R. V.**, Radiation sterilization of food. II. Some aspects of the growth, sporulation, and radiation resistance of spores of *Clostridium botulinum* Type E, *J. Food Sci.*, 27, 77, 1962.
55. **Goldblith, S. A. and Nickerson, J. T. R.**, The effect of gamma rays on haddock and clams inoculated with *Clostridium botulinum* Type E, Rep. from MIT, Cambridge, No. MIT-3325, to U.S. Atomic Energy Commission, 1965.
56. **Learson, R. J., Schwartz, M. S., and Ronsivalli, L. J.**, Study of Irradiated-Pastuerized Fishery Products, Final rep. by National Marine Fisheries Service, Gloucester, Mass., to U.S. Atomic Energy Commission, No. TID25544, Contract No. AT(49-11)-1889, October 1970.
57. **Eklund, M. W., Poysky, F. T., and Wieler, D. I.**, Growth and Toxin Production of *Clostridium botulinum*, Types E, Nonproteolytic B and F in Nonirradiated and Irradiated Fishery Products in the Temperature Range 36 — 72°F, Final summary rep. by National Marine Fisheries Service, Seattle, Wash., to U.S. Atomic Energy Commission, Contract No. AT(49-7)-2442, November 1967.
58. **Zak, J.**, Some Factors in the Outgrowth of *Cl. botulinum* Type E, Ph.D. thesis, Massachusetts Institute of Technology, Cambridge, 1970.
59. **Ronsivalli, L. J., Schwartz, M. S., and Nickerson, J. T. R.**, A proposed method for testing the possibility of a botulism hazard in radiation-pasteurized fish, *Isot. Radiat. Technol.*, 8(2), 211, 1971.
60. **Licciardello, J. J., Nickerson, J. T. R., Goldblith, S. A., Shannon, C. A., and Bishop, W. W.**, Development of radiation resistance in *Salmonella* cultures, *Appl. Microbiol.*, 18, 24, 1969.
61. **Stapleton, G. E.**, Variations in the Radiosensitivity of *Escherichia coli* During the Growth Cycle, Ph.D. thesis, University of Tennessee, Knoxville, 1952.
62. **Liston, J. and Matches, J. R.**, Single and multiple doses in the radiation pasteurization of seafoods, *Food Technol. (Chicago)*, 22, 893, 1968.
63. **Hannesson, G.**, Objectives and present status of irradiation of fish and seafoods, *Food Irradiat. Inf.*, 1, 39, 1972.
64. **Ampola, V. G. and Ronsivalli, L. J.**, Effect of preirradiation quality of eviscerated haddock on postirradiation shelf life of fillets, *J. Food Sci.*, 34(1), 27, 1969.
65. **Laycock, A. and Regier, L. W.**, The effect of gamma irradiation on the microbial spoilage patterns on fish in relation to initial quality, in *Preservation of Fish by Irradiation*, IAEA, Vienna, 1970.
66. **Teeny, F. M. and Miyauchi, D.**, Irradiation of Pacific coast fish at sea, *J. Milk Food Technol.*, 33(8), 330, 1970.
67. **Novak, A. F., Ramachandra Rao, M. R., Grodner, R. M., and Liuzzo, J. A.**, Commercial Irradiation of Shellfish with a Portable Shipboard Irradiator, Final rep. U.S. Atomic Energy Commission Rep. ORO-663, Louisiana State University, Baton Rouge, June 30, 1968.
68. **Nickerson, J. T. R. and Licciardello, J. J.**, Storage of eviscerated haddock in refrigerated sea water treated with gamma radiation, *J. Milk Food Technol.*, 28(7), 223, 1965.
69. **Power, H. E., Fraser, D. I., Dyer, W. J., Neal, W., and Castrell, C. H.**, Effect of pasteurizing doses of gamma radiatin on storage life of cooked lobster meat, *J. Fish. Res. Board Can.*, 24(2), 221, 1967.
70. **Kumta, U. S., Mavinkurve, S. S., Gore, M. S., Sawant, P. L., Gangal, S. V., and Sreenivasan, A.**, Radiation pasteurization of fresh and blanched tropical shrimps, *J. Food Sci.*, 35, 360, 1970.
71. **Wood, T. H.**, Influence of temperature and phase state on X-ray sensitivity of yeast, *Arch. Biochem. Biophys.*, 52, 157, 1954.
72. **Licciardello, J. J.**, Effect of temperature on radiosensitivity of *Salmonella typhimurium*, *J. Food Sci.*, 29(4), 469, 1964.
73. **Pallas, J. E., III, and Hamdy, M. K.**, Effects of thermoradiation on bacteria, *Appl. Environ. Microbiol.*, 32(2), 250, 1976.
74. **Goldblith, S. A. and Nickerson, J. T. R.**, Simultaneous Radiation-Heating Treatment of Precooked Marine Products, Final Rep. TID 4500 for Division of Isotopes Development, U.S. Atomic Energy Commission on Contract No. AT(30-1)-3343, MIT, Cambridge, February 1, 1965 to February 28, 1966.
75. **Nickerson, J. T. R. and Goldblith, S. A.**, Simultaneous Radiation-Heating of Haddock, U.S. Atomic Energy Commission, Rep. TID-21932, MIT, Cambridge, 1965.
76. **Shiflett, M. A.**, Preservation of Dover Sole by Low-Dose Radiation and Antimicrobial Agents, Master's thesis, Oregon State University, Corvallis, 1965.
77. **Tomiyama, T., Jeng, S. S., Shiraishi, E., Hamada, M., and Yamada, Y.**, Complementary effect of food preservative and gamma radiation of whale and horse mackerel flesh, in *Freezing and Irradiation of Fish*, Kreuzer, R., Ed., Fishing News (Books), London, 1969.
78. **Lee, J. S., Shiflett, M. A., and Sinnhuber, R. O.**, Radiation pastuerization of seafood. I. The combined effect of sodium benzoate and irradiation for retarding microbial growth in Dover sole *(Microstomus pacificus)*, *Int. J. Appl. Radiat. Isot.*, 16, 221, 1965.

79. **Shewan, J. M. and Liston, J.**, Experiments on the irradiation of fish with 4 Mev. cathode rays and cobalt-60 gamma rays, in 2nd Int. Conf. Peaceful Uses Atomic Energy (United Nations), Geneva, 1958.
80. **Lerke, P. A., Farber, L., and Huber, W.**, Preservation of fish and shellfish by relatively low doses of beta radiation and antibiotics, *Food Technol. (Chicago)*, 15(3), 145, 1961.
81. **Awad, A. A., Sinnhuber, R. O., and Anderson, A. W.**, Radiation pasteurization of raw and chlortetracycline treated shrimp, *Food Technol. (Chicago)*, 19(5), 182, 1965.
82. **Roy, A. N. and Kaylor, J. D.**, Low level ionizing radiation and spice treatment of raw, headless, white shrimp, *Mar. Fish. Rev.*, 37(4), 16, 1975.
83. **Spinelli, J. and Miyauchi, D.**, Irradiation of Pacific coast fish and shellfish. V. The effect of 5′ inosine monophosphate on the flavor of irradiated fish fillets, *Food Technol., (Chicago)*, 22(6), 123, 1968.
84. **Ronsivalli, L. J. and Peters, J. A.**, Packaging requirements for irradiated fishery products, *Fish. Gaz.*, 81(13), 134, 1964.
85. **Ronsivalli, L. J., Steinberg, M. A., and Seagran, H. L.**, Radiation preservation of fish of the northwest Atlantic and the Great Lakes, in Radiation Preservation of Foods, Publ. 1273, National Research Council, National Academy of Sciences, Washington, D.C., 1965.
86. **Tinker, B. L., Ronsivalli, L. J., and Slavin, J. W.**, Suitability of flexible plastics as packaging materials for radiopasteurized seafoods, *Food Technol. (Chicago)*, 20(10), 122, 1966.
87. **Astrack, A., Brasch, O., and Huber, W.**, Effects of high-intensity electron bursts upon various vegetable and fish oils, *Food Res.*, 17, 571, 1952.
88. **Spinelli, J., Eklund, M., Stoll, N., and Miyauchi, D.**, Irradiation preservation of Pacific coast fish and shellfish. III. Storage life of petrale sole fillets at 33° and 42°F, *Food Technol. (Chicago)*, 19(6), 126, 1965.
89. **Jorgensen, B. V. and Hansen, P.**, Storage life of vacuum-packed iced trout. II. Influence of radiation pasteurization, *J. Sci. Food Agric.*, 17, 140, 1966.
90. **Ampola, V. G., Connors, T. J., and Ronsivalli, L. J.**, Preservation of fresh unfrozen fishery products by low-level radiation. VI. Optimum radiopasteurization dose studies on ocean perch, pollock and cod fillets, *Food Technol. (Chicago)*, 23(3), 357, 1969.
91. **Hansen, P. and Jorgensen, B. V.**, The effect of ionizing radiation and antioxidant treatment on the quality and storage life of vacuum-packed trout at 0°C, in *Microbiological Problems in Food Preservation by Irradiation*, IAEA, Vienna, 1967.
92. **Kazanas, N., Emerson, J. A., Seagran, H. L., and Kempe, L. L.**, Effect of gamma irradiation on the microflora of freshwater fish. I. Microbial load, lag period and rate of growth on yellow perch *(Perca flavescens)* fillets, *Appl. Microbiol.*, 14(2), 261, 1966.
93. **Pelroy, G. A. and Eklund, M. W.**, Changes in the microflora of vacuum-packaged irradiated petrale sole *(Eopsetta jordani)* fillets stored at 0.5°C, *Appl. Microbiol.*, 14(6), 921, 1966.
94. **Sinskey, R. J., Pablo, I. S., Silverman, G. J., and Ronsivalli, L.**, Effect of packaging on the major microbial flora of irradiated haddock, *Nature (London)*, 213(5074), 425, 1967.
95. **Castell, C. H. and MacCallum, W. A.**, The value of temperatures close to freezing on the storage of fish, *J. Fish. Res. Board Can.*, 8(2), 111, 1950.
96. **Spencer, R. and Baines, C. R.**, The effect of temperature on the spoilage of wet white fish. I. Storage at constant temperatures between −1 and 25°C, *Food Technol. (Chicago)*, 18(5), 175, 1964.
97. **Masurovsky, E. B., Goldblith, S. A., and Nickerson, J. T. R.**, Effects of substerilization doses of cobalt-60 gamma radiation on the cold-storage life extension of shucked soft-shelled clams and haddock fillets, *Appl. Microbiol.*, 11(3), 220, 1963.
98. **Slavin, J. W., Steinberg, M. A., and Ronsivalli, L. J.**, Radiation preservation of New England seafoods, *Isot. Radiat. Tech.*, 1(4), 317, 1964.
99. **Miyauchi, D., Eklund, M., Spinelli, J., and Stoll, N.**, Irradiation preservation of Pacific coast shellfish. I. Storage life of king crab meats at 33 and 42°F, *Food Technol. (Chicago)*, 18(6), 138, 1964.
100. **Hannesson, G. and Dagbjartsson, B.**, Report on a survey project in Iceland on the use of radiation pasteurization of various seafoods, in *Preservation of Fish by Irradiation*, IAEA, Vienna, 1970.
101. **Elliot, R. P. and Michener, H. D.**, Factors affecting the growth of psychrophilic microorganisms in foods, Tech. Bull. No. 1320, U.S. Department of Agriculture, Washington, D.C., 1965.
102. **Corlett, D. A., Jr.**, Microbial selection due to ionizing radiation, *Food Technol. (Chicago,)* 21, 755, 1967.
103. **Harris, N. D.**, The influence of the recovery medium and the incubation temperature on the survival of damaged bacteria, *J. Appl. Bacteriol.*, 26(3), 387, 1963.
104. **Dyer, F. E.**, Microorganisms from Atlantic cod, *J. Fish. Res. Board Can.*, 7(3), 128, 1947.
105. **Liston, J.**, The bacterial flora of fish caught in the Pacific, *J. Appl. Bacteriol.*, 23(3), 469, 1960.
106. **Shewan, J. M. and Hobbs, G.**, The bacteriology of fish spoilage and preservation, in *Progress in Industrial Microbiology*, Vol. 6, Hockenhull, D. J. D., Ed., Lliffe Books, London, 1967.

107. **Castell, C. H., MacCallum, W. A., and Power, H. E.**, Spoilage of fish in the vessels at sea. II. Treatment on the deck and in the hold, *J. Fish. Res. Board Can.*, 13(1), 21, 1956.
108. **Shewan, J. M.**, The microbiology of seawater fish, in *Fish as Food*, Vol. 1, Borgstrom, G., Ed., Academic Press, London, 1961.
109. **Gillespie, N. C. and Macrae, I. C.**, The bacterial flora of some Queensland fish and its ability to cause spoilage, *J. Appl. Bacteriol.*, 39, 91, 1975.
110. **Kazanas, N.**, Effect of gamma irradiation on the microflora of freshwater fish. II. Generic identification of aerobic bacteria from yellow perch fillets, *Appl. Microbiol.*, 14(6), 957, 1966.
111. **Niven, C. F., Jr.**, Microbiological aspects of radiation preservation of food, *Ann. Rev. Microbiol.*, 12, 507, 1958.
112. **Masurovsky, E. B., Voss, J. S., and Goldblith, S. A.**, Changes in the microflora of haddock fillets and shucked soft-shelled clams after irradiation with cobalt-60 gamma rays and storage at 0°C and 6°C, *Appl. Microbiol.*, 11, 229, 1963.
113. **Laycock, R. A. and Regier, L. W.**, Pseudomonads and achromobacters in the spoilage of irradiated haddock of different pre-irradiation quality, *Appl. Microbiol.*, 20(3), 333, 1970.
114. **Corlett, D. A., Jr., Lee, J. S., and Sinnhuber, R. O.**, Application of replicate plating and computer analysis for rapid identification of bacteria in some foods. II. Analysis of microbial flora in irradiated Dover sole *(microstomus pacificus)*, *Appl. Microbiol.*, 13(5), 818, 1965.
115. **Shiflett, M. A., Lee, J. S., and Sinnhuber, R. O.**, Microbial flora of irradiated Dungeness crabmeat and Pacific oysters, *Appl. Microbiol.*, 14(3), 411, 1966.
116. **Kazanas, N. and Emerson, J. A.**, Effect of gamma irradiation on the microflora of freshwater fish. III. Spoilage patterns and extension of refrigerated storage life of yellow perch fillets irradiated to 0.1 and 0.2 megarad, *Appl. Microbiol.*, 16(2), 242, 1968.
117. **Pelroy, G. A., Seman, J. P., Jr., and Eklund, M. W.**, Changes in the microflora of irradiated Petrale sole *(Eopsetta jordani)* fillets stored aerobically at 0.5°C, *Appl. Microbiol.*, 15(1), 92, 1967.
118. **Eklund, M. W., Spinelli, J., Miyauchi, D., and Dassow, J.**, Development of yeast on irradiated Pacific crab meat, *J. Food Sci.*, 31(3), 424, 1965.
119. **Pelroy, G. A. and Eklund, M. W.**, Changes in the microflora of vacuum-packaged, irradiated Petrale sole *(Eopsetta jordani)* fillets stored at 0.5°C, *Appl. Microbiol.*, 14(6), 921, 1966.
120. **Shewan, J. M.**, Food poisoning caused by fish, in *Fish as Food*, Vol. 2, Borgstrom, G., Ed., Academic Press, New York, 1962.
121. **Angelotti, R., Foter, M. J., and Lewis, K. H.**, Time-temperature effects on salmonellae and staphylococci in foods. I. Behavior in refrigerated foods, *Am. J. Pub. Health*, 51(1), 76, 1961.
122. **Matches, J. R. and Liston, J.**, Growth of salmonellae on irradiated and nonirradiated seafoods, *J. Food Sci.*, 33, 406, 1968.
123. **Matches, J. R. and Liston, J.**, Radiation destruction of *Vibrio parahaemolyticus*, *J. Food Sci.*, 36, 339, 1971.
124. **Quinn, D. J., Anderson, A. W., and Dyer, J. F.**, The inactivation of infection and intoxication micro-organisms by irradiation in seafood, in *Microbiological Problems in Food Preservation by Irradiation*, IAEA, Vienna, 1967.
125. **Anderson, A. W.**, Irradiation inactivation of food infection microorganisms in seafoods, in *Freezing and Irradiation of Fish*, Kreuzer, R., Ed., Fishing News (Books), London, 1969.
126. **Slabyj, B. M., Dollar, A. M., and Liston, J.**, Post-irradiation survival of *Staphylococcus aureus* in sea foods, *J. Food Sci.*, 30, 344, 1965.
127. **Van der Schaaf, A. and Mossel, D. A. A.**, Gamma radiation sanitation of fish and blood meals, *Int. J. Appl. Radiat. Isot.*, 14, 557, 1963.
128. **Ostovar, K., Pereira, R. R., and Gallop, R. A.**, Effects of gamma irradiation on salmonella spp. in smoked lake whitefish *(Coregonus clupeaformis)*, *J. Fish. Res. Board Canada*, 28, 643, 1971.
129. **Hobbs, G.**, *Clostridium botulinum* and its importance in fishery products, in *Advances in Food Research*, Vol. 22, Academic Press, New York, 1976.
130. **Segner, W. P. and Schmidt, C. F.**, Radiation resistance of spores of *Clostridium botulinum* type E, in *Food Irradiation*, International Atomic Evergy Agency, Vienna, 1966, 956.
131. **Wheaton, E. and Pratt, G. B.**, Radiation survival curves of *Clostridium botulinum* spores, *J. Food Sci.*, 27, 327, 1962.
132. **Roberts, T. A. and Ingram, M.**, Radiation resistance of spores of clostridium species in aqueous suspension, *J. Food Sci.*, 30, 879, 1965.
133. **Hobbs, G.**, Prospects for the elimination of *Clostridium botulinum* from fish and fishery products by irradiation, in *Elimination of Harmful Organisms from Food and Feed by Irradiation*, IAEA Vienna, 1968, 101.
134. **Eklund, M. W. and Poysky, F. T.**, The significance of non-proteolytic *Clostridium botulinum* types B, E and F in the development of radiation pasteurized fishery products, in *Preservation of Fish by Irradiation*, IAEA, Vienna, 1970, 125.

135. **Shewan, J. M. and Hobbs, G.**, The botulism hazard in the proposed use of irradiation of fish and fishery products in the United Kingdom, in *Preservation of Fish by Irradiation*, IAEA, Vienna, 1970, 117.
136. **Ajmal, M.**, Growth and toxin production of *Clostridium botulinum* type E, *J. Appl. Bacteriol.*, 31(1), 120, 1968.
137. **Hobbs, G.**, Toxin production by *Clostridium botulinum* type E in fish, in *Microbiological Problems in Food Preservation by Irradiation*, IAEA, Vienna, 1967, 148.
138. **Goldblith, S. A. and Nickerson, J. T. R.**, The effect of gamma rays on ocean perch inoculated with type E *Clostridium botulinum*, Final rep. for Division of Biology and Medicine, U.S. Atomic Energy Commission on contract number AT(30-1)-3343, 1967.
139. **Goldblith, S. A. and Nickerson, J. T. R.**, The effect of gamma rays on marine finfish fillets inoculated with type E, *Clostridium botulinum*, Final rep. for Division of Isotopes Development, U.S. Atomic Energy Commission on contract number AT(30-1)-3343, 1968.
140. **Cann, D. C., Wilson, B. B., Shewan, J. M., and Hobbs, G.**, Incidence of *Clostridium botulinum* type E in fish products in the United Kingdom, *Nature (London)*, 211, 205, 1966.
141. **Nickerson, J. T. R., Goldblith, S. A., DiGioia, G., and Bishop, W. W.**, The presence of *Cl. botulinum* type E in fish and mud taken from the Gulf of Maine, in *Botulism 1966*, Ingram, M. and Roberts, T. A., Eds., Chapman and Hall, London, 1967.
142. **Licciardello, J. J., Nickerson, J. T. R., Ribich, C. A., and Goldblith, S. A.**, Thermal inactivation of type E botulinum toxin, *Appl. Microbiol.*, 15, 249, 1967.
143. **Hobbs, G., Cann, D. C., and Wilson, B. B.**, An evaluation of the botulinum hazard in vacuum packaged smoked fish, *J. Food Technol.*, 4, 185, 1969.
144. **Daget, J.**, Insect infection of African dried or smoked fish and the possibility of its control by irradiation, in *Application of Food Irradiation in Developing Countries*, IAEA, Vienna, 1966.
145. **Boisot, M. H. and Gauzit, M.**, Disinsectization of African dried and smoked fish by means of irradiation, in *Application of Food Irradiation in Developing Countries*, IAEA, Vienna, 1966.
146. **Kumta, U.S. and Sreenivasan, A.**, Preservation by gamma radiation of Bombay duck, shrimps and white pomfrets, in *Preservation of Fish by Irradiation*, IAEA, Vienna, 1970.
147. **Van Mameren, J. and Houwing, H.**, Effect of irradiation on *Anisakis* larvae in salted herring, in *Elimination of Harmful Organisms from Food and Feed by Irradiation*, IAEA, Vienna, 1968.
148. **Gould, E. and Peters, J. A.**, *On Testing the Freshness of Frozen Fish*, Fishing News (Books), London 1971, 9.
149. **Shewan, J. M. and Hobbs, G.**, Quality assessment and consumer acceptance of irradiated seafoods, in *Freezing and Irradiation of Fish*, Kreuzer, R., Ed., Fishing News (Books), London, 1969.
150. **Spinelli, J., Eklund, M., and Miyauchi, D.**, Irradiation preservation of Pacific coast shellfish. II. Relation of bacterial counts, trimethylamine and total volatile base to sensory evaluation of irradiated king crab meat, *Food Technol. (Chicago)*, 18(6), 143, 1964.
151. **Spinelli, J., Pelroy, G., and Miyauchi, D.**, Quality indices that can be used to assess irradiated seafoods, in *Freezing and Irradiation of Fish*, Kreuzer, R., Ed., Fishing News (Books), London, 1969.
152. **Power, H. E., Fraser, D. I., Neal, W., Dyer, W. J., and Castell, C. H.**, Gamma irradiation as a means of extending the storage life of haddock fillets, *J. Fish. Res. Board Can.*, 21, 827, 1964.
153. **Van Mameren, J., Van Spreekens, K. J. A., Houwing, H., and Mossel, D. A. A.**, Effect of irradiation on the keeping quality of packaged cod flesh, in *Freezing and Irradiation of Fish*, Kreuzer, R., Ed., Fishing News (Books), London, 1969, 446.
154. **Castell, C. H., Anderson, G. W., and Pivick, H.**, Relation of bacterial counts to quality of cod fillets, *J. Fish, Res. Board Can.*, 7(6), 378, 1948.
155. **Shewan, J. M. and Hobbs, G.**, The bacteriology of fish spoilage, in *Progress in Industrial Microbiology*, Vol. 6, Gordon and Breach, Science Publishers, London, 1965.
156. **Kumta, U.S. and Mavinkurve, S. S.**, Radiation sensitivity and biochemical characteristics of microflora of Bombay duck *(Harpodon nehereus)*, *J. Food Sci.*, 36, 63, 1971.
157. **Carver, J. H., Connors, T. J., Ronsivalli, L. J., and Holston, J. A.**, Shipboard Irradiation Studies, Bureau of Commercial Fisheries, Technological Laboratory, for U.S. Atomic Energy Commission, Contract No, AT(49 11) 1889, Gloucester, Mass., 1967.
158. **Anon.**, Irradiation extends storage life of Bombay duck, *Food Irradiat.*, 5, A12, 1965.
159. **Mavinkurve, S. S., Gangal, S. Y., Sawant, P. L., and Kumta, U. S.**, Bacterial studies on irradiated tropical fish, *J. Food Sci.*, 32, 711, 1967.
160. **Sawant, P. L., Mavinkurve, S. S., Saroja, S., Savagaon, K. A., and Kumta, U. S.**, Radiation pasteurization of tropical fish, Bombay duck *(Harpodon nehereus)*, *Food Technol. (Chicago)*, 21, A126, 1967.
161. **Aiyar, A. S.**, Fish, *Food Irradiat. Inf.*, 6, 35, 38, 1976.
162. **Carver, J. H. and Steinberg, M. A.**, Effect of radiation pasteurization on the storage life and acceptability of some North Atlantic fish, *Commer. Fish. Rev.*, 21, 1, 1959.

163. **Shewan, J. M.**, The present status of radiation preservation of food products — fish, *Int. J. Appl. Radiat. Isot.*, 6, 143, 1959.
164. **Coleby, B. and Shewan, J. M.**, The radiation preservation of fish, in *Fish as Food*, Vol. 4, Borgstrom, G., Ed., Academic Press, New York, 1965, 419.
165. **Shewan, J. M. and Liston, J.**, Experiments on the irradiation of fish with 4 MeV cathode rays and cobalt-60 gamma rays, *Proc. 2nd U.N. Int. Conf. PUAE*, 27, 377, 1958.
166. **Slavin, J. W. and Ronsivalli, L. J.**, Study of Irradiated-Pasteurized Fishery Products, Bureau of Commercial Fisheries Technological Laboratory, for U.S. Atomic Energy Commission, Contract No, AT(49-11)-1889, Gloucester, Mass., 1964.
167. **Rhodes, D. N.**, Pasteurization of Fish by Ionizing Radiation. A Study of Feasibility in the United Kingdom, *Food Irradiat.*, 4, A8, 1964.
168. **Steinberg, M. A.**, The atom preserves seafoods, in Proc. Gulf Caribbean Fish. Inst., Key Biscayne, Fla., 1965, 56.
169. **Slavin, J. W., Ronsivalli, L. J., Kaylor, J. D., and Carver, J. H.**, Study of Irradiated-Pasteurized Fishery Products, Bureau of Commercial Fisheries Technological Laboratory for U.S. Atomic Energy Commission, Contract No. AT(49-11)-1889, 1965.
170. **Slavin, J. W., Nickerson, J. T. R., Goldblith, S. A., Ronsivalli, L. J., Kaylor, J. D., and Licciardello, J. J.**, The quality and wholesomeness of radiation-pasteurized marine products with particular reference to fish fillets, *Isot. Radiat. Technol.*, 2, 365, 1966.
171. **Wittfagel, H.**, Ionizing irradiation of marine fish and marine fish products in Germany, *Arch. Lebensmittelhyg.*, 16, 227, 1965.
172. **Slavin, J. W., Carver, J. H., Connors, T. J., and Ronsivalli, L. J.**, Shipboard Irradiator Studies, Bureau of Commercial Fisheries Technological Laboratory, for U.S. Atomic Energy Commission, Contract No. AT(49-11)-1889, Gloucester, Mass., 1966.
173. **Ronsivalli, L. J., Ampola, V. G., King, F. J., and Holston, J. A.**, Study of irradiated-pasteurized fishery products. Bureau of Commercial Fisheries Technological Laboratory, for U.S. Atomic Energy Commission, Contract No. AT(49-11)-1889, Gloucester, Mass., 1968.
174. **Metlitskii, L. V., Rogachev, V. N., and Krushchev, V. G.**, Radiation Processing of Food Products, Oak Ridge National Laboratory, U.S. Atomic Energy Commission, Oak Ridge, Tenn., 1968.
175. **Ronsivalli, L. J., Kaylor, J. D., and Slavin, J. W.**, Status of research on irradiated fish and shellfish in the United States, in *Freezing and Irradiation of Fish*, Kreuzer, R., Ed., Fishing News (Books), London, 1969, 482.
176. **Ronsivalli, L. J., King, F. J., Ampola, V. G., and Holston, J. A.**, Study of Irradiated-Pasteurized Fishery Products. Maximum Shelf Life Study. B. Radiation Chemistry, Bureau of Commercial Fisheries Technological Laboratory, for U.S. Atomic Energy Commission Contract No. AT(49-11)-1889, Gloucester, Mass., 1970.
177. **Miyauchi, D.**, Application of Radiation Pasteurization Processes to Pacific Coast Fishery Products, National Marine Fisheries Service for U.S. Atomic Energy Commission Contract No. AT(49-11)-2058, Seattle, Wash., 1970.
178. **Miyauchi, D.**, Irradiation preservation of Pacific northwest fish. 1. Cod fillets, *Food Technol. (Chicago)*, 14, 379, 1960.
179. **Miyauchi, D., Spinelli, J., Pelroy, G., Teeny, F., and Seman, J.**, Application of Radiation-Pasteurization Process to Pacific Crab and Flounder, Bureau of Commercial Fisheries Technological Laboratory for U.S. Atomic Energy Commission Contract No. AT(49-11)-2058, Seattle, Wash., 1967.
180. **Miyauchi, D., Spinelli, J., Pelroy, G., and Stoll, N.**, Application of Radiation-Pasteurization Processes to Pacific Crab and Flounder, Bureau of Commercial Fisheries Technological Laboratory for U.S. Atomic Energy Commission Contract No. AT(49-11)-2058, Seattle, Wash., 1965.
181. **Miyauchi, D., Spinelli, J., and Pelroy, G.**, Application of Radiation-Pasteurization Processes to Pacific Crab and Flounder, Bureau of Commercial Fisheries Technological Laboratory for U.S. Atomic Energy Commission Contract No. AT(49-11)-2058, Seattle, Wash., 1966.
182. **Spinelli, J. Pelroy, G., and Miyauchi, D.**, Irradiation of Pacific coast fish and shellfish. VI. Treatment with sodium tripolyphosphate, *Fish. Ind. Res.*, 4, 37, 1967.
183. **Miyauchi, D., Spinelli, J., Pelroy, G., and Steinberg, M. A.**, Radiation preservation of Pacific coast fishery products, *Isot. Radiat. Technol.*, 5, 136, 1968.
184. **Spinelli, J., Dollar, A. M., Wedemeyer, G. A., and Gallagher, E. C.**, Irradiation of fish fillets. Relation of vapor phase reactions to storage quality, *Int. J. Appl. Radiat. Isot.*, 20, 167, 1969.
185. **Liston, J., Dollar, A. M., and Matches, J. R.**, Effects of multiple dose irradiation on the bacterial flora of seafoods, *Freezing and Irradiation of Fish*, Kreuzer, R., Ed., Fishing News (Books), London, 1969, 463.
186. **Miyauchi, D. and Teeny, F.**, Application of Radiation-Pasteurization Processes to Pacific Coast Fishery Products, National Marine Fisheries Service Technological Laboratory for U.S. Atomic Energy Commission Contract No. AT(49-11)-2058, Seattle, Wash., 1970.

187. **Eukel, W. W. and Huber, W.**, Tooling up for low-dose irradiation processing of fresh foods, *Food Technol. (Chicago)*, 14, 198, 1960.
188. **Sieling, D. H.**, Preservation of Food by Low-Dose Ionizing Energy, Quartermaster Research and Engineering Center, Natick, Mass., 1961.
189. **Miyauchi, D., Eklund, M., Spinelli, J., and Stoll, N.**, Application of Radiation-Pasteurization Processes to Pacific Crab and Flounder, Bureau of Commercial Fisheries Technological Laboratory for U.S. Atomic Energy Commission Contract No. AT(49-11)-2058, Seattle, Wash., 1963.
190. **Dassow, J. A. and Miyauchi, D. T.**, Radiation preservation of fish and shellfish of the northwest Pacific and Gulf of Mexico, in Radiation Preservation of Foods, Publ. No. 1273, National Research Council, National Academy of Sciences, Washington, D.C., 1965.
191. **Miyauchi, D., Spinelli, J., Pelroy, G., and Stoll, N.**, Application of Radiation-Pasteurization Processes to Pacific Crab and Flounder, Bureau of Commercial Fisheries Technological Laboratory, U.S. Atomic Energy Commission Contract No. AT(49-11)-2058, Seattle, Wash., 1965.
192. **Anon.**, Radiation of Fish, Bureau of Commercial Fisheries Technological Laboratory for U.S. Atomic Energy Commission Contract No. AT(49-11)-2058 and U.S. Bureau of Commercial Fisheries Contract No. 14-17-001-523, Seattle, Wash., 1962.
193. **Miyaushi, D., Eklund, M., Dassow, J., Spinelli, J., and Stoll, N.**, Application of Radiation-Pasteurization Processes to Pacific Crab and Flounder, Bureau of Commercial Fisheries Technological Laboratory for U.S. Atomic Energy Commission Contract No. AT(49-11)-2058, Seattle, Wash., 1963.
194. **Slavin, J. W., Ronsivalli, L. J., and Connors, T. J.**, Status of research and development studies on radiation-pasteurization of fish and shellfish in the United States, in *Food Irradiation*, IAEA, Vienna, 1966.
195. **Slavin, J. W. and Ronsivalli, L. J.**, Study of Irradiated-Pasteurized Fishery Products, U.S. Fish and Wildlife Technological Laboratory for U.S. Atomic Energy Commission Contract No. AT(49-11)-1889, Gloucester, Mass., 1963.
196. **Slavin, J. W., Steinberg, M. A., and Connors, T. J.**, Low-Level Radiation Preservation of Fishery Products, U.S. Fish and Wildlife Technological Laboratory for U.S. Atomic Energy Commission Contract No. AT(49-11)-1889, Gloucester, Mass., 1963.
197. **Nickerson, J. T. R.**, The storage life extension of refrigerated marine products by low-dose radiation treatment, in *Exploration in Future Food Processing Techniques*, MIT Press, Cambridge, 1963.
198. **Ronsivalli, L. J. and Slavin, J. W.**, Pasteurization of fishery products with gamma rays from a cobalt-60 source, *Commer. Fish. Rev.*, 27, 1, 1965.
199. **Connors, T. J. and Steinberg, M. A.**, Preservation of unfrozen fishery products by low-dose level radiation. IV. Organoleptic studies on radiation-pasteurized skinless haddock fillets, *Food Technol.(Chicago)*, 20, 117, 1966.
200. **Hembree, H. W. and Burt, T. B.**, Troop acceptance tests of irradiated foods, in Radiation Preservation of Foods, Publ. No. 1273, National Research Council, National Academy of Sciences, Washington, D.C., 1965.
201. **Ampola, V. G. and Ronsivalli, L. J.**, Effects of special handling of haddock on the post irradiation shelf life of haddock fillets, *Fish. Ind. Res.*, 4, 109, 1968.
202. **Stansby, M. E. and Kudo, G.**, Effect of Ionizing Radiation on the Lipids of Fish, Fish and Wildlife Service Technological Laboratory for U.S. Atomic Energy Commission Contract No. AT(49-11)-2058, Seattle, Wash., 1964.
203. **Schonborn, W., Kinkel, H. J., and Hafferl, W.**, Investigations on pasteurization of cold marinades by cobalt-60 gamma rays, in *Food Irradiation*, IAEA, Vienna, 1966.
204. **Goresline, H. E.**, The potential of ionizing radiation for food preservation, *Food Irradiat. Inf.*, 2, 20, 1973.
205. **Amano, K. and Yamanaka, H.**, Meat color regeneration in frozen tuna by irradiation, in *Freezing and Irradiation of Fish*, Kreuzer, R., Ed., Fishing News (Books), London, 1969.
206. **Massa, D., Perna, A., and Beliomini, A.**, Further reserach on whitings pasteurized with ionizing radiation, Abstr. in *Preservation of Fish by Ionizing Radiation*, An Annotated Bibliography, Weiss, F. J., McKinney, F. E., and Frye, L. M., Eds., US. Atomic Energy Commission, 1969.
207. **Emerson, J. A., Kazanas, N., Gnaedinger, R. H., Krzeczkowski, R. A., Seagran, H. L., Markakis, P., Nicholas, R. C., Schweigert, B. S., and Kempe, L. L.**, Irradiation Preservation of Fresh Water Fish and Inland Fruits and Vegetables, Bureau of Commercial Fisheries Technological Laboratory, University of Michigan and Michigan State University for U.S. Atomic Energy Commission Contract No. AT(11-1)-1283, Ann Arbor, and East Lansing, Mich., 1966.
208. **Emerson, J. A., Kazanas, N., Greig, R. A., and Seagran, H. L.**, Irradiation Preservation of Fresh Water Fish, Bureau of Commercial Fisheries Technological Laboratory for U.S. Atomic Energy Commission Contract No. AT(11-1)-1283, Ann Arbor, Mich., 1965.
209. **Emerson, J. A., Kazanas, N., Greig, R. A., and Seagran, H. L.**, Irradiation Preservation of Fresh Water Fish, Bureau of Commerical Fisheries Technological Laboratory for U.S. Atomic Energy Commission Contract No. AT(11-1)-1283, Ann Arbor, Mich., 1964.

210. **Emerson, J. A., Kazanas, N., Gnaedinger, R. H., Krzeczkowski, R. A., and Seagran, H. L.**, Irradiation of Fresh Water Fish, Bureau of Commercial Fisheries Technological Laboratory for U.S. Atomic Energy Commission Contract No. AT(11-1)-1283, Ann Arbor, Mich., 1965.
211. **Emerson, J. A., Kazanas, N., Greig, R. A., Seagran, H. L., and Kempe, L. L.**, I. Extension of the refrigerated storage life of fresh yellow perch, *Food Technol. (Chicago)*, 20, 108, 1966.
212. **Graikowski, J. T., Kazanas, N., Watz, J., Ducharme, S., Emerson, J. A., and Seagran, H. L.**, Irradiation Preservation of Fresh Water Fish, Bureau of Commercial Fisheries Technological Laboratory for U.S. Atomic Energy Commission Contract No. AT(49-11)-2954, Ann Arbor, Mich., 1968.
213. **Graikowski, J. T., Ducharme, S., Jackson, M., and Dudley, S.**, Preparatory Efficacy Studies for the Application of Ionizing Radiation to the Extension of the Refrigerated Storage of Fresh Water Perch *(Perca flavescens)*, Bureau of Commercial Fisheries Technological Laboratory for U.S. Atomic Energy Commission Contract No. AT(49-11)-2954, Ann Arbor, Mich., 1970.
214. **Hansen, P.**, The effect of ionizing radiation and antioxidant treatment on the quality and storage life of vacuum-packed trout at 0°C, in *Microbial Problems in Food Preservation by Irradiation*, IAEA, Vienna, 1967.
215. **Ostovar, K., Slusar, M., and Vaisey, M.**, Preservation of fresh whitefish with gamma radiation, *J. Fish. Res. Board Can.*, 24, 9, 1967.
216. **Gadbois, D. F., Mendelsohn, J. M., and Ronsivalli, L. J.**, Effects of radiation, heating and storage on volatile compounds in clam meats, *J. Food Sci.*, 32, 511, 1967.
217. **Ronsivalli, L. J., King, F. J., Mendelsohn, J. M., Gadbois, D. F., Brooke, R. O., and Holston, J. A.**, Chemistry of Radiopasteurized Foods, Bureau of Commercial Fisheries Technological Laboratory for U.S. Atomic Energy Commission Contract No. AT(49-11)-2443, Gloucester, Mass., 1968.
218. **Mendelsohn, J. M. and Brooke, R. O.**, Radiation processing and storage effects on the head space gas components in clam meats, *Food Technol. (Chicago)*, 22, 1163, 1968.
219. **Novak, A. F., Liuzzo, J. A., Grodner, R. M., and Lovell, R. T.**, Radiation pasteurization of Gulf coast oysters, *Food Technol. (Chicago)*, 20, 201, 1966.
220. **Novak, A. F. and Liuzzo, J. A.**, Radiation Pasteurization of Shrimp and Oysters, Louisiana State University for U.S. Atomic Energy Commission Rep. No. ORO-636, Baton Rouge, La., 1966.
221. **Novak, A. F. and Liuzzo, J. A.**, Radiation Pasteurization of Gulf Shellfish, Louisiana State University for U.S. Atomic Energy Commission Rep. No. ORO-654, Baton Rouge, La. 1967.
222. **Novak, A. F. and Liuzzo, J. A.**, Radiation Pasteurization of Gulf Shellfish, Louisiana State University for U.S. Atomic Energy Commission Rep. No. ORO-660, Baton Rouge, La., 1968.
223. **Gardner, E. A. and Watts, B. M.**, Effect of ionizing radiation on southern oysters, *Food Technol. (Chicago)*, 11, 329, 1957.
224. **Schultz, H. W. and Lee, J. S.**, Food preservation by irradiation. Present status, *Food Technol. (Chicago)*, 20, 38, 1966.
225. **Power, H. E., Fraser, D. I., Neal, W., Dyer, W. J., Castell, C. H., Freeman, H. C., and Idler, D. R.**, Gamma radiation for the preservation of scallop meat, *J. Fish. Res. Board Can.*, 21, 813, 1964.
226. **Scholz, D. J., Sinnhuber, R. O., East, D. M., and Anderson, A. W.**, Radiation pasteurized shrimp and crabmeat, *Food Technol. (Chicago)*, 16, 118, 1962.
227. **Miyauchi, D., Spinelli, J., Stall, N., Pelroy, G., and Eklund, M.**, Irradiation of Pacific coast fish and shellfish. IV. Storage life of Dungeness crab meat at 33°F and 42°F, *Int. J. Appl. Radiat. Isot.*, 17, 137, 1966.
228. **Dagbjartsson, B. and Solberg, M.**, Texture changes in precooked lobster meat *(Homarus americanus)* resulting from radurization followed by refrigerated storage, *Food Sci.*, 38, 165, 1973.
229. **Kaylor, J. D., Murphy, E. J., Green, J. H., and Hill, W. S.**, Marine Products Development Irradiation Program, National Marine Fisheries Service, Fishery Products Technology Laboratory for U.S. Atomic Energy Commission Contract No. AT(49-11)-3005, Gloucester, Mass., 1970.
230. **Lievens, L.**, Progress in food irradiation, *Food Irradiat. Inf.*, 6, 6, 1976.
231. **Novak, A. F. and Liuzzo, J. A.**, Radiation Pasteurization of Shrimp, Louisiana State University for U.S. Atomic Energy Commission Rep. No. ORO-601, Baton Rouge, La., 1962.
232. **Novak, A. F., Liuzzo, J. A., and Amant, L.**, Radiation pasteurization with cobalt-60 of fresh shrimp prior to freezing, *Quick Frozen Foods*, 25, 40, 1963.
233. **Novak, A. F.**, Radiation Pasteurization of Shrimp, Louisiana State University for U.S. Atomic Energy Commission Rep. No. ORO-615, Baton Rouge, La., 1964.

Chapter 3

RADURIZATION AND RADICIDATION: FRUITS AND VEGETABLES

James H. Moy

TABLE OF CONTENTS

I. INTRODUCTION

All fruits and vegetables are perishable due to physiological changes, postharvest fungal diseases, other pathological breakdown, and insect infestation. A good preservation technique for fresh fruits and vegetables should have the following characteristics:

1. It should be effective and efficient as a postharvest treatment.
2. It should retain the qualities and nutrients of the commodity.
3. It should have a proven capacity for controlling insect larvae and eggs if insect infestation is a problem.
4. It should have a synergistic effect on the commodity if and when combined with other preservation techniques.

Various preservation techniques such as dehydration, canning, freezing, or refrigerating the fruits and vegetables with controlled atmosphere serve their purposes as a means of extending the marketable life of these commodities but do not retain their original qualities in most instances. The recent concept of using subatmosphere, better known as "hypobaric" storage, for preserving fresh commodities, appears to be a sensible approach because fruits and vegetables stored in a partial vacuum have a lower respiration rate and retard bacterial and fungal growth. There might even be adverse effects on insects and their eggs. This remains to be studied, however, and verified.

Ionizing radiation, commonly in the form of gamma radiation, as a postharvest, preshipment treatment probably meets the above-mentioned requirements more completely than any other known technique including hypobaric storage. Hence, low-dose gamma radiation has been suggested and studied as a practical means of treating fresh fruits and vegetables for: (1) disinfestation, (2) shelf-life extension, and (3) pasteurization.

The first two applications have been extensively studied and have a great deal of potential and promise, especially for tropical fruits. The third application, radiopasteurization, also known as radurization and defined as "exposure of a fruit or vegetable to limited doses of ionizing radiation to effect a reduction in spoilage and disease-causing microorganisms", has the objective of controlling postharvest spoilage and diseases that affect fruits and vegetables without impairing their sensory qualities.

The applied dose level for radurization has been as high as 5 to 6 kGy. Various vegetative microorganisms are usually controlled at this dose level but some of the more resistant ones such as streptococci and yeasts will probably remain in the treated foods and some of the bacterial spores might still survive.[1] In many cases, a dose level of 2 to 3 kGy is recommended for fungal decay control.

Another application of ionizing radiation to foods is "radicidation". Ingram and Roberts[1] pointed out that the aim of radicidation is to remove microorganisms hazardous to public health by the application of low-dose gamma radiation. A number of microorganisms may still survive the treatment. The purpose of radicidation is to exert bacteriocidal effects on foods by ionizing radiation and it has been proposed to apply it in particular to frozen or dried foods. But the problems are (1) to determine the degree of microbial inactivation desired and hence the radiation dose necessary and (2) to control the efficacy of the treatment in commercial applications. On fruits and vegetables, studies have been carried out for the purpose of radurization rather than radicidation.

Taking full account of the available literature, this chapter will be devoted primarily to the discussion of problems and the state-of-the-art of radurization of fruits and vegetables. These problems are depicted in a summary chart as shown in Table 1.

Table 1
PROBLEMS OF RADURIZATION OF FRUITS AND VEGETABLES

Problems	State-of-the-art in using ionizing radiation to control postharvest decay-causing fungi
Radiobiological effects	Radiation treatments cause different morphological, genetic, and physiological changes in fungi; higher dose required to cause death than to inhibit germination than to retard growth
Pretreatment condition of host	Host-pathogen specificity and suitability of host to support fungal growth could vary
Environmental effects	High moisture and temperature favor spore germination; varying degrees of field infection and lesion development affect dose requirement
Fungal sensitivity to radiation	Inherent resistance to radiation damage varies with genetic makeup; apparent sensitivity depends on fungal cell population
Technical effects desired	Types of storage conditions, shipment requirements, and physiological life of host could change dose requirement
Effects of other interacting factors	Controlling effects on fungi vary according to dose rate, oxygen level, chemical protection of host (in vivo), repair and recovery mechanisms, and mutation of fungi

II. RADIOBIOLOGICAL AND RELATED PROBLEMS OF RADURIZATION

Applying ionizing radiation to fruits and vegetables at the pasteurization dose level is a means of extending the shelf life of these commodities. At the dose level where the growth of fungi on a commodity can be controlled, a number of problems may be encountered because the dose levels used are always higher than those used for delaying ripening of some fruits and inhibiting sprouting of some tuber-type vegetables. The problems are usually related to the physiological response of the commodity being irradiated and the dose requirements to achieve the desired effect plus some other interrelated factors.

Maxie and Sommer[2] discussed criteria that should be established to assure the beneficial application of ionizing radiation to any perishable commodity. They listed the following:

1. Less decay in the product as a distinct benefit and lower susceptibility to postirradiation infections by decay organisms
2. Retention of nutritional quality
3. Minimal changes in texture and flavor even after transport and storage
4. Attractive appearance
5. Wholesomeness and safety for human consumption as determined by government regulatory authority
6. Cost effective in terms of handling, radiation treatment, postirradiation refrigeration, and final marketing

Maxie and Abdel-Kader[3] discussed in depth the physiology of fruits as related to the feasibility of the application of ionizing radiation. They felt in 1966 that there probably was no product that had met all the requirements or criteria mentioned above.

The physiological response of fruits to gamma radiation, a very important aspect of applying this technology to delay postharvest ripening and senescence of a commodity, will be covered in Chapter 5 entitled "Delay in Postharvest Ripening and Senescence of Fruits".

A. Radiobiological Effects on Fungi

Ionizing radiation may cause morphological, genetic, and physiological changes in pathogens.[3-7] The morphological changes in irradiated fungi are sometimes very pronounced.[5,8-10] Rapid and extensive germination in irradiated spores often produces abnormally large germ tubes and excessive growth. A point is reached where all development stops and death follows. The time between irradiation and the occurrence of lethal damage may vary from hours to days depending on the fungi, the dose level, and the environment under which irradiated spores are stored. The phenomena of stimulated germination and growth following irradiation are believed to be related to postirradiation changes in cellular activities.[11,12]

For controlling fungal growth, the ability of fungi spores to germinate in relation to radiation dose is less important than their ability to form a colony with continual growth.[13] A much lower dose is required to cause the fungi to lose their capacity for indefinite growth than that for germination.[14,15] For example, a dose of 500 krad is needed to reduce the ability of *Rhizopus stolonifer* to form a colony to less than 1% but would hardly affect its ability to germinate. Reducing the capacity to germinate to 1% would require a dose of 1500 krad.[16]

When fungal colonies are irradiated at a dose insufficient to be permanently inactivated, growth is temporarily halted but is resumed after a delay of several days.[14,15] It is believed that some mycelia have been irreversibly injured but certain portions of hyphae may have recovered and then continued to grow.

B. Parameters Related to Dose Requirements for Controlling Fungal Growth in Fruits and Vegetables

1. Postharvest and Pretreatment Conditions of Fruits and Vegetables

Most fruits and vegetables grow in an environment subject to attack by a number of natural, biological factors such as insect invasion, bacterial and fungal infection, and imbalanced weather conditions. Although thousands of species of fungi and bacteria might be in contact with fruits and vegetables, very few are, in fact, capable of invading sound, living tissue.[17] Fungi and bacteria known to have serious, pathological effects on fruits and vegetables have been studied.[18-21] Many other postharvest pathogens are of minor importance because of their slow growth or control by such methods as refrigeration. For example, strawberries grown in California are usually invaded by about six species of fungi.[17] Only two of these, *Botrytis cinerea* Pers. ex Fr. and *R. stolonifer* (Ehrenb ex Fr.) Lind., would extensively develop after harvesting of the berries. However, growth of *R. stolonifer* can be effectively controlled by holding the berries below 10°C,[22,23] but the growth of *B. cinerea* still continues to a lesser extent at a temperature about 0°C. Therefore, once *B. cinerea* has already colonized the berries extensively, gray mold rot development might be rapid in spite of adequate refrigeration.

Why living tissues of fruits and vegetables are attacked by some microorganisms but not by others is a complex phenomenon not yet fully understood.[17] One of the explanations, which is not generally applicable, is that some living tissues are not a good medium for fungi growth. In some other cases, failure of the fungi to synthesize necessary enzymes may minimize pathogenicity on the living host tissue. In most cases, host-pathogen specificity depends upon host resistance which may be biochemical, mechanical, or physiological.

2. Environmental Conditions Surrounding the Fruits and Vegetables

The degree of postharvest disease is closely related to environmental conditions. High moisture content in the air or its availability is favorable to spore germination. Though reducing the humidity around the fruits and vegetables would decrease spreading of diseases, a dry environment is not necessarily desirable because it will promote dehydration and shrivel the fruits and vegetables.[17] Therefore, storage temperature is an interrelated, major environmental factor influencing spore germination and growth. If the fruit or vegetable could tolerate 0°C, many rot-causing fungi, except for a few cold-tolerant species, would not grow.[22-24]

Infections occurring in the field or orchard are very difficult to control since fungal lesions already exist in fruits and vegetables at harvest. Spores of *B. cinerea* frequently invade senescing floral parts of strawberries, thus the developing fruits are infected with gray mold disease when the fruits are ripening. Even if only one fruit is infected, the rest of the healthy fruits would be affected because of the contact between fruits or what is known as "nesting". Therefore, one infected fruit in a container may eventually result in a complete loss of a batch or container of fruits.[17]

3. Sensitivity of Various Disease-Causing Fungi

Effective treatment of fruits and vegetables by ionizing radiation for fungal disease control depends on the relative sensitivity of the host and the fungi to radiation. The host must be able to tolerate the treatment without any apparent injury or other undesirable side effects.

The sensitivity of a fungus to radiation is governed by several factors. One is the inherent resistance of the fungus to radiation damage. The degree of resistance, a complex phenomenon, varies greatly among species and is determined genetically. Sommer and Fortlage[17] discussed the relationship between radiation resistance and cellular composition. For example, a diploid possesses greater resistance because of the presence of a second set of chromosomes. Also, multinuclearity imparts radiation resistance analogous to the effect of ploidy. Spores with multinuclei can survive radiation treatment if one nucleus escapes inactivation. Multicellular spores would presumably escape lethal injury as long as one constituent cell survives to retain the ability to form a colony.

The degree of damage to cytoplasm is also important in relation to radiation treatment. However, nuclear damage may not be all genetic damage. The apparent sensitivity to ionizing radiation of a species of fungus depends upon the amount or size of fungus "tissue" present.[17] The minimum dose required to inactivate all the cells is related to cell population.[25] Lethal doses for inactivating mycelium would vary if the size of the colonies or the amount of mycelium is different. This presents a problem in the application of ionizing radiation to control fungal diseases in that there might not be a definite limited dose range for controlling a certain species of fungus. Even radiation-sensitive fungi may require a relatively large dose if the population is large.

4. Technical Effects Desired

The physiological or shelf life of various fruits and vegetables differs according to types and other storage conditions and shipment requirements. A related problem is whether or not complete inactivation of lesions is required. For fruits such as strawberry or sweet cherry which have a relatively short physiological life, a temporary halt in lesion growth in localized, infected areas may provide adequate control for marketing purposes. Thus a low dose would suffice. Fungal cells treated with sublethal doses are apparently injured for a length of time before some growth is resumed.

Infected commodities that normally have a longer physiological life would require a higher dose to inactivate the fungal lesions. This is related to the postharvest fungal population on the commodity.

5. Other Interacting Factors

a. Dose Rate

Varying the dose rate or a total absorbed dose applied in several fractional doses apparently produces different effects on the fungi in fruits and vegetables. Whiting[26] suggested a rapidly applied dose to achieve greater radiobiological effect on fungi because opportunities would not be available for repair.

Beraha et al.[27] reported in 1959 that within the range of 137 to 182 krad, *Pythium debaryanum* Hesse inoculated in potato tubers was better controlled at a dose rate of 7 krad/min than the same dose applied at 3 krad/min. When *Penicillium italicum* was inoculated in oranges and irradiated at 137 krad with a dose rate of 40 krad/min or 182 krad at 20 krad/min, the fruits could be kept in marketable conditions for 12 days.[28] Comparable results of a high-dose rate were found for *B. cinerea* Pers. ex Fr. in pears and *Monilinia fructicola* (Wint) Honey in peaches.

b. Atmosphere Around Radiation Site

The presence or absence of oxygen in the atmosphere around the radiation site where fruits and vegetables are radurized results in different degrees of effectiveness and dose requirement for inhibiting fungal growth. Peroxides formed in the presence of oxygen are apparently injurious to the fungi and increases lethality.[29,30] For example, Sommer and co-workers[31] found the dose required to reduce postharvest fungi survivors to the 1% level (in the presence of oxygen) to be 60% of that in anoxia.

Under some conditions, when fruits to be irradiated are sealed in plastic bags or other containers, the amount of oxygen may be depleted and carbon dioxide may accumulate due to the respiration of the fruit and the fungi or bacteria which have been parasitizing the fruit tissues.[17] Fungi prepared for irradiation in vitro may create near-anaerobic conditions by consuming all the oxygen in a closed system, thereby requiring a higher dose to control their growth.

Ozone produced in the radiation site in a closed container may reach concentrations toxic to both host and fungi. A more difficult problem is the ozone accumulation within host tissues which may account for up to 20% air space in the fruit.[3] Circulating air through the irradiation chamber could minimize these problems.

c. Host Protection

In vivo irradiation of disease-causing fungi and bacteria is not as effective as in vitro irradiation at the same dose level. The lower effectiveness by in vivo irradiation is believed to be due to some degree of chemical protection provided by the constituents in the host.[17] In some cases, the protective effect is due to the removal of oxygen by the chemical compounds or the promotion of oxygen removal by the cell's oxidation. Stapleton[32] found certain compounds to exert greater protection than the removal of oxygen. These compounds are believed to protect specific radiosensitive sites or enhance recovery.

d. Recovery of the Decay-Causing Microorganisms

In vitro studies have shown that under certain conditions, some microorganisms are able to repair their radiation damage, depending, of course, on the postirradiation environment.[33,34] Thus the dose effect is influenced by the ability of the organism to recover. Generally, recovery is preceded by their ability to be in the inactive stage through slowing down of metabolism and growth.

The significance of the recovery mechanism which affects the dose required for disease control has not been completely evaluated.[17] Whether or not the recovery mechanism would operate in vivo is even less certain. More fungal spores tend to recover when growth has been slowed or prevented. A cold environment is unfavorable to fungal growth in a host, but low temperature would limit metabolic recovery appreciably. Another approach is the use of a modified or controlled atmosphere where the oxygen content is depressed while the level of carbon dioxide is elevated. Under this atmosphere, growth would be slowed and the storage condition might still permit metabolic repair. For storage and transport of fruit, the combination of controlled atmosphere and low temperature will be unfavorable to fungal growth as well as negating the repair and recovery of the fruit.

e. Mutants

In terms of treating fruits and vegetables by ionizing radiation for fungal disease control, the concern is whether mutants might become more vigorous pathogens or exhibit higher radioresistance. Sommer and Fortlage[17] cited studies that demonstrated variation to radioresistance and suggested the possibility that radioresistant strains of phytopathogenic fungi could appear after a radiation technology is adopted. Therefore, a potential problem exists which cannot yet be adequately evaluated.

Generally, mutants are less vigorous and less pathogenic than their parents.[35] The occasional occurrence of increased pathogenicity from mutation has been demonstrated in plants with a resistance introduced through a breeding program. Usually, such plants are resistant to certain fungus strains but not others.[17] Gaumann[36] found that parasitism exhibited by rust fungi to be an advanced type which depends upon a delicate physiological balance between host and pathogen. On the other hand, fruit and vegetable rot disease organisms exhibit a primitive type of parasitism in which the physiological balance between host and parasite is relatively less important. Therefore, even though increased pathogenicity could appear among postharvest disease mutants, the increased pathogenicity is probably less dramatic than in the case of rusts.

III. TECHNOLOGY OF RADURIZATION

The application of ionizing radiation to fruits and vegetables for decay control is more complicated and difficult than for other purposes such as for delaying ripening or extending the shelf life when applied to other types of foods such as meat and meat products. This is because of several interrelated problems:

1. The tolerance dose level of the host to radiation is usually lower than the optimal dose required for pathogen control.
2. Fruits and vegetables are naturally grown in an environment where disease-causing microorganisms are free to move around — resulting in varying degrees of infection and different population sizes of the infecting microorganisms.
3. The condition mentioned above also results in the impossibility of finding a sufficient number of naturally grown fruits and vegetables (as compared to those grown in germ-free, controlled-atmosphere chamber) which are free of contamination and at the same stage of ripeness to serve as a control for determining dose requirements for inhibiting microbial growth.
4. There is a lack of any convenient, efficient method of estimating the population sizes of infecting microorganisms.

Experimental results from the studies of radiation effects on individual organisms by using in vitro techniques can probably be interpreted with a higher degree of confidence than those from in vivo studies. To discuss the technology of radurization, it is therefore logical to divide this section into experimental studies in which have been carried out: (1) in vitro radiation treatment of microorganisms, especially fungi, (2) single radiation treatment of various fruits and vegetables, and (3) various combined treatments.

A. In Vitro Treatment of Decay-Causing Microorganisms

Sommer and Fortlage[17] presented a comprehensive discussion of the results of their experiments using in vitro irradiation of decay-causing microorganisms. It was pointed out that previous studies by some workers were made whereby fungal populations were represented by all or parts of mycelial colonies or by uniform loops of a spore suspension. A common defect of these techniques was the use of only one fungus population of a size which might not have been well defined. Colonies of different species were sometimes grown on agar media in petri dishes but the relative sensitivities between species or growth conditions could only be approximated. This is complicated by the fact that: (1) density of the mycelial mat varies with species; the same block size may not represent comparable mycelia; (2) when the same species is grown under different environments or media, block size also may not provide comparable mycelia; and (3) the medium constituents might influence the dose response by providing a protective action during irradiation — the degree of protection might vary with the medium used, or the medium constituents might change with age or as a consequence of fungal growth.

Since the major obstacle of in vitro studies of radiation effect on fungi is the lack of a convenient method of estimating population size accurately, spores are considered the most convenient fungus structure available for dose-response studies, provided the fungus sporulates abundantly. Being discrete individuals, spores can usually be quantified by the counting procedure. There are however, a few drawbacks in using spores for dose-response studies. These are (1) a sufficient number of spores is not always obtainable; (2) some spores are large, of indefinite multicellularity, and difficult to quantify on a cellular basis; (3) sometimes more than one type of spore may be present in a culture; uniform spore suspension cannot be readily obtained; and (4) more often than not it is the mycelia rather than the spores that are inactivated in a fruit.

Dose-response curves for spores of postharvest disease fungi can be established by plotting percent survivors vs. absorbed dose in kilorad. This is somewhat similar to the survivor curve for thermal sterilization of microorganisms in the canning process.

Another technique to determine the doses required to inactivate a large percentage of each size of fungus population is called the "endpoint" technique. This technique allows infection levels in "orchard-run" fruit to be estimated and compared with experimental end-point results from known populations. Sommer et al.[37,38] used a variation of the endpoint technique to determine experimentally the doses required to inactivate every fungus cell, in vivo and in vitro, in populations of inocula varying from about 10 to 10^6 cells or spores. Doses were selected for every population size so that the highest dose would inhibit all growth, the intermediate dose would halt some growth, while the lowest dose would not have any effect.

Figure 1 illustrates the doses required to inactivate every cell in populations of various sizes of conidia of *Monilinia fructicola* where 20 replicates were used for each population and every dose.[17] From such a plot, the approximate dose required to inactivate every cell in 80% (colony inactivation dose 80) or other percentage of various sized populations of a selected number of postharvest disease-causing fungi spores in vitro can be established (Figure 2).

B. Single Radiation Treatment of Fruits and Vegetables

1. In Vivo Technique

For in vivo studies with spores, the technique is to inject a portion of a suspension containing the desired number of spores into each fruit with a hypodermic syringe.[17] In a hole on the fruit about 5 mm deep and 5 mm in diameter, 1/10 or 1/100 m*l* of spore suspension of a concentration that delivers the required number of spores was placed. The hole was immediately covered with a film, usually cellulose acetate. To

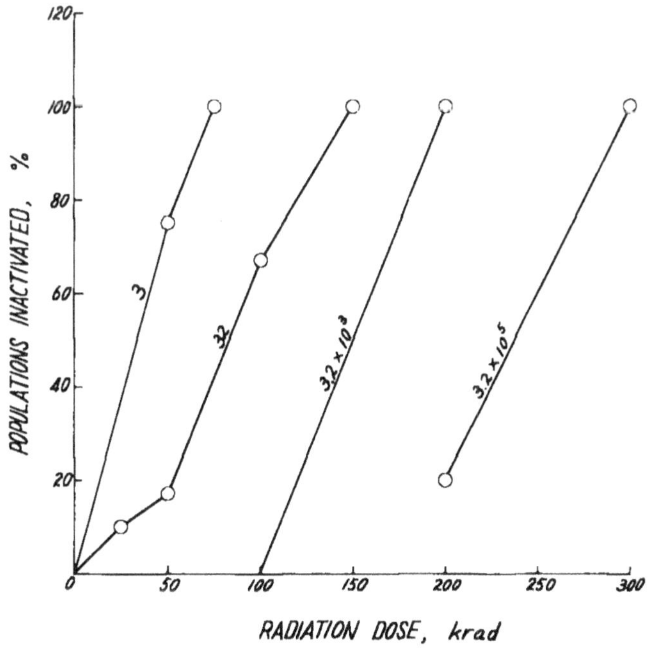

FIGURE 1. Inactivation doses for various sized populations of conidia of *Monilinia fructicola*, 20 replicates of which were used for each population and every dose. (Adapted from Sommer, N. F. and Fortlage, R. J., in *Advances in Food Research*, Vol. 15, Chichester, C. O., Mrak, E. M., and Stewart, G. F., Eds., Academic Press, New York, 1966, 147. With permission.)

prevent contamination, each fruit was placed in a paper cup of appropriate size which was tightly covered with cellulose acetate film. To detect possible development of low oxygen or high carbon dioxide conditions, the atmosphere around certain fruits should be tested by gas chromatography.

In general, germinated spores can provide young mycelia in the form of germ tubes; otherwise, mycelial inoculum cannot be used conveniently for in vivo experiments. Fungal colonies growing in fruit could be sized indirectly by comparing the radiation dose required for inactivation of the lesion with the spore population size similarly affected by the same dose. The in vivo mycelial colony size could be expressed as "spore equivalents". The colonies must, however, be reasonably uniform. For measurement of colony inactivation or lesion delay, the end-point technique seems to be useful and practical.

2. Radiation Treatment of Various Fruits and Vegetables

To achieve the objective of controlling postharvest disease in fruits and vegetables by ionizing radiation at the radurization dose level, the radiation dose used must be one that the host can tolerate. In many cases, the radiation dose used will be determined by the host rather than by the pathogen. Complicating the matter are the different host species and varieties within a species which have different degrees of radiation tolerance. Also, stage of ripeness at the time of treatment and the availability of extended storage will also influence the dose tolerance by the host. Maxie and Abdel-Kader[3] suggested that fruits destined for storage are usually limited to a lower radiation dose.

A sizable number of fruits and vegetables has been irradiated for several purposes.

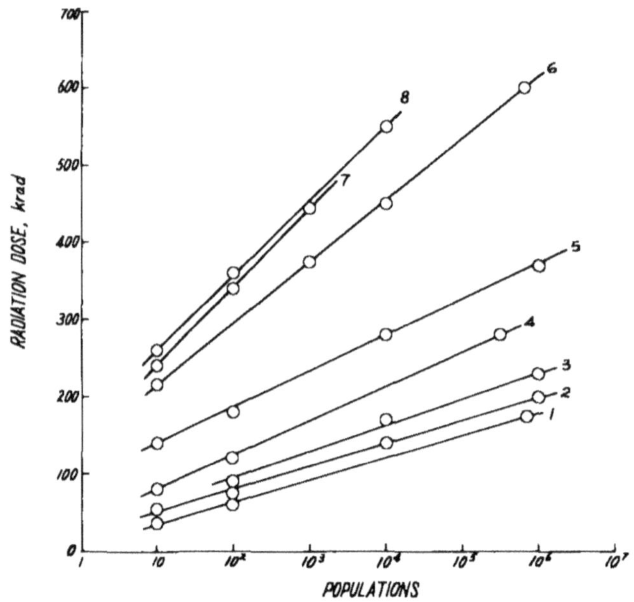

FIGURE 2. Approximate dose required to inactivate every cell in 80% (colony inactivation dose 80) of various sized postharvest disease-causing fungi. Data are for spores in vitro. 1. *Penicillium italicum*, 2. *P. expansum*, 3. *P. digitatum*, 4. *Monilinia fructicola*, 5. *Botrytis cinerea*, 6. *Rhizopus stolonifer*, 7. *Alternaria citri*, 8. *Cladosporium herbarum*. (Redrawn from Sommer, N. F. and Fortlage, R. J., in Advances in Food Research, Vol. 15, Chichester, C. O., Mrak, E. M., and Stewart, G. F., Eds., Academic Press, New York, 1966, 147. With permission.)

Some were treated to test the effect of ionizing radiation on the host or to control specific host responses such as ripening or sprouting. Others were irradiated to study the relationship between treatment and disease control.

Those that have been studied rather extensively or have been shown to be potential candidates for radiation treatment for disease control will be summarized here. Discussion, surveys, and reports in this subject area have also been reported by various authors.[39-45]

a. Berries

Strawberries (*Fragaria chiloensis* L.) are the berries that have been studied extensively for various aspects of preservation by ionizing radiation.

Strawberries are frequently attacked by *B. cinerea* Pers ex Fr. which causes the disease called "gray mold".[17] Since small lesions are almost always present at harvest, decay may occur without the time delay for infection and colonization.[46] The fungus is capable of spreading vigorously from fruit to fruit, its growth only being slowed, not stopped, under refrigerated transit conditions.

In vivo and in vitro studies have determined that treatment of about 200 krep (186 krad) could halt the growth of *B. cinerea* mycelial colonies.[17] At 5°C, a normal rate of growth resumed after 10 to 14 days. For a similar period, radiation at that dose level prevented contact infection and nesting.

Maxie et al.[47] irradiated strawberries (var. Shasta) at 200 krad and conducted simulated and rail transit tests. When the fruits were cooled immediately after harvest and kept cool, irradiation significantly reduced the amount of decayed fruits in transit although *B. cinerea* was not completely inhibited.

Lovell and Flick[48] reported irradiating Gulf coast area strawberries at 150, 250, and 350 krad and found 99% mold inactivation with 350 krad treatment, but the fruits were softened. At the lower dose, gray mold was controlled for 10 to 12 days when stored at 4.4 and 15.6°C, but there was slight softening of the fruits.

Barkai-Golan et al.[49] studied the effect of 100- to 300-krad treatments on *B. cinerea* and *R. nigricans* on strawberries. The incubation period of both fungi was extended for 2 to 7 days by the 200 to 300-krad treatments. Dose requirements were much higher to slow *B. cinerea* mycelia growth in vivo than in vitro. Treatment of 300 krad delayed in vitro growth of *R. nigricans* but only delayed the in vivo incubation period by 2 days.

Another major disease of strawberries is "leak", usually caused by *R. stolonifer* (Ehr. ex Fr.) Lind. although other mucroraceous species may sometimes be associated with the disease. Wounds inflicted during harvesting and handling are frequently infected by *R. stolonifer*, which may also infect by contact from fruit to fruit, resulting in an extremely rapid growth throughout the contents of a fruit container. However, *R. stolonifer* does not grow below 10°C. Proper refrigeration at below 10°C can prevent this disease.

A third fungus, *Cladosporium herbarum* Lk. ex Fr., is commonly found in strawberries. It is radiation resistant as well as capable of growing at refrigerated temperature. However, its growth is slow and the organism is apparently only weakly pathogenic.

Rots caused by *Phytophthora* sp. and *Rhizoctonia* sp. rarely develop vigorously on strawberries after harvest,[17] but Skou[50] has shown that *Aureobasidium pullulans* (de By.) Arnaud is radiation resistant and might be a limiting factor in irradiated fruits and vegetables.

In general, results of laboratory studies have indicated that treatment of ripe strawberries at 200 krad combined with refrigerated storage was the optimal condition for shelf-life extension through decay control in storage (Plate 1)*. Simulated and actual shipments of irradiated strawberries verified these findings.[47]

b. Citrus

Microorganisms causing severe postharvest losses in citrus fruits are *Penicillium digitatum* Sacc. and *P. italicum* Wehmer.[21] Infection by the former results in green mold while the latter causes blue mold. Deposits of colored spores on neighboring fruits in a container give an unsightly appearance. Conventional control measures include (1) chemical fumigation with nitrogen trichloride or ammonia; (2) use of in-package chemicals such as biphenyl; and (3) exercising care to avoid fruit injuries.

Beraha et al.[14] applied ionizing radiation to control *Penicillium* blue and green mold and found that fruits inoculated with *P. digitatum* could be protected against rotting with treatment of 150 to 200 krep (140 to 186 krad) for 12 days at 24°C and 17 days at 13°C. Similar results were obtained when *P. italicum* was isolated on sample fruits The effect of irradiation at this dose level on other postharvest citrus disease was not reported.

A higher incidence of stem-end rot in citrus fruits caused by *Alternaria citri* following irradiation was observed by Beraha et al.[14] and confirmed by Maxie et al.[3,51] Apparently *A. citri* is very radiation resistant[52] and might escape inactivation by radiation treatment if hidden in lesions in calyx tissue. With irradiation inducing death of calyx tissue, the fungus grows into the fruit proper and causes rotting.

Studies using gamma radiation to control molds and stem-end rots in citrus, especially oranges (*Citrus sinensis* L.), have been reported by various researchers.[53-60] In

* Plate 1 appears after page 144.

most cases, the pasteurizing doses varied from a low of 45 krad to a high of 200 krad. Work conducted by Dennison and Ahmed[61] (summarized by McKinney) perhaps best represents the overall findings of using gamma radiation to control decay in citrus fruits. It was stated that no single radiation dose can be given to protect citrus fruits against postharvest spoilage in the forms of stem-end rots and green and blue molds. The dose required to retard established infections is higher than that to retard incipient infections. The peel of citrus fruits is very sensitive to radiation damage, especially at the dose level required for decay control.

Dennison and Ahmed[61] pointed out that there was considerable variation in the control of postharvest decay in oranges by irradiation. In some tests the amount of spoilage in irradiated lots was as great as or greater than in the control lots. Some of the variables were time of harvest, fruit maturity, waxing or no waxing, and variety of fruit. Dipping the fruit in hot water (53°C for 5 min) prior to irradiation reduced the decay incidence in most cases.

c. Stone Fruits

Stone fruits represented by apricots (*Prunus armeniaca* L.), nectarines (*P. persica* L.), peaches (*P. persica* L.), and plums (*P. cerasifera* L.) are infected by two major fungi: *M. fructicola* (Wint.) Honey and *R. stolonifer* (Ehr. ex Fr.) Lind. which cause brown rot both before and after harvest. Wet weather conditions make fruit rot in orchards quite extensive even with a protective spray.[17] Postharvest sprays are less effective because the mycelium of lesions is internal. Chemical treatments applied to reduce postharvest infections are also ineffective owing to the prolific contamination through wounds and lesions in fruits by fungal spores.

Maxie and Abdel-Kader[3] observed that 200 to 250 krad, the maximum permissible radiation dose for stone fruits, was inadequate to inactivate a large proportion of established colonies. Sommer et al.[62] confirmed this by allowing inoculated stone fruits to incubate at ambient temperature for 48 hrs before irradiation. Treatment at 200 to 250 krad failed to inactivate a high proportion of infected sites.

Zehnder[63] reported that fully ripe apricots irradiated at 180 krad retain good appearance and firm texture for 3 weeks at 4°C and 90% relative humidity, but the fruit lost its typical taste.

Dose level of gamma radiation applied to various *Prunus* fruits such as peaches, cherries, and nectarines has varied from 50 to 400 krad[64-68] with 250 krad the most effective for controlling brown rot in cherries. Whereas brown rot was controlled in peaches and nectarines when treated at 200 to 300 krad, there was softening and loss of flavor in the fruits.[69,70]

Sommer and Fortlage[17] pointed out that if stone fruits are harvested and marketed only partially ripe, as is common with peaches and nectarines, *Rhizopus* rot should not be a problem. Prompt and proper cooling of fruits would also reduce the incidence of *Rhizopus* rot.[23,24]

d. Pome Fruits

A major postharvest disease called blue mold rot occurs in pome fruits such as apples (*Pyrus malus* L.) and pears (*P. communis* L.) caused by *Penicillium expansum* LK. ex Thom. and other species of *Penicillium*. Growing saprophytically and sporulating profusely, the fungus contaminates wounds on the fruit and gains entry into the fruit.[17]

Beraha et al.[71] reported a study of irradiated Jonathan apples inoculated with *P. expansum* and incubated for 24 or 96 hr prior to irradiation. Growth of the fungus was suppressed for 10 days at 21 to 24°C by treatment of 200 krep (186 krad). In vitro determination of the sensitivity of *P. expansum* as well as in hosts other than apples suggests that gamma radiation could effectively control the blue mold disease.[17] This was based on the assumption that disease lesions were not already present in the fruit at harvest and that the fruit could tolerate a dose of about 200 krad. Gray mold could be similarly controlled provided infection was due primarily to harvest contamination and handling wounds.

There were other studies of gamma-radiation treatment of pome fruits for decay control by Phillips,[72] Maxie et al.[73], and Barkai-Golan et al.[74] reported the effective dose level to be 200 to 300 krad. However, the main problem was softening of the fruit at that dose level. This led Sommer and Maxie[75] to suggest that interest in irradiation of pome fruits for control of storage disease would appear to be limited by the availability of reasonably satisfactory and cheap chemical treatments, on the one hand, and the possible development of delayed irradiation injuries of the treated fruits in storage, on the other.

e. Tropical Fruits

Tropical fruits that have received the most attention for the potential application of gamma radiation are papayas (*Carica papaya* L.) and mangoes (*Mangifera indica* L.).

Papaya is one of a few fruits that can tolerate gamma radiation quite well. Studies by Moy et al.[76-78] and by Moy[79] have shown that irradiation for fruit fly disinfestation and for shelf-life extension are well suited for papayas. Among the postharvest fungal diseases, anthracnose is a major problem; it could not be controlled by irradiation. Akamine and Wong[80,81] used hot water treatment (49°C for 20 min) to control anthracnose on papayas and found it to be more effective than irradiation. Doses of 75 to 100 krad brought about only a slight control of storage decay. For effective decay control, Akamine and Wong[81] found doses of 600 krad and higher were required, which were damaging to the fruits.

Additional work by Akamine and Goo[82] showed that if papayas were hot water treated and irradiated at 75 krad, storing the fruits at 15.6°C would simultaneously control storage decay, delay ripening and senescence, and extend the shelf life of the fruits.

Seven microorganisms that would cause stem-end rot and eight that would cause surface rot and spots in papayas have been identified by Buddenhagen and Kojima.[83,84] However, a radiation dose level that could control these decay-causing organisms has not been established. The recommended procedure for decay control on papayas is still to apply the hot water treatment which could be in the form of a dip or a high-temperature short-time system of sprays.[85]

Hatton et al.[86] studied irradiation of Florida grown mangoes (var. Irwin and Sensation) and found anthracnose control by irradiation erratic and inconclusive.

Brodrick et al.[87] studied the irradiation of mangoes grown in South Africa (var. Kent, Peach and Sabre) for disease control. They reported results on the irradiation control of anthracnose caused by *Colletotrichum gloeosporioides* to be inconclusive, and Diplodia stem-end rot caused by *Physalospora rhodina* (Syn. *Diplodia natalensis*) could not be successfully controlled with gamma radiation, even at a dose level of up to 200 krad.

Akamine and Moy report in Chapter 5 that there is considerable discrepancy in the findings on the effects of gamma irradiation on mangoes, caused by or due to varietal differences, varying pre and postirradiation treatments used, and observations on the different parameters of irradiation effects by various workers. At the tolerant dose level of 10 to 100 krad depending on the variety studied, there was no significant storage decay control by gamma-radiation treatment.

Bananas (*Musa sapientum* L) are variously classified as a subtropical and a tropical fruit. Most of the studies concerned with the gamma-radiation treatment of bananas have as their objective achievement of the technical effect of delayed ripening and, additionally, in some instances, disinfestation.

Chowdhury and Hamid[88] and Ferguson et al.[89] studied microbiological aspects of irradiated bananas, but Chowdhury and Hamid used a dose level up to only 60 krad. Ferguson et al. applied doses of 25 to 500 krad to bananas with an inoculum of *Gloeosporium musarum* at 1.5×10^5 spores per milliliter. It was found that there was a 50% survival of the organisms after treatment of 100 krad. Doses of 250 to 300 krad were effective in achieving complete inhibition of growth.

f. Other Fruits

Bramlage and Covey[90] studied the reduction of decay incidence in figs (*Ficus carica* L.) by irradiating the figs at 200 to 400 krad; they found that when stored at 3.3 to 21°C after irradiation, decay was appreciably reduced. The shelf life of the figs was thereby extended by several days.

Maxie et al.[91] studied the effects of gamma irradiation on table grapes (*Vitis vinifera* L.) and found that irradiating fresh Emperor and Tokay grapes at 100 to 300 krad did not control fungal decay on these grapes. There was a loss of firmness in the grapes. It was concluded that irradiation at the level used was not a practical substitute for standard SO_2 treatment. This was in agreement with findings by Bramlage and Covey.[90]

Donini and Pansolli[92] irradiated table grapes of the "hoanes" variety when the grapes were 50 to 80% ripe. It was reported that when irradiated at 200 krad, shelf life of the grapes was increased by 3 weeks.

g. Tomatoes (Lycopersicon esculentum Mill.)

Postharvest diseases on tomatoes are Alternaria rot caused by *A. tenuis* Nees ex Corda and Rhizopus rot caused by *R. stolonifer.* While most of the studies on irradiated tomatoes were on delayed ripening, Bramlage and Lipton[93] reported that when tomatoes were irradiated at 300 krad, storage decay of some lots was reduced, but the treatment caused mottling (an uneven red pigmentation) and softening of the fruit.

Studies by Abdel-Kader et al.[94] showed that shelf life of two varieties of tomatoes, "Early Pak No. 7" and "Ace", irradiated at 250 to 300 krad and stored at 12 to 15°C, was extended 4 to 12 days as a result of decay control, but the fruits must be at the pink or riper stage when irradiated. Doses above 300 krad caused physiological breakdown resulting in more decay. Lower doses were ineffective for decay control. Because of the adverse sensitivity to gamma radiation, there was no shelf-life extension in irradiated mature green or breaker stage tomatoes.

The tamarillo (*Cyphomandra betacea* [Car.] Sendt syn. *C. crassifolia*), is a woody shrub in the same family (Solanaceae) as the tomato (*L. esculentum*). A native of South America, it is found in Brazil and Peru and is widely grown in frost-free areas in New Zealand. Storage life of the fruit is limited to a few weeks because of field infection of several diseases such as stem-end rot. Sutton and Strachan[95] reported that irradiation of the tamarillo with attenuated 2 MeV electrons at 70 to 150 krad resulted in some delay in the appearance of stem-end rot, but the resultant increase in storage life was only marginal.

h. Tubers, Root Crops, and Vegetables

Shelf-life extension in the form of sprout inhibition through gamma-radiation treatment has been studied extensively. Approval of commercial application of this technology has been granted in at least nine countries. Aspects of this technology are discussed in Chapter 6 of this book.

On the other hand, limited work has been done with using gamma radiation to control storage disease of tubers and root crops. Beraha[27] reported that treatment of potatoes (var. Red Pontiac) at 18 to 477 krad did not prevent soft rot caused by preinoculated *Erwinia carotovora* (Jones) Holland. Extensive discoloration and softening resulted when higher doses were applied. Treatment with 46 krad prevented late blight rot in potatoes caused by *Phytophthora infestans* (Mont.) de By., but did not control natural infections of *Fusarium* sp. in the same tubers. Beraha further reported that treatment with 137 krad almost completely controlled preinoculated *Pythium debaryanum* Hesse, but a slight softening of the tubers occurred.

Taro (*Colocasia esculenta* L. Schott), a tropical root crop, is prone to postharvest fungal disease during storage, but no reported work has been carried out on the radurization of taro for decay control.

Farkas et al.[96] reported on the effect of ionizing radiation on the after-ripening of red pepper *(Capsicum annuum)* which is prone to postharvest decay caused by *Fusarium, Cephalosporium, Alternaria,* and *Stemphylium* spp. Freshly picked, sweet red peppers were irradiated with X-rays at 0 to 800 krad. It was found that a dose of 100 to 300 krad was needed to achieve substantial inhibition of mold growth and a dose of 400 to 800 krad was necessary for complete control of the organisms. Farkas et al. concluded that since there was a concurrent deleterious effect on the quality of the red pepper at the several-hundred-kilorad level, the application of irradiation to decrease storage losses in fresh red pepper was not advisable.

Staden has reported experiences with irradiation of vegetables in the Netherlands.[97] These included head lettuce, melons, cucumbers, onions, mushrooms, and some cut vegetables such as red cabbage, onion, and leek. The purpose of irradiation was twofold: a combination of shelf-life extension through sprout inhibition in some cases and decay control in others. Most of these vegetables were irradiated up to 200 krad. Susceptibility to radiation injury, resulting in impaired quality at this dose level, was observed. No detailed report was made as to the exact effect of radiation on the control of fungi on these vegetables.

i. Fruit Juices and Purees

Kaindl[98] reported on the international project on the irradiation of fruit juices. Since yeast was the main spoilage organism in fruit juices, Kaindl found that the most resistant strain of yeast, *Saccharomyces cerevisiae* Hansen required 1.8 Mrad to stop its colony-forming process over a period of 20 days. Studies also showed that when apple juice was heated to 50°C and irradiated with 0.3 Mrad, it could be stored without fermentation for more than a year at room temperature. For grape juice, a dose of about 1 Mrad was needed to delay fermentation in the filtered state.

Brekke et al.[99] studied the effects of gamma radiation and refrigeration on the spoilage and microbial population in guava puree concentrate (30.5% total soluble solids). Guava puree was inoculated with small portions of spoiled guava puree which contained yeasts. Microbial population was determined before and after irradiation at 0, 100, 500, and 1000 krad. Canned, irradiated puree samples were stored at 7.2 and 24°C and assayed every 3 days by series dilution and plate culture.

It was found that unirradiated, canned guava puree spoiled after 2 days at 24°C and 4 to 6 days at 7.2°C. Storage life of guava puree irradiated with 100 krad was extended 4 to 6 days when stored at 24°C, and still longer when stored at 7.2°C. Irradiation with 500 and 1000 krad increased the cell count in samples stored at 24°C but decreased it with storage at 7.2°C. The disadvantage of irradiation at 500 and 1000 krad was that the guava puree developed off-aroma. There was, however, no color change.

C. Combined Treatments

1. With Heat

Studies by Sommer et al.[100] have shown that some postharvest pathogens could be sensitized to irradiation by proper heat treatments without injuring the host. Based on

Table 2
EFFECTS OF GAMMA RADIATION AND COMBINED HEAT-RADIATION TREATMENTS ON GRAY MOLD (*BOTRYTIS CINEREA* PERS.) ROT ON STRAWBERRIES

Treatment		
Hot vapor	Gamma radiation (krad)	Gray mold rot (%)
0	0	44 ± 6
0	100	15 ± 6
0	200	8 ± 6
39°C/10 min	100	5 ± 1
40°C/10 min	100	2 ± 1
41°C/10 min	100	2 ± 2

Note: Data obtained after a simulated transit period of 7 days at 5°C following treatments. In combined treatments, heating preceded irradiation.

their studies on the disease-causing fungi of stone fruits, the observation was that a synergistic effect from the combined gamma radiation and heat treatment was obtained for the inactivation of the fungi.

Stehlik et al.[101] studied the influence of combined process of heat and irradiation on the survival of *S. cerevisiae* var. ellipsoideus and found that simultaneous heating and irradiation was more effective than sequential application for the control of the yeast growth.

In a study of the control of gray mold of strawberries by gamma irradiation, Sommer et al.[102] found that heating sensitizes *B. cinerea* Pers. to irradiation more than irradiation sensitized to heating. Synergism was greater in controlling this fungus when heating preceded irradiation. Above a dose of 200 krad, irradiation causes appreciable softening of the berries. However, with preheating to sensitize the pathogens, radiation dose could be reduced to 150 or 100 krad (Table 2). The time and temperature of heat treatment for sensitizing the fungus must be provided for different microorganisms.

"Leak" caused by *R. stolonifer* (Ehernb ex Fr.) Lind. does not usually develop under refrigeration but the fungus is resistant to heat and radiation. Sommer et al.[102] suggested that the potential exists for the use of combined heat and radiation treatments in the control of gray mold disease in strawberries.

In a study of the inactivation (loss of colony forming ability) of *R. stolonifer* sporangiospores by single and combined treatments of heating, chilling, and gamma irradiation, Buckley et al.[103] applied hot water treatment at 39 to 46°C (samples reaching that temperature then rapidly cooled), irradiation at 50 to 200 krad, or chilling rapidly to 0°C. It was found that although the nongerminating spores were markedly resistant to single applications of heat or radiation within these limits, a strong inactivation effect (<1% survival) was obtained when radiation plus heat (125 krad + 46°C/5 min) were applied in sequence. The interaction was less pronounced with treatments in reverse sequence. With germinating spores (6-hr incubation) an even greater synergistic effect (<0.1% survival) resulted from heating (39°C/5 min) followed by irradiation (125 krad). Thus, maximum synergism was obtained in nongerminating spores with a radiation-heat sequence whereas in germinating spores maximum synergism was obtained with a heat-radiation sequence. With both nongerminating and germinating spores, heating and radiation had a stronger interaction than heating and chilling.

Dennison and Ahmed[61] treated mangoes of the Zill variety with gamma irradiation up to 210 krad in combination with a hot water dip at 49°C for 7 min. The fruits were then stored at 20°C for up to 17 days. Significant reduction in anthracnose infection was obtained by the combined treatment especially at the highest dose level of 157 and 210 krad.

In a parallel study, Dennison and Ahmed[61] applied a hot water dip to peaches at 49°C for 7 min plus gamma radiation a 75, 150, and 225 krad. The fruits were stored for 10 days at 2°C and 4 days at 20°C. At the end of the storage, the control batch had 53% decay, those receiving radiation only at 150 and 225 krad had 2.2 and 8.9% decay, respectively, whereas those receiving hot water dip only 4.4% decay. Those receiving the combined treatment at 150 and 225 krad had no decay.

Other studies of combining gamma radiation with heat for the control of fungal decay include those by Salunkhe,[104] Pierson et al.,[105] Ben-Arie et al.,[106] Akamine and Goo,[82] Moy et al.,[85] Barkai-Golan and Padova,[107] Chou et al.,[108] and Sreenivasan.[109] In most cases, hot water treatment at 45 to 55°C was applied to fruits or vegetables (tubers) for 4 to 5 min followed by gamma radiation at various doses: 10 krad for potatoes, 25 to 35 krad for five varieties of bananas, 50 krad for mangoes, 75 krad for papayas, 100 krad for table grapes, and 150 krad for figs. In general, some synergistic effects were obtained in delaying incidence of rot or stem rot caused by *Rhizopus* sp., *Aspergillus* sp., *Botryodiplodia theobromae, Colletotrichum gloeosporioides* and *Penicillium digitatum.*

2. With Chemicals

Lapidot[110] reported the use of diphenyl and gamma radiation on oranges (var. Shamouti) inoculated in vivo with fungal spores of *P. digitatum* and *P. italicum.* When irradiation was applied at 100 krad and above, there were peel injuries and other harmful effects on the qualities of the fruit. When pears (var. Spadona) were dipped in diphenyl, irradiated at 50 to 400 krad and then stored at −1°C, the irradiated fruits did not ripen after removal from storage and kept at ambient temperature for 10 days.

Roy and Mokewar[111] studied the combined gamma radiation and chemical treatment in the control of *Aspergillus niger* van Tieghem and *Fusarium coeruleum* (Lib.) Sacc. Treatment with 250 krad had little fungicidal effect in vitro on the growth of *A. niger,* the black mold rot of apples, and *F. coeruleum,* the cause of dry rot in potatoes. The combination of 1000 ppm Aureofungin or captan as a preirradiation dip with 200 krad could protect about 80 to 85% of the apples at 21°C for 3 weeks. Benomyl alone or combined with Aureofungin singly or in combination with 200 krad delayed fungi growth. Captan with 200 krad prevented growth or sporulation of *A. niger* on apples.

3. With Controlled Atmosphere

Chalutz et al.[112] studied the effectiveness of gamma irradiation and controlled atmospheres, alone and in combination, in checking the development of botrytis rot on "Shasta" strawberries. It was found that controlled atmospheres (CO_2 concentrations varying from 0.035 to 10.46%; O_2 varying from 21.5 to 4.79%, the balance being nitrogen) inhibited lesion development of botrytis rot caused by *Botrytis cinerea* Pers. ex Fr., but were generally less effective than 200 krad of gamma radiation. The combined effect of controlled atmospheres plus gamma irradiation at 200 krad did not confer enough benefit over irradiation alone to justify the cost.

4. With Ultraviolet Radiation

Moy et al.[113] made an exploratory study to determine if combining UV and gamma radiation treatment at selective doses would decrease the fungal decay incidence in papaya (*Carica papaya* var. Solo). Also studied was the effect of the combined treatment on the spores of *Ascochyta* spp., *Colletotrichum* spp., and *Phytophthora* spp., commonly found on papaya grown in Hawaii. The use of this combined treatment was to test the postulate that such treatment would prevent photoreactivation of the disrupted nuclei in microorganisms and lead to eventual death.

The UV energy received by the fruits (total of 4 rotating positions facing the "Uviarc®", germicidal, mercury vapor lamp) varied from 1.22×10^4 erg/mm^2 to 19.1×10^4 erg/mm^2 whereas gamma-radiation dose varied from 25 to 75 krad on the papayas but was up to 250 krad on the three fungi.

In most cases, combining UV and gamma radiation treatment did not exhibit any noticeable improvement in the marketable life of the fruits. Gamma radiation at 25 krad plus UV dose of 3.58×10^4 erg/mm^2 was sufficient to prevent growth and colony formation of *Phytophthora* spp. while the combined doses of 150 krad and 1.19×10^4 erg/mm^2 were needed to control the growth of *Colletotrichum* spp. Of the three fungi treated, *Ascochyta* spp. were the most resistant to radiation. Gamma-radiation doses of up to 250 krad and UV doses of up to 7.33×10^4 erg/mm^2 alone or combined were not sufficient to prevent complete growth and colony formation. Thus, the experimental results did not indicate a conclusive trend to effectiveness but suggested a number of problems to be resolved before this type of combined treatment could become useful in decay control of fruits.

IV. POTENTIAL OF RADURIZATION OF FRUITS AND VEGETABLES

During the mid 1960s, researchers working on radiation treatment of the strawberry considered it to be "the fruit most likely to benefit in the near future from irradiation as a fungicidal treatment".[17] Sommer and Fortlage[17] pointed out that this was especially true in California where benefits would be high because of several factors: relatively long producing periods (6 to 8 months) in a single location, long distances (3000 to 6000 mi) to market, the presence of destructive diseases not adequately controlled by other means, high market value of the fruit, and important benefits to be derived from only a few days' delay in disease development.

The potential of using a pasteurization dose to control postharvest diseases in fruits and vegetables has been reported and discussed by Salunkhe,[104] Dharkar and Sreenivasan,[114] Maxie,[115] and Maxie et al.[116,117] The consensus was that most fruits and vegetables could not tolerate the irradiation doses (200 to 300 krad) needed for disease control. The host usually suffers quality damage, especially softening of the texture, at the radurization dose level.

Shallow irradiation has been suggested as one means of reducing host damage.[115,117] With diseases in which the lesions are near the surface, a fungicidal radiation dose applied to a depth of 2 mm or so might be nearly as effective as complete penetration. With limited penetration, the energy absorbed per fruit would be relatively small and host injury might be minimized. However, shallow treatment would be quite ineffective if lesions were deep within the fruit. Obtaining a uniform dose distribution with gamma radiation is a formidable problem.

Heat has shown promise of appreciably or greatly improving the effectiveness of irradiation. The synergistic effect of heat and ionizing radiation on postharvest disease-causing fungi could be used to increase the fungicidal activity without noticeable damage to the host. Fruits and vegetables that cannot tolerate a radurization dose can tolerate combined treatments since the radiation dose can be reduced. The sensitivity to heat and ionizing radiation varies among hosts and pathogens. The combined treatments, once determined for various host-pathogen combinations to be treated, could hold promise to extend the marketable life of many fruits and vegetables.

Based on available data, dose requirements and effects of radurization as a single treatment on selected fruits and vegetables for postharvest control are shown in Table 3.

V. SUMMARY

A major problem in applying ionizing radiation at the radurization dose level (mostly 200 to 300 krad, sometimes as high as 500 to 600 krad) to fruits and vegetables is that the host is often more sensitive to quality change such as softening than the pathogens for which the treatment is intended.

In addition to the relatively low tolerance of host to radiation, the technology of radurization of fruits and vegetables is complicated by several interrelated problems, namely: (1) varying degrees of field infection, (2) lack of contamination-free samples to determine dose requirement, and (3) lack of any convenient, efficient methods to estimate fungal populations. Therefore, results of various studies have indicated that it is not practical to prescribe a single radiation dose as being adequate to control a given decay-causing organism in a specific host. It also implies that radurization is a technology not too likely to be applied singly.

The more promising application is to combine ionizing radiation with heat treatment so as to take advantage of the synergistic effect between the two in order to use a lower dose and lesser amount of heat to achieve the purpose of pasteurization.

A good example is the inactivation (loss of colony forming ability) of *R. stolonifer* sporangiospores by single and combined treatments of heating, chilling, and gamma irradiation. It was found by Buckley et al.[103] that although nongerminating spores were markedly resistant to single applications of heat (hot water at 39 to 46°C) or radiation (50 to 200 krad), a strong inactivation effect (<1% survival) was obtained when radiation plus heat (125 krad + 46°C/5 min) were applied in sequence. The interaction was less pronounced with treatments in reverse sequence. With germinating spores (6-hr incubation) an even greater synergistic effect (0.1% survival) resulted from heating (39°C/5 min) followed by irradiation (125 krad). Thus, maximum synergism was obtained in nongerminating spores with a radiation-heat sequence whereas in germinating spores maximum synergism was obtained with a heat-radiation sequence. With both nongerminating and germinating spores, heating and radiation had a stronger interaction than heating and chilling.

Results of studies combining radiation treatment with other techniques such as chemicals, controlled atmosphere, and UV radiation have shown these treatments to be less effective than with heat as a means of postharvest disease control.

Table 3
DOSE REQUIREMENT AND EFFECTS OF RADURIZATION ON SELECTED FRUITS AND VEGETABLES FOR POSTHARVEST DISEASE CONTROL

Fruit/vegetable type	Disease	Causative pathogen	Dose required (krad)	Effects of radiation treatment	Ref.
Berries					
Strawberry	Gray mold	*Botrytis cinerea*	186—200	Pathogen growth halted 10—14 days at 5°C	17
		Rhizopus nigricans	300—400	Pathogen growth retarded 2—7 days	48, 49
	Leak	*R. stolonifer*	200	No growth under 10°C	17
	Rot	*Cladosporium herbarum*	200	Pathogen grows slowly at refrigerated temperature	17
		Aureobasidium pullulans	>300	Pathogen quite radiation resistant	50
Citrus Orange, etc.	Green mold	*Penicillium digitatum*	140—186	Rotting retarded for 12 days at 24°C and 17 days at 13°C	14, 21, 61
	Blue mold	*P. italicum*	140—186	Rotting retarded for 12 days at 24°C and 17 days at 13°C	14, 21, 61
	Stem-end rot	*Alternaria citri*	>300	Pathogen very radiation resistant	3, 14, 51
	Citrus brown rot	*Phytophthora spp.*	200	Inconsistent control	53—61
Stone Apricot, cherry, peach, plum,and nectarine	Brown rot	*Monilinia fructicola*	250—300	Some softening and loss of fruit flavor	3, 62, 63—70
	Brown rot	*R. stolonifer*	250—300	Some softening and loss of fruit flavor	3, 62, 63—70
	Gray mold	*B. cinerea*	200—300	Some softening and loss of fruit flavor	3, 17
	Blue mold	*P. expansum*	200—300	Some softening and loss of fruit flavor	62, 64
	Rot (sweet cherry)	*C. herbarum*	200—300	Some softening and loss of fruit flavor	62, 64
Pome Apple, pear	Blue mold	*P. expansum*	186—200	Pathogen growth suppressed 10 days at 21—24°C	17, 71
	Gray mold	*B. cinerea*	186—200	Pathogen growth suppressed 10 days at 21—24°C	17, 71
	Black mold	*Aspergillus niger*	200—300	Some softening of fruit	72, 75

Tropical					
Papaya, mango, banana	Stem-end rot	*Ascochyta caricae*	300—600	Damage to fruit texture and flavor; irradiation of 75 krad combined with hot water (49°C/20 min) more effective	80—85, 86—89, 113
		R. stolonifer and 5 others	300—600		
	Surface rot	*Colletotrichum gloeosporioides* and 7 others	300—600	Damage to fruit texture and flavor; irradiation of 75 krad combined with hot water (49°C/min) more effective	80—85, 113
Tomato	Alternaria rot	*Alternaria tenuis*	250—300	Some reduction in storage decay, but mottling and softening occur	93, 94
	Rhizopus rot	*R. stolonifer*	200—300	Some reduction in storage decay, but mottling and softening occur	93, 94
Tuber					
Potato	Soft rot	*Erwinia carotovora*	18—477	Not effective in controlling rot; tuber softened and discolored	17
	Late blight rot	*Phytophthora infestans*	46	Did not control *Fusarium* sp.	17
	Pythium soft rot	*Pythium debaryanum*	137	Slight softening of the tubers	27

REFERENCES

1. **Ingram, M. and Roberts, T.A.**, Microbiological principles in food irradiation, in Proc. Symp. Food Irradiation, STI/PUB/127, International Atomic Energy Agency (IAEA), Vienna, 1966.
2. **Maxie, E.C. and Sommer, N.F.**, Status of gamma irradiation as a technology for perishable commodities, in Proc. Fruit Veg. Handling Conf., University of California, Davis, 1964, 96.
3. **Maxie, E. C. and Abdel-Kader, A.**, Food irradiation — physiology of fruits as related to feasibility of the technology, in *Advances in Food Research*, Vol. 15, Chichester, C. O., Mrak, E. M., and Stewart, G. F., Eds., Academic Press, New York, 1966, 105.
4. **Atwood, K. C. and Pittenger, T. H.**, X-ray-induced mutations in growing mycelium of Neurospora, *Radiat. Res.*, 3, 212, 1955.
5. **Brace, K. C.**, Effects of X-rays on size of yeast cells, *Proc. Soc. Exp. Biol. Med.*, 74, 751, 1950.
6. **Burns, V. W.**, X-rays-induced division delay of individual yeast cells, *Radiat. Res.*, 3, 217, 1955.
7. **Johnson, F. H.**, Effects of electromagnetic waves on fungi, *Phytopathology*, 22, 277, 1932.
8. **Berk, S.**, The effects of ionizing radiation from polonium on the spores of *Aspergillus niger.*, *Mycologia*, 45, 488, 1953.
9. **Diamond, A. E. and Duggar, B. M.**, Effects of ultraviolet radiation on the germination and morphology of spores of *Rhizopus suinus*, *J. Cell. Comp. Physiol.*, 16, 55, 1940.
10. **Luyet, B. J.**, The effects of ultra-violet, X-rays and cathode rays on the spores of mucoraceae, *Radiology*, 18, 1019, 1932.
11. **Buchwald, C. E. and Wheldon, R. M.**, Stimulation of growth in *Aspergillus niger* under exposure to low velocity cathode rays, *Am. J. Bot.*, 26, 778, 1939.
12. **Vasudeva, R. S., Bajaj, B. S., Chatrath, M. S., and Granju, P. J.**, Stimulating effect of ionizing radiation on certain microorganisms, *Indian Phytopathol.*, 12, 19, 1959.
13. **Uber, F. M. and Goddard, D. R.**, Influence of death criteria on the X-ray survival curves of the fungus *Neurospora*, *J. Gen. Physiol.*, 17, 577, 1934.
14. **Beraha, L., Ramsey, G. B., Smith, M. A., Wright, W. R.**, Factors influencing the use of gamma radiation to control decay of lemons and oranges, *Phytopathology*, 49, 91, 1959.
15. **Beraha, L., Ramsey, G. B., Smith, M. A., and Wright, W. R.**, Effects of gama radiation on brown rot and Rhizopus rot of peaches and the causal organisms, *Phytopathology*, 49, 354, 1959.
16. **Sommer, N. F., Creasy, M., Romani, R. J., and Maxie, E. C.**, Recovery of gamma irradiated Rhizopus stolonifer sporangiospores during autoinhibition of germination, *J. Cell. Comp. Physiol.*, 61, 93, 1963.
17. **Sommer, N. F. and Fortlage, R. J.**, Ionizing radiation for control of postharvest diseases of fruits and vegetables, in *Advances in Food Research*, Vol. 15, Chichester, C. O., Mrak, E. M., and Stewart, G. F., Eds., Academic Press, New York, 1966, 147.
18. **Wiant, J. S. and Brately, C. O.**, Spoilage of Fresh Fruits and Vegetables in Rail Shipments Unloaded in New York City, U.S. Department of Agriculture Circ., 1948, 773.
19. **Walker, J. C.**, *Diseases of Vegetable Crops*, McGraw-Hill, New York, 1952, 529.
20. **Anderson, H. W.**, *Diseases of Fruit Crops*, McGraw-Hill, New York, 1956, 501.
21. **Klotz, L. J.**, *Color Handbook of Citrus Diseases*, University of California, Berkeley, 1961, 75.
22. **Muller, W. H.**, Influence of temperature on growth and sporulation of certain fungi, *Bot. Gaz.*, 117, 336, 1956.
23. **McClure, T. T.**, Brown and Rhizopus rots of peaches as affected by hydrocooling, fungicides, and temperature, *Phytopathology*, 48, 322, 1958.
24. **Smith, W. L., Jr. and McClure, T. T.**, Rhizopus rot of peaches as affected by postharvest temperature and moisture, *Phytopathology*, 50, 558, 1960.
25. **Gouey, H. M. and Bramlage, W. J.**, Effect of spore population and age of infection on the response of *Botrylis cinerea* to gamma radiation, *Phytopathology*, 55, 1013, 1965.
26. **Whiting, A. R.**, Protection and recovery of the cell from radiation damage, in *Radiation Protection and Recovery*, Hollaender, A., Ed., Pergamon Press, Elmsford, N.Y., 1960, 117.
27. **Beraha, L., Ramsey, G. B., Smith, M. A., and Wright, W. R.**, Effects of gamma radiation on some important potato tuber decays, *Am. Potato J.*, 36, 333, 1959.
28. **Beraha, L.**, Influence of gamma radiation dose rate on decay of citrus, pears, peaches, and on *Penicillium italicum* and *Botrytis cinerea in vitro*, *Phytopathology*, 54, 755, 1964.
29. **Laser, H.**, The "Oxygen effect" in ionizing irradiation, *Nature (London)*, 174, 753, 1954.
30. **Bridges, B. A. and Horne, T.**, The influence of environmental factors on the microbial effect of ionizing radiations., *J. Appl. Bacteriol.*, 22, 96, 1959.
31. **Sommer, N. F., Creasy, M., Romani, R. J., and Maxie, E. C.**, An oxygen-dependent postirradiation restoration of *Rhizopus stolonifer* sporongiospores, *Radiat. Res.*, 22, 21, 1964.
32. **Stapleton, G. E.**, Protection and recovery of bacteria and fungi, in *Radiation Protection and Recovery*, Hollaender, A., Ed., Pergamon Press, Elmsford, N.Y., 1960, 87.

33. Sobels, F. H., *Repair from Genetic Radiation Damage*, Macmillan, New York, 1963.
34. Bernard, D. D., Dulbecco, R., Eisen, H. N., Ginsberg, H. S., Wood, W. B., Jr., and McCarty, M., Radiation damages and their repair, in *Microbiology — Including Immunology and Molecular Genetics*, 2nd ed., Davis, B. D., Ed., Harper & Row, New York, 1973, chap. 2.
35. Beraha, L., Garber, E. D., and Stromaes, O., Genetics of phyto pathogenic fungi. Virulence of color and nutritionally deficient mutants of *Penicillium italicum* and *Penicillium digitatum*, *Can. J. Bot.*, 42, 429, 1964.
36. Gaumann, E., *Principles of Plant Infection*, Hafner Publishing, New York, 1950, 543.
37. Sommer, N. F., Maxie, E. C., and Fortlage, R. J., Quantitative dose-response of *Prunus* fruit decay fungi to gamma irradiation, *Radiat. Bot.*, 4, 309, 1964.
38. Sommer, N. F., Maxie, E. C., Fortlage, R. J., and Eckert, J. W., Sensitivity of citrus fruit decay fungi to gamma irradiation, *Radiat. Bot.*, 4, 317, 1964.
39. Clarke, I. D., Possible applications of ionizing radiations in the fruit, vegetable and related industries, *Int. J. Appl. Radiat. Isot.*, 6, 175, 1959.
40. DeZeeuw, D., Experiments on the preservation of fruit by irradiation, *Food Irradiat.*, 1(3), A5, 1961.
41. Heeney, H. B., Rutherford, W. M., and MacQueen, K. F., Some effects of gamma radiation on the storage life of fresh strawberries, *Can. J. Plant Sci.*, 44, 188, 1964.
42. Salunkhe, D. K., Gamma radiation effects on fruits and vegetables, *Econ. Bot.*, 15, 28, 1961.
43. Saravacos, G. D., Hatzipetrou, L. P., and Georgiadou, E., Lethal doses of gamma radiation of some fruit spoilage microorganisms, *Food Irradiat.*, 3(1—2), A6, 1962.
44. Truelsen, T. A., Radiation pasteurization of fresh fruits and vegetables, *Food Technol. (Chicago)*, 17, 336, 1963.
45. Willison, S. R., Ionizing radiation for the control of plant pathogens: a review, *Can. Plant Dis.* (Survey), 43, 39, 1963.
46. Powelson, R. L., Initiation of strawberry fruit rot caused by *Botrytis cinerea*, *Phytopathology*, 50, 491, 1960.
47. Maxie, E. C., Sommer, N. F., and Rae, H. L., Effects of gamma irradiation on Shasta strawberries under marketing conditions, *Isot Radiat. Technol.*, 2, 50, 1964.
48. Lovell, R. T. and Flick, G. J., Irradiation of gulf coast area strawberries, *Food Technol.*, *(Chicago)*, 20, 99, 1966.
49. Barkai-Golan, R., Temkih-Gorodeiski, N., and Kahan, R. S., Effect of gamma irradiation on development of fungi, *Botrytis cinerea* and *Rhizopus nigricans*, causing rot in strawberry fruits, *Food Irradiat.*, 8(1—2), 34, 1967.
50. Skou, J. P., Aureobasidium pullulans (de By Arnaud) — a common and a very radio-resistant fungus on fresh fruits and vegetables, in *Radiation Preservation of Foodstuffs*, Kinnell, P. O. and Runnstrom-Reio, V., Eds., Second Scand. Meet. for Preserv. Radiation, 1964.
51. Maxie, E. C., Eaks, I. L., and Sommer, N. F., Some physiological effects of gamma irradiation on lemon fruit, *Radiat. Bot.*, 4, 405, 1964.
52. Sommer, N. F., Maxie, E. C., Fortlage, R. J., and Eckert, J. W., Sensitivity of *Citrus* fruit decay fungi to gamma irradiation, *Radiat. Bot.*, 4, 317, 1964.
53. Kahan, R. S. and Monselise, S. P., Extension of storage life of citrus fruit by irradiation, *Food Technol. (Chicago)*, 19, 122, 1965.
54. Dennison, R. A. and Ahmed, E. M., Review of the status of irradiation effects on citrus fruits, Paper IAEA/SM-73/24, in Proc. Symp. Food Irradiat., IAEA, Vienna, 1966.
55. Benn, T. R., Shallow irradiation of oranges by electrons. A feasibility study for in-line processing, *Advances in Chemistry* Series 65, Gould, R. R., Ed., American Chemical Society, Washington D.C., 1967, 126.
56. Macfarlane, J. J. and Roberts, E. A., Some effects of gamma radiation on Washington Naval and Valencia oranges, *Aust. J. Exp. Agric. Anim. Husb.*, 8(34), 625, 1968.
57. Klotz, L. J. and Harding, P. R., Control of postharvest decay of citrus fruits, *Calif. Lithograph*, 55(7), 259, 1970.
58. Saiz de Bustamante, C., Garcia de Mateos, A., Hernandez Gimenez, F., and Safont Thomas, A., Use of irradiation in the preservation of Spanish oranges, I. Influence of irradiation on microbial agents and on external appearance of fruits, *Rev. Agroquimic Technol. Alimento*, 10(3), 371, 1970.
59. Sato, T., Present status of food irradiation in Japan, Paper IAEA/A/CONF. 49/P/256, in Proc. 4th Int. Conf. Peaceful Uses Atomic Energy, Vol. 12, IAEA, Vienna, 1972.
60. Matsuyama, A., Present status of food irradiation research in Japan with special reference to microbiological and entomological aspects, Paper IAEA SM-166/30, in Proc. Symp. Radiat. Preservation Food, IAEA Vienna, 1973, 261.
61. Dennison, R. A. and Ahmed, E. M., Effects of low-level irradiation on the preservation of fruit: a 7-year summary, *Isot. Radiat. Technol.* 9(2), 194, 1971.
62. Sommer, N. F., Maxie, E. C., and Fortlage, R. J., Quantitative dose-response of *Prunus* fruit decay fungi to gamma irradiation, *Radiat. Bot.*, 4, 309, 1964.

63. **Zehnder, H. J.,** Food irradiation in Switzerland, Paper IAEA/A/CONF. 49/P/120, in Proc. 4th Int. Conf. Peaceful Uses Atomic Energy, Vol. 12, IAEA, Vienna, 1972.
64. **Sommer, N. F., Maxie, E. C., and Fortlage, R. J.,** Quantitative dose-response of *Prunus* fruit decay fungi to gamma irradiation, *Radiat. Bot.*, 4, 309, 1964.
65. **Massey, L. M., Jr., Robinson, W. B., Spaid, J. F., Splittstoesser, D. F., van Buren, J. P., and Kertesz, Z. I.,** Effect of gamma radiation upon cherries, *J. Food Sci.*, 30, 759, 1965.
66. **Maxie, E. C., Johnson, C. F., Boyd, C., Roe, H. L., and Sommer, N. F.,** Effect of gamma irradiation on ripening and quality of nectarines and peaches, *Proc. Am. Soc. Hortic. Sci.*, 89, 91, 1966.
67. **Kuhn, G. D., Merkley, M. S., and Dennison, R.,** Irradiation inactivation of brown rot infections on peaches, *Food Technol. (Chicago)*, 22, 91, 1968.
68. **Larmond, E. and Hamilton, H. A.,** The effect of low level gamma irradiation on peaches, *Food Irriadiat.*, 8(4), 2, 1968.
69. **Somogyi, L. P. and Romani, R. J.,** Irradiation-induced textural change in fruits and its relation to pectin metabolism, *J. Food Sci.*, 29, 366, 1964.
70. **Dhalival, A. S. and Salunke, D. K.,** Ionizing radiation and packaging effects on respiratory behavior, fungal growth, and storage-life of peaches *Prunus persica*, *Radiat. Bot.*, 3, 75, 1963.
71. **Beraha, L., Ramsey, G. B., Smith, M. A., and Wright, W. R.,** Gamma radiation for possible control of postharvest diseases of apples, strawberries, grapes and peaches, *Phytopathology*, 47, 4, 1947.
72. **Phillips, W. R., Poapst, P. A., and MacQueen, K. F.,** *The Effect of Irradiation on the Development of Storage Disorders of Apples*, presented at Int. Inst. of Refrigeration, Copenhagen Conf., August 1959, Atomic Energy of Canada, Ltd. Publ. 1120, 35.
73. **Maxie, E. C., Sommer, N. F., Muller, C. J., and Rae, H. L.,** Effect of gamma radiation on the ripening of Bartlett pears, *Plant Physiol.*, 41, 437, 1966.
74. **Barkai-Golan, R., Ben-Arie, R., Reich, S. G., and Kahan, R. S.,** Sensitivity to gamma irradiation of fungi pathogenic to pears, *Int. J. Appl. Radiat. Isot.*, 20(8), 577, 1969.
75. **Sommer, N. F. and Maxie, E. C.,** Recent research on the irradiation of fruits and vegetables, Paper IAEA/SM-73/27, in Proc. Symp. Food Irradiat., IAEA, Vienna, 1966.
76. **Moy, J. H., Akamine, E. K., Brewbaker, J. L., Buddenhagen, I. W., Ross, E., Spielmann, H., Upadhya, M. D., Wenkam, N., Helber, D., Dollar, A. M., Hanaoka, M., and McGlish, G. A.,** Dosimetry, tolerance and shelf-life extension related to disinfestation of tropical fruits by gamma irradiation, in *Disinfestation of Fruit by Irradiation*, STI/PUB/299, IAEA, Vienna, 1971, 43.
77. **Moy, J. H.,** Problems and prospects of radiation disinfestation and shelf-life extension of tropical fruits and vegetables, in *Proc. Aspects of the Introduction of Food Irradiation in Developing Countries*, STI/PUB/362, IAEA, Vienna, 1973, 13.
78. **Moy, J. H., Akamine, E. K., Wenkam, N., Dollar, A. M., Hanaoka, M., Kao, H. Y., Liu, W. L., and Revetti, L. M.,** Tolerance quality and shelf life of gamma-irradiated papaya grown in Hawaii, Taiwan, and Venezuela, in *Radiation Preservation of Food*, STI/PUB/317, IAEA, Vienna, 1973, 375.
79. **Moy, J. H.,** Potential of gamma irradiation of fruits: a review, *J. Food Technol.*, 12, 449, 1977.
80. **Akamine, E. K. and Wong, R.,** Shelf-life Extension of Papayas with Gamma Irradiation; In Dosimetry Tolerance, and Shelf Life Extension Related to Disinfestation of Fruits and Vegetables by Gamma Irradiation, Rep. 1964-1965, Contract No. AT(043)-235 Proj. Agreement No. 5, Div. Isot. Dev., U.S. Atomic Energy Commission, Washington, D.C., 1965, 55.
81. **Akamine, E. K. and Wong, R. T. F.,** Extending the shelf life of papayas with gamma irradiation, *Hawaii Farm. Sci.*, 15, 4, 1966.
82. **Akamine, E. K. and Goo, T.,** Effects of gamma irradiation on shelf life extension of fresh papaya (*Carica papaya* L. var. Solo) *Hawaii Agric. Exp. Stn. Res. Bull.*, 165, 1, 1977.
83. **Buddenhagen, I. W. and Kojima, E. S.,** Biology of Postharvest Disease Fungi of Papaya in Relation to Disinfestation of Fruits and Vegetables by Gamma Irradiation, Rep. 1965—1966, UH-235-P5-2, U.S. Atomic Energy Commission, Washington, D.C., 1966, 30.
84. **Kojima, E. S. and Buddenhagen, I. W.,** Studies on Gamma Irradiation in Relation to Postharvest Diseases of Papayas and Mango, Rep. 1966—1967, UH235-P5-3, U.S. Atomic Energy Commission, Washington, D.C., 1967, 54.
85. **Moy, J. H., Dollar, A. M., Hanaoka, M., and Hundtoff, E. B.,** High-temperature short-time hot-water treatment for surface disinfestation of papaya, presented at Annu. Meet. Inst. Food Technol., Chicago, June 1975.
86. **Hatton, T. T., Jr., Beraha, L., and Wright, W. R.,** Preliminary trials of gamma radiation on mature green Irwin and Sensation mangoes, in *Proc. 21st Annu. Meet.*, Fla. Mango Forum, 1961, 15.
87. **Brodrick, H. T., Jacobs, C. J., Kok, I. B., Milne, D. L., Basson, R. A., and Thomas, A. C.,** Application of ionizing radiation to the preservation of subtropical fruit in South Africa, in *Proc. Natl. Conf. Technol. Appl. Nucl. Tech.*, Pelindanba, South Africa, South Africa Atomic Energy Board, Pretoria, 1972, 1.

88. **Chowdhury, S. U. and Hamid, M. A.**, Some aspects of the rate of appearance and type of microbiological damage in irradiated banana, *Food Irradiat.*, 10(1—2), 18, 1969.
89. **Ferguson, W. E., Yates, A. R., MacQueen, K. F., and Robb, J. A.**, The effects of gamma radiation on bananas, *Food Technol. (Chicago)*, 20, 105, 1966.
90. **Bramlage, W. J. and Covey, H. M.**, Gamma radiation of fruits to extend market life, Marketing Res. Rep., No. 717, Agricultural Research Service, U.S. Department of Agriculture, Washington, D.C., 1975, 1.
91. **Maxie, E. C., Nelson, K. E., and Johnson, C. F.**, Effect of gamma irradiation on table grapes, *Proc. Am. Soc. Hortic. Sci.*, 84, 263, 1964.
92. **Donini, M. L. B. and Pansolli, P.**, Gamma irradiation of table grapes of the "hoanes" variety, *Food Irradiat.*, 10(4), 15, 1970.
93. **Bramlage, W. J. and Lipton, W. J.**, Gamma radiation of vegetables to extend market life, Marketing Res. Rep. No. 703, Agricultural Research Service, U.S. Department of Agriculture, Washington, D.C., 1965, 1.
94. **Abdel-Kader, A. S., Morris, L. L., and Maxie, E. C.**, Physiological studies of gamma irradiated tomato fruits. II. Effects on deterioration and shelf life, *Proc. Am. Soc. Hortic. Sci.*, 94, 831, 1968.
95. **Sutton, H. C. and Strachan, G.**, An attempt to control *Botyris* rot in tamarillos (*Cyphomandra betacea* (Cav.) Sendt) by electron irradiation, *N.Z. J. Sci.*, 14(4), 1097, 1971.
96. **Farkas, J., Kiss, I., and Andrassy, E.**, After-ripening of red pepper *(Capsicum annuum)* as affected by ionizing radiation, Paper IAEA/SM-73/60, in Proc. Symp. Food Irradiat., IAEA, Vienna, 1966.
97. **Staden, O. L.**, Experiences with the irradiation of vegetables in the Netherlands, Paper IAEA/SM-73/68, in Proc. Symp. Food Irradiat., IAEA, Vienna, 1966.
98. **Kaindl, K.**, International project on the irradiation of fruit and fruit juices, Paper IAEA/SM-73/in Proc. Symp. Food Irradiat., IAEA, Vienna, 1966.
99. **Brekke, J. E., Tonaki, K., and Moy, J. H.**, The effects of gamma irradiation and refrigeration on spoilage and microbial population in guava puree concentrate, *Food Irradiat.*, 8(4), 10, 1968.
100. **Sommer, N. F., Fortlage, R. J., Buckley, P. M., and Maxie, E. C.**, Radiation-heat synergism for inactivation of market disease fungi of stone fruits, *Phytopathology*, 57, 428, 1967.
101. **Stehlik, G. and Kaindl, K.**, Microbiological studies on the influence of combined processes of heat and irradiation on the survival of *Saccharomyces cerevisiae* var. ellipsoideus, in *Food Irradiation*, IAEA, Vienna, 1966, 299.
102. **Sommer, N. F., Buckley, P. M., Fortlage, R. J., Coon, D. A., Maxie, E. C., and Mitchell, F. G.**, Heat sensitization for control of grey mold of strawberry fruits by gamma irradiation, *Radiat. Bot.*, 8, 441, 1968.
103. **Buckley, P. M., Sommer, N. F., Coon, D. A., Dolly, M., and Maxie, E. C.**, Inactivation of *Rhizopus stolonifer* sporangiospores by single and combined treatments of heating, chilling and gamma-irradiation, *Radiat. Bot.*, 40, 26, 1969.
104. **Salunkhe, D. K.**, Gamma radiation effects on fruits and vegetables, *Econ. Bot.*, 15, 28, 1961.
105. **Pierson, N. W., Wilcox, E. B., and Salunkhe, D. K.**, Radiation preservation of fruits, *Utah Farm Home Sci.*, 23, 4, 1962.
106. **Ben-Arie, R. and Barkai-Golan, R.**, Combined heat-radiation treatments to control storage rots of spadona pears, *Int. J. Appl. Radiat. Isot.*, 20(10), 687, 1969.
107. **Barkai-Golan, R. and Padova, R.**, Combined irradiation, heat and biphenyl treatments for the control of *Penicillium digitatum* in inoculated citrus fruits, *Isr. J. Agric. Res.*, 20(3), 129, 1970.
108. **Chou, T. W., Singh, B., Salunkhe, D. K., and Campbell, W.**, Effects of gamma radiation on *Penicillium expansum* L. I. Some factors influencing the sensitivity of the fungus, *Radiat. Bot.*, 10, 511, 1970.
109. **Sreenivasan, A.**, Compositional and quality changes in some irradiated foods, Paper IAEA/PL-561/10, in Proc. Panel Improvement Food Qual. Irradiat., IAEA, Vienna, 1974.
110. **Lapidot, M.**, The Israel food irradiation program 1965-1970: application of a concepted approach to the commercialization of irradiated agricultural products, Paper IAEA/A/CONF. 49/P/012, in Proc. 4th Int. Conf. Peaceful Uses Atomic Energy, Vol. 12, IAEA, Vienna, 1972.
111. **Roy, M. K. and Mukewar, P.**, Combined gamma-irradiation and chemical treatment in the control of *Aspergillus niger* van Tieghem and *Fusarium coeruleum* (Lib.) Sacc., in *Radiation Preservation of Food*, STI/PUB/317, IAEA, Vienna, 1973, 193.
112. **Chalutz, E., Maxie, E. C., and Sommer, N. F.**, Interaction of gamma irradiation and controlled atmospheres on Botrytis rot of strawberry fruit, in *Proc. Am. Soc. Hortic. Sci.*, 88, 365, 1965.
113. **Moy, J. H., McElhaney, T., Matsuzaki, C., and Piedrahita, C.**, Combined treatment of UV and gamma-radiation of papaya for decay control, in *Proc. Int. Symp. Food Preservation Irradiat* Vol. 1, STI/PUB/470, IAEA, Vienna, 1978, 361.
114. **Dharkar, S. D. and Sreenivasan, A.**, Irradiation of tropical fruits and vegetables, Paper IAEA/SM-73/64, in Proc. Symp. Food Irradiat., IAEA, Vienna, 1966.

115. **Maxie, E. C.,** Radiation technology in conjunction with postharvest procedures as a means of extending the shelf life of fruits and vegetables, in Proc. Conf. Contractors, U.S. Atomic Energy Commission, Washington, D.C., 1968, 123.
116. **Maxie, E. C., Sommer, N. F., and Mitchell, F. G.,** Infeasibility of irradiating fresh fruits and vegetables, *HortScience*, 6, 202, 1971.
117. **Maxie, E. C., Sommer, N. F., Mitchell, F. G., and Bringhurst, R. S.,** Radiation Technology in Conjunction with Postharvest Procedures as a Means of Extending the Shelf Life of Fruits and Vegetables, Rep. UCD-34P80, U.S. Atomic Energy Commission, Washington, D.C., 1971.

Chapter 4

RADURIZATION AND RADICIDATION: SPICES

József Farkas

TABLE OF CONTENTS

I. MICROBIAL CONTAMINATION OF SPICES AND PROBLEMS ARISING FROM THEIR USE

Spices are used in virtually all categories of the food industry — meat, fish, vegetable products, bakery products, and other prepared and convenience foods. The meat industry and soup manufacturers use the largest quantities of spices. There is an increasing demand around the world for spices in connection with the increasing amount of processed food, which must be flavored.[1,2]

With regard to whole and ground natural spices, both the producing countries and the food industries that use spices as ingredients of various manufactured foods face the problem of the high microbial count of spices, especially the contamination by molds[3,4] and heat-resistant bacterial spores.[5-17]

In general, black pepper, caraway, coriander, capsicum peppers, ginger, marjoram, and turmeric are the most highly contaminated spices. The total viable count may reach the 80 to 100 million/1g level.[10,18] On the basis of this contamination it can be appreciated that the addition of spices at a level of 0.1 to 1% in meat products can lead to a contamination of 10^5 to 10^6 organisms per gram of products arising from the spices alone. The viable cell counts of spice samples of various origin might show deviation of several log cycles as shown in Figure 1.

Mold counts vary between the 10^2 to 10^5/g range and do not correlate with the total viable counts. There is no apparent correlation between the country of origin and the quality of spice with respect to numbers of microorganisms or predominant flora.[34]

The majority of the microbial flora of spices consists of aerobic spore-forming bacteria.[5,9,15,20,23,30,34,40-42] Powers et al.[43] tested 110 samples of various spices for incidence and levels of *Bacillus cereus*. The organism was found in 53% of the spices and counts ranged from 50 to 8500/g. This incidence was more than double the incidence found in selected dry products.[42] Of the isolates, 89% were enterotoxigenic. The frequent occurence of *B. cereus* in spices, which is reported also by other authors,[44] is significant from the point of view of food safety and sanitation.

A relatively high incidence of *Clostridium perfringens* is also reported.[28,38,42,45,46] Since spores of these organisms may survive cooking temperatures and will grow in foods at temperatures between 20 to 50°C, spices harboring these spores must be considered a potential health hazard.

Although spices may not be suitable substrates for the growth or long survival of *Salmonellae*,[47] which are usually absent from spices,[3,28,34] it is not known if this is due to inadequate methodology. Occasional *Salmonella* contamination of spices, however, cannot be excluded.[48,49] Black pepper and white pepper have both been implicated as vehicles for the spread of *S. weltevreden* resulting in several cases of salmonellosis.[50]

According to several authors, spices play a major etiological part in mold contamination of meat products.[3,4,8,13-15,17] Spices examined for mold counts have revealed that *Penicillium* species, *Aspergillus glaucus* (group), *A. niger*, and *A. flavus* are frequently found to varying extents.[3,39,51-53] Relative high incidence of toxigenic molds has been found in spices[3,53-55] and several authors have detected aflatoxins in a range of spices[56-58] although the levels of aflatoxins recorded are generally low.

In many cases contaminated seasoning is responsible for the spoilage of canned meats or may cause defective sausage products.[8,13,15,20,23,59,65] The thermoduric bacterial spores present in seasonings often indicate a heat treatment which ensures the microbial stability only at the cost of substantial reduction of nutritional and sensory values of the spiced product. Logically, the practical control of microbial contamination of food products by organisms from canning ingredients lies in the selection for the use of materials relatively free from significant spoilage organisms. The work involved in carrying out a routine bacteriological examination of all ingredients is, how-

FIGURE 1. Contamination of various spices by mesophylic aerobic bacteria.[3,4,8,10,19-40]

ever, hardly within the capacity of all but a few of the largest laboratories. This accounts for the world-wide attempts on the one hand to reduce the viable cell count of spices by appropriate sterilization treatment, and on the other hand to use spice oleoresins and essential oils which are free from microbial contamination.[23,66]

II. THE ADVANTAGES AND DISADVANTAGES OF THE USE OF SPICE EXTRACTS

In the convenience food sectors, spice extracts offer certain advantages over natural spice powders, such as consistency of quality, freedom from enzymes and microorganisms,[23,27,60,64,66] uniform dispersion in the products,[67] and easy handling and storage.[68,69] But meat packers in several countries (for instance Italy and Hungary) do not prefer oleoresins to whole spices. In salami, hams, and sausages they make the point of showing the spices, particularly pepper and cloves used in their meat products. The "bolder" the pepper so exhibited, the greater apparently the appreciation of the spice-quality of the meat products displayed. Another problem in the use of spice extracts is that their flavor quality generally does not equal that of the whole or ground spices.[70,72] Comparing the spicing power of ground pepper and pepper oil in bologna sausage, Körmendy[72] found that by using ground pepper a more intensive taste is obtained even if the concentrations of the spicing agents were equal according to the profilograms of the flavor dilutions. The reason for this fact is that in contrast to the homogeneous distribution of pepper extract in the sausage, ground pepper is not homogeneously distributed, forming "hot spots" in the product. When these "spots" are crushed by the teeth, they release a pungent taste sensation. As spice extracts are volatile and no longer encapsulated by the plant cells, their loss of flavor during any cooking process, unless fully enclosed as in a can, is greater than that of natural spice.[27] Another disadvantage of oleoresins arises from the need for a solvent in their production. Unless great care is exercised in ensuring the purity of the solvents used, high boiling trace solvents may well remain. These are of concern from the standpoint of unpleasant flavor and toxicity.

III. CHEMICAL TREATMENT OF SPICES FOR REDUCING MICROBIAL CONTAMINATION

A. Fumigation with Ethylene Oxide or Propylene Oxide

Gassing with ethylene oxide (EO) or propylene oxide (PO) is used commercially in several countries to kill insect pests in seasonings and to reduce the viable cell count.[14,24,73-80] It has been established that at a temperature of 20 to 25°C, fumigation for 6 to 7 hr is required, though the requirement varies according to the resistance of the microflora present.[81,82] But EO does not kill bacterial spores very effectively[83] and slight changes in color and flavor of fumigated spices are unavoidable.[14,18,84,85] A significant decrease in the volatile and nonvolatile oil content of pepper and allspice by EO fumigation was reported by Vajdi and Pereira.[35,103] It is now recognized that mustard is unsuitable for sterilization with EO since it develops an unpleasant aroma and quickly becomes rancid.[86] What is a more serious disadvantage, fumigation, a chemical treatment, cannot be excluded from the danger of leaving toxic chemical residues.

EO may be physically retained for some time by the spices. More important, it may undergo, however, chemical reaction with food components (formation of alkylated derivatives of sugars, amino acids, proteins, and vitamins).[81,87]

According to Kröller,[88] the amount of the adsorptive bound gas remained fairly high after gassing, even after repeated evacuation. For example, pepper treated with EO at a gas concentration of 934 g/m^3 in the atmosphere for 3 hr at 22°C contained 570 ppm residual EO, and even after evacuation had been repeated three times a residual concentration of 384 ppm was observed. The adsorptive bound EO residues decreased continuously during storage although this was due to further chemical reactions rather than to additional loss of the gas.

Wesley et al.[89] pointed out that during treatment with EO or PO in the presence of water and inorganic chlorides, chlorohydrin from a part of EO, and chloropropanol isomers from PO are formed which are persistent toxic compounds. In commercially sterilized ground spice mixtures they found up to 1030 ppm ethylene chlorohydrin (ECH) with an average of 805 ppm. Similar observations were reported by other laboratories.[90,91] Ragelis et al.[92] found residues of ECH ranging from 45 ppm in paprika to 110 ppm in pepper after commercial treatments with EO. Recently Stijve et al.[93] measured 680 to 1130 ppm and 880 to 3400 ppm ECH levels in fumigated black pepper and turmeric, respectively. The persistence of ECH residues during storage varied considerably from one substrate to another. No decrease in residue was observed within 4 months in black pepper and whole turmeric.

Much data are available showing the mutagenicity of EO in a variety of plants and lower animals. Recent investigations[94,95] indicate that EO is capable of inducing mutagenic effect in rats.

Both epychlorohydrin and PO were considered to be carcinogenic in animal tests.[96] Recent investigations have shown that ECH in high occupational doses has a detectable mutagenic effect on the lymphocyte chromosomal structure in exposed workers.[97]

B. Reduction of Microbial Contamination by *In Situ* Salt Formation

Another chemical treatment, reduction of viable cell count of spices by acidification or *in situ* salt formation[98,99] has not yet gained much attention. This method seems to be applicable only over a limited range, and it is therefore highly desirable to find physical methods for decontamination of seasonings.

IV. PHYSICAL METHODS FOR DECONTAMINATION OF SPICES

Because of the heat sensitivity of the delicate flavor and odor of spices,[100-102] regular heat sterilization cannot be applied. Microwave treatment has also been applied in an attempt to control microorganisms in selected spices. Since spices have a low moisture content, sufficient heat to achieve the desired reduction in numbers of organisms cannot be generated in the product. Microwave treatment was therefore considered to be unsuited to bring about decontamination.[35,103]

Initially, UV radiation was thought to hold promise for reducing spice contamination.[62] However, experiments carried out by Coretti[104] and Bergmann and Pradella[105] showed that the very low penetration of UV radiation is an obstacle to the technical realization of the method. In some cases even a stimulation of microbial growth, particularly that of molds, could be observed after UV treatment.[3,59]

A. Reduction of Viable Count by Ionizing Radiation

Scientists at the Massachusetts Institute of Technology (MIT) who did pioneering work in radiation sterilization of food, found during their early investigations[19,106] that high doses of ionizing radiations are suitable for the sterilization of some spices and other dry food materials. These results were confirmed and expanded by additional experiments on a large variety of spices in various laboratories. It was established that to achieve nearly total sterility (i.e., a reduction of total viable count to less than 10/g) a radiation dose of 0.8 to 2.0 Mrad is required, depending on the initial level of contamination.[21,22,23,26,30,107-112] Electron bombardment techniques, X-rays, and gamma rays have been employed with success for the sterilization of spices. However, because of the limited penetration of the electron beam, only thin packages can be successfully treated with this type of ionizing radiation. More penetrating radiation, such as gamma rays, can feasibly be applied to bulky materials and larger packages.

Comparative investigations of EO and gamma-irradiation treatment in selected spices made by Vajdi[35] and Vajdi and Pereira[103] showed that radiation was more effective than EO in destroying bacterial contamination. According to another comparative investigation by Szabad,[113] a radiation treatment of ground paprika with 0.5 Mrad reduced the viable count of mesophilic aerobic bacteria and aerobic bacterial spores to the same extent as an EO treatment of 6 hr at 24°C in a gas mixture of 90% EO and 10% CO_2 in a Degesch apparatus. However, in relation to fungi this EO treatment of paprika was practically ineffective, whereas irradiation with 0.5 Mrad destroyed the molds very effectively (Figure 2).

On studying the total viable cell count of various spices and mixed seasonings for meat, it was found that a reduction of 2 to 3 log cycles of the viable cell count could be achieved with a dose of 0.25 to 0.75 Mrad.[22,29,112,114,115] The efficiency of the radiation treatment as measured by the survivors directly after treatment was not significantly affected by varying the water activity of the paprika powder.[33] However, several observations point to an aftereffect of irradiation in the dry spices, viz., the occurrence

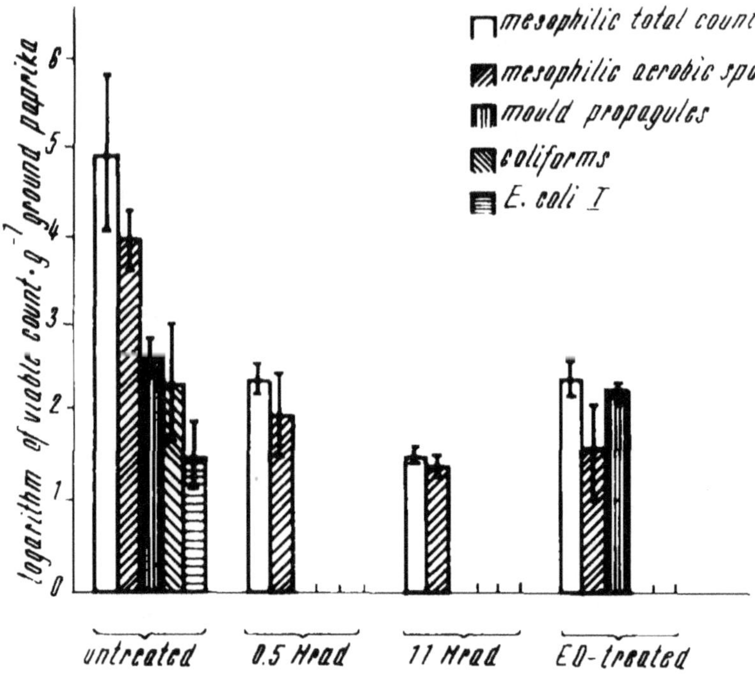

FIGURE 2. Effect of gamma irradiation and ethylene oxide on the microflora of ground paprika. Treatment with ethylene oxide was carried out in a chamber of 600 g/m^3 ethylene oxide concentration.[113]

of a further reduction of the viable cell count during postirradiation storage.[112,116] This aftereffect is probably due to some long-lived free radicals formed by irradiation in a low-moisture system.

It was also established[29,33,117] that the surviving microbial flora of spices treated with a "pasteurizing" radiation dose of 0.3 Mrad became more sensitive to environmental effects. The heat and salt tolerances of the surviving microbial flora of irradiated seasonings were significantly lower than those of untreated seasonings (Table 1). This observation is in good agreement with the investigations carried out with pure cultures of microorganisms which prove that the bacterial spores surviving irradiation become more sensitive to heat[118] and reduced water activity of the growth medium.[117,119,120] On the other hand, it seems unnecessary to achieve full sterility as the product in most cases is incorporated into unsterilized food. A radiation dose of 0.5 to 0.8 Mrad seems to be sufficient to reduce the microbial contamination of spices, even of the heavily contaminated ones, to below a level of 10^3 to 10^4/g.[22,26,29,30-33,35,37,103,111-113,115,116] This level of viable count is generally accepted as a maximal count of the decontaminated spices (entkeimte Gewürze) desired in the spice trade.[18,23,121]

B. The Sensory Quality of Irradiated Spices

According to several authors, the flavor characteristics of black pepper,[21,107,108] cinnamon,[21,107] and cloves[21] are slightly or noticeably changed by radiation doses of 1 to 2 Mrad. No significant changes were found in the sensory quality of powdered paprika,black pepper, and cumin, respectively, after exposure to 1.0-Mrad dose level.[37] These spices were mixed into yogurt for the triangle taste test.

The threshold doses required to produce a change in flavor of various spices were reported by Bachman and Gieszczynska[112] and are shown in Table 2.

In experiments with radiation decontamination of onion powder, the characteristic

Table 1
HEAT AND SALT SENSITIVITY IN THE MICROBIAL FLORA OF MIXED SEASONING SURVIVING RADIATION TREATMENT

Heat treatment 80°C (min)	Incubation temperature (°C)	Log number of cells per gram capable of colony formation					
		0 Mrad			0.3 Mrad		
		0.5% NaCl	5% NaCl	10% NaCl	0.5% NaCl	5% NaCl	10% NaCl
		Water Activity (a_w)					
		0.99	0.97	0.94	0.99	0.97	0.94
0	30	6.0	6.3	6.0	3.1	2.0	1.0
	10	5.5	5.5	3.6	2.6	2.2	<1.0
10	30	6.0	6.1	—[a]	2.4	1.3	<1.0
	10	4.6	4.5	<2.0	<1.0	<1.0	<1.0

[a] Not investigated.

From Farkas, J., Beczner, J., and Incze, K., in *Radiation Preservation of Food*, IAEA, Vienna, 1973, 389. With permission.

Table 2
THRESHOLD DOSES REQUIRED TO PRODUCE A CHANGE IN FLAVOR OF SPICES.

Spice	Dose (Mrad)
Black pepper	1.25
Caraway	1.25
Coriander	0.75
Cardamon	0.75
Charlock	1.0
Juniper	1.50
Marjoram	0.75—1.25
Pimento	1.50
Spice mixtures	>1.50

From Bachman, S. and Geiszczynska, J., *Aspects of the Introduction of Food Irradiation in Developing Countries*, IAEA, Vienna, 1973, 33. With permission.

odor of onion powder decreased significantly above 0.8 Mrad. Caramel-like off-odor appeared and the color of the dried onions darkened.[32] Organoleptic tests of nutmeg, ginger, and white pepper indicated that irradiation with doses of 0 to 4.5 Mrad did not significantly alter the taste or odor of these spices.[30]

Vajdi[35] found less damage to the color and volatile oil content of various spices decontaminated by gamma irradiation than in those treated with EO.

Luncheon meat prepared with spices that had been given a dose of 2 Mrad could not be differentiated by a flavor profile panel from luncheon meat prepared with the corresponding nonirradiated spices. On the other hand, luncheon meat prepared with a number of different brands of spice extracts or with chemically decontaminated spices showed detectable off-flavors, or flavors deviating from those of luncheon meat containing natural spices.[111] In other taste tests of brawn (head cheese) seasoned with a mixture consisting of ground marjoram, pepper, and paprika, no significant differences could be observed between the samples prepared with the irradiated seasoning mixture and the nonirradiated control.[33]

C. Chemical Changes in Irradiated Spices

The capsaicin content of the pungent paprika did not change even after a treatment with 1.6 Mrad. Neither the vitamin C nor the linoleic and linolenic acid content of ground paprika was affected to a significant extent even by this high sterilization dose.[33] No decrease in benzene soluble pigment content (carotenoid content) and reducing sugar content was observed following doses up to 0.4 Mrad. A 90% retention of pigment was found after treatment with 1.6 Mrad.[22] No significant changes in the chemical composition and organoleptic quality were found in various spices irradiated with a dose of up to 2 Mrad at the Bulgarian Meat Research Institute.[26]

Changes in the yield of essential oils by steam distillation have been investigated in marjoram, black pepper, and juniper by Bachman and Gieszczynska.[112] An increased yield of volatile oil of marjoram has been observed after exposure to ionizing radiation

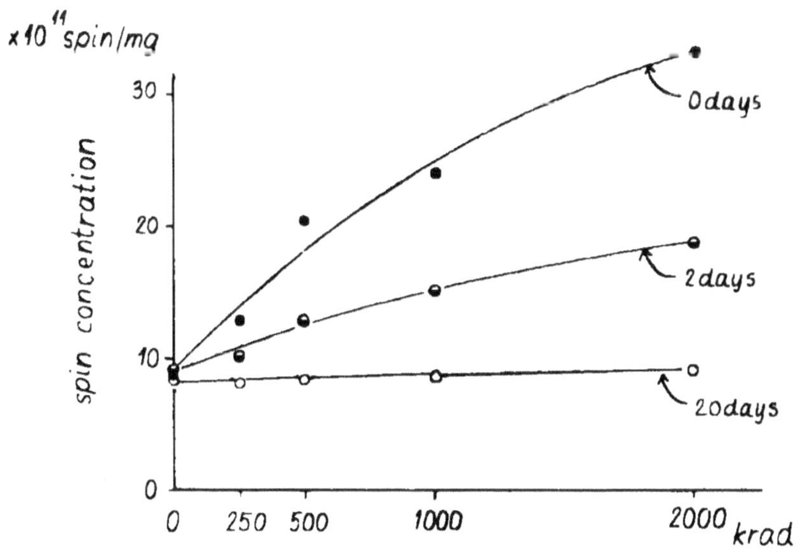

FIGURE 3. The spin concentration in paprika powder as a function of irradiation dose and storage time at room temperature. (From Farkas, J., *Aspects of the Introduction of Food Irradiation in Developing Countries*, IAEA, Vienna, 1973, 43. With permission.)

over the 0.5- to 2.0-Mrad dose range. This was accompanied by changes in the composition of marjoram oil. The content of terpenes decreased; that of the alcoholic fraction increased. The yield of essential oils in black pepper and juniper decreased at a dose of 0.5 Mrad.

According to Vajdi,[35] gamma irradiation of 6 spices resulted in no change in the volatile oil content and only insignificant changes in the nonvolatile oils after irradiation. Similar results were reported by Inal et al.[37] after 1.0-Mrad radiation treatment of black pepper, cumin, and paprika, respectively.

Production of H_2 and CO_2 was observed as an effect of gamma irradiation of various spices.[30] The yield of H_2 with doses up to 4.5 Mrad was proportional to the dose. The G-values (the amount of gas produced per 100 eV absorbed dose) varied between 0.5 and 1.7.

Formation and disappearance of free radicals in irradiated spices were investigated in Hungary.[122,123] The irradiation-induced free radicals diminished rapidly during storage (Figure 3). According to these investigations, a certain amount of stable free radicals is always present even in the unirradiated material.[123] The Electron Spin Resonance (ESR) measurements by Tjaberg et al.[30] indicated that the amount of permanent free radicals present 3 weeks after irradiation was negligible.

D. Storage Stability of Irradiated Spices

1. Chemical Aspects

The storage stability of carotenoids in ground paprika was not affected by a dose of 0.5 Mrad.[33,124] However, the surface color of ground paprika depends not only on the carotenoid content but also on the browning reactions at higher storage temperatures, owing to the high reducing sugar content of paprika. To determine color changes, samples stored for 120 days were scored by a panel of nine panelists according to a 5-point scale. Since moisture content of ground paprika influences the surface color, samples with the same moisture content stored in closed aluminum tubes were tested. The sensory scores and the percentage retention of benzene soluble pigment are shown in Figures 4 and 5.

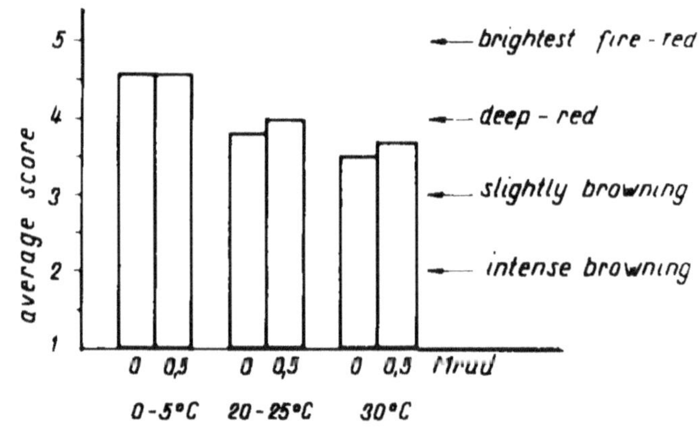

FIGURE 4. Visual scoring of ground paprika after 120 days storage in closed aluminium tubes as affected by storage temperature. (From Farkas, J., Beczner, J., and Incze, K., *Radiation Preservation of Food*, IAEA, Vienna, 1973, 389. With permission.)

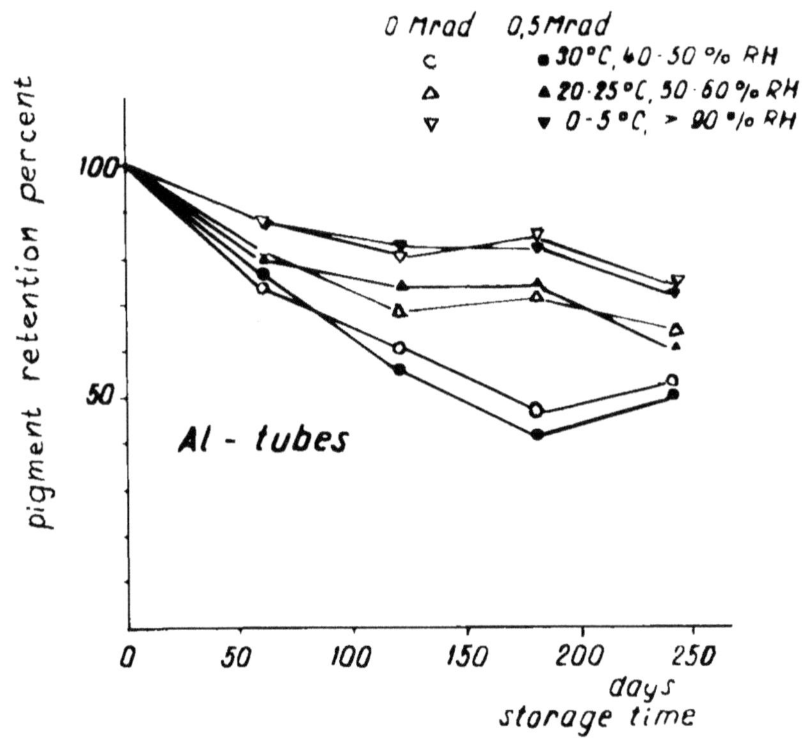

FIGURE 5. The retention of the benzene soluble pigment content of paprika samples in closed aluminium tubes as affected by storage time and temperature. (From Farkas, J., Beczner, J., and Incze, K., *Radiation Preservation of Food*, IAEA, Vienna, 1973, 389. With permission.)

2. Microbiological Aspects

From the point of view of microbial stability, the mold contamination of spices is highly important. Because of the microbial flora of spices, only molds are capable of growth in storage at ambient temperature at a minimum relative humidity of 72 to 75%. Thus, the ability of surviving microorganisms in irradiated ground paprika to

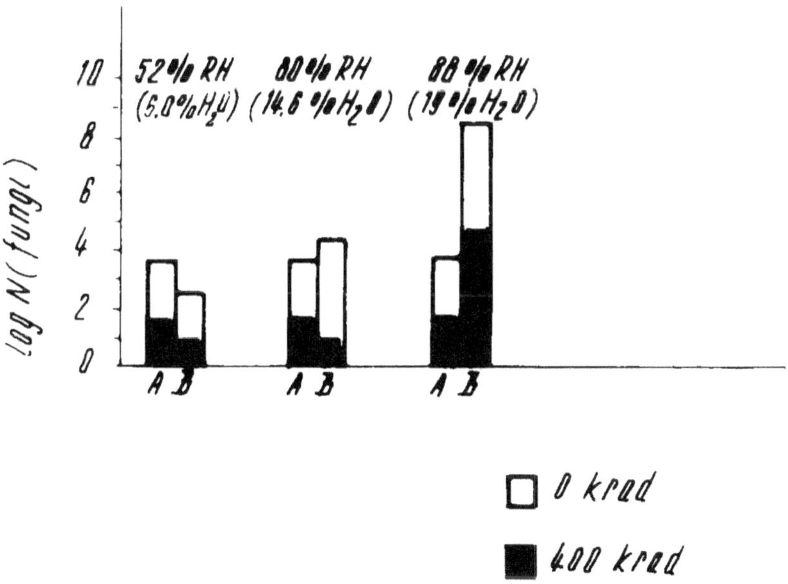

FIGURE 6. Viable mold counts of paprika powder as affected by the relative humidity of the storage space. (A) immediately upon radiation treatment; (B) after 4 months' storage subsequent to radiation. (From Farkas, J., *Aspects of the Introduction of Food Irradiation in Developing Countries*, IAEA, Vienna, 1973, 43. With permission.)

propagate was investigated by Farkas et al.[116] with particular attention to the humidity prevailing during storage. Figure 6 shows the viable mold counts immediately upon radiation treatment and after 4 months of storage. As seen in the figure, the dose of 0.4 Mrad reduced the mold count by 99%. Furthermore, the surviving molds in paprika treated with 0.4-Mrad dose were not capable of growth even at 80% relative humidity and 25°C storage temperature. The mold count of radiation-pasteurized paprika powder reached the present level prior to irradiation only after 4 months' storage in a chamber of 88% relative humidity. The unirradiated samples, when stored under identical conditions, spoiled within a few weeks.

In another storage experiment[33] ground paprika samples were stored at 0 to 5°C and a relative humidity higher than 90%. During 120 days storage the control paprika faded completely and got visibly moldy (Figure 7). The color of the irradiated sample remained bright and its mold count was 2.3×10^2/g.

These experiments established that the surviving mold flora of the radurized ground paprika is unable to propagate to an undesirable extent even in very humid storage space. This supports the frequent observation that the surviving, but damaged microorganisms are not capable of reproduction in the medium in which they were exposed to radiation treatment even under conditions favorable to the growth of undamaged microorganisms, and are more sensitive than the undamaged cells.[117,119,120,125,126] Because this experience with bacterial spores seems to be valid in relation to the mold flora of radiation treated seasonings, it may become important in the prevention of the growth of toxigenic molds, too.

E. The Use of Irradiated Spices in the Processing of Food

Another microbiological aspect with great practical importance is the contamination of spices and other dry food ingredients with highly heat-resistant bacterial spores. In some canned foods, the microbial flora of the major components consists of relatively heat-sensitive organisms. If, however, a certain minor ingredient contributes a com-

FIGURE 7. Ground paprika after 120 days storage at 0 to 5°C, >90% RH packaged in paper bags: (a) control; (b) sample treated with 0.5 Mrad. (From Farkas, J., Beczner, J., and Incze, K., *Radiation Preservation of Food,* IAEA, Vienna, 1973, 389. With permission.)

pletely heat-resistant microflora in terms of the process being used, then this item becomes the single important determinant of the bacteriological quality of the product. Control over the finished product involves control over the spore load of this ingredient.[127]

Spices and cereals are potentially prolific sources of heat-resistant bacteria, including thermophilic flat sours, putrefactive anaerobes, and sulfide stinkers.[28,128,129] Therefore, the effect of radiation at a 1.5-Mrad dose on the additive mixture used in the manufacture of canned pork liver paste, consisting of ground rice, common salt, "French seasoning", and powdered onion, was studied on samples treated at about one tenth (F_o = 1.47) and one hundredth (F_o = 0.14) of the heat treatment applied in the industry (F_o = 13 to 15 sterilization equivalent).[122] The total viable aerobic cell count of the product prior to the heat treatment was 7.0×10^7 per can, of which the total aerobic cell count introduced with the untreated seasoning amounted to 2.9×10^6 per can. The additive mixture formed 6.6% of the weight of the product and the total viable cell count introduced with the untreated seasoning formed about 4% of the total viable cell count of the material prior to sterilization. In spite of this, the results of incubating (at 30°C) the samples exposed to reduced heat treatment described above showed that the effectiveness of the heat treatment depended to a great extent on the microbial contamination of the additive mixture. The number of blown cans plotted against time over the 80-day storage period at 30°C is given in Figure 8.

The distribution of the individual symptoms of spoilage was also evaluated. These results are shown in Figure 9.

As seen in Figure 9, the spoilage percentage of pork liver paste made with irradiated additive mixture was not quite one seventh of that prepared with untreated additives even if given a heat-treatment equivalent of F_o = 1.47. This permits the conclusion that by using a radicidated additive mixture a heat treatment of F_o = 2.5 would be sufficient to achieve sterility even with respect to *C. botulinum*. This heat treatment is just above one sixth of the treatment as applied in the industry. Such a substantial reduction of the heat requirement with unchanged microbiological safety would result in a substantial improvement of the product, saving of energy, and increasing the production capacity.

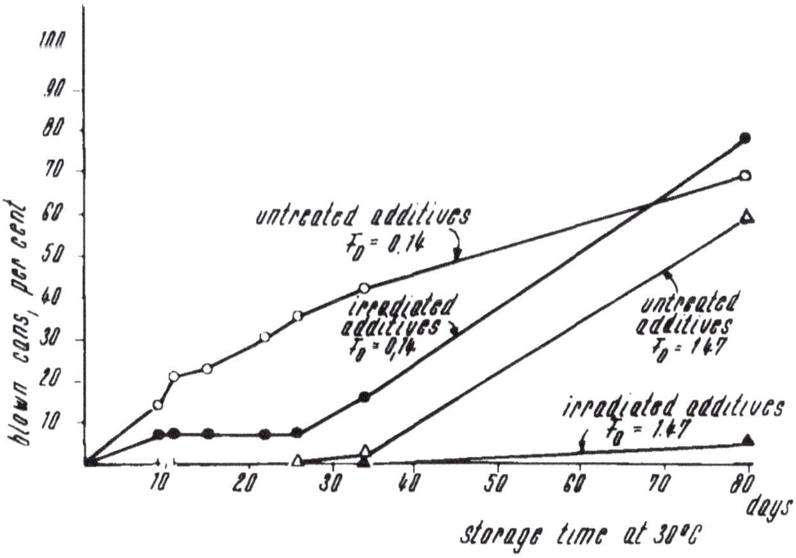

FIGURE 8. The percentage of blown cans of canned pork liver paste as a function of heat treatment, decontamination of the additives and storage time. (From Farkas, J., *Aspects of the Introduction of Food Irradiation in Developing Countries,*, IAEA, Vienna, 1973, 43. With permission.)

In another experiment[33] the viable count of brawn loaves (head cheese) manufactured with irradiated seasonings was substantially lower after cooking than those prepared with untreated seasonings. The difference in viable cell count of samples made with seasonings treated with 0.3 and 1.5 Mrad was not significant. It was found that the use of irradiated seasonings reduced the heat-treatment requirement during brawn manufacture by about half, while the microbiological quality of the product was improved.

In a Russian study introduction of radiation-sterilized spice mixtures into sprat preserves resulted in a considerable reduction in spoilage and in an increase in shelf life from 14 to 20 days.[130]

F. Economic Feasibility of Radiation Treatment of Spices

Since the cost of irradiation is directly proportional to the dose requirement, it is desirable to aim at a significant reduction in the cell count instead of complete sterilization. According to the results described earlier, the heat treatment and the salt content of the products as applied in the food processing industry complement the effect of radiation treatment. Since it is not necessary to aim at complete sterility, the dose requirement and the cost of treatment can be reduced considerably.

For a detailed analysis of economic feasibility, actual experience with commercial facilities would be required. However, in view of the several factors outlined below it seems possible that radiation treatment would be able to compete favorably with other methods of decontamination. These factors that support the economic feasibility of radicidation and radurization are

1. Unlike fumigation, the radiation treatment is easy to automate and the process is a continuous one.
2. Radiation can be applied to prepacked materials in their final packaging.
3. The irradiation of spices need not be seasonal but could be done practically the whole year round, thus avoiding seasonal production peaks.

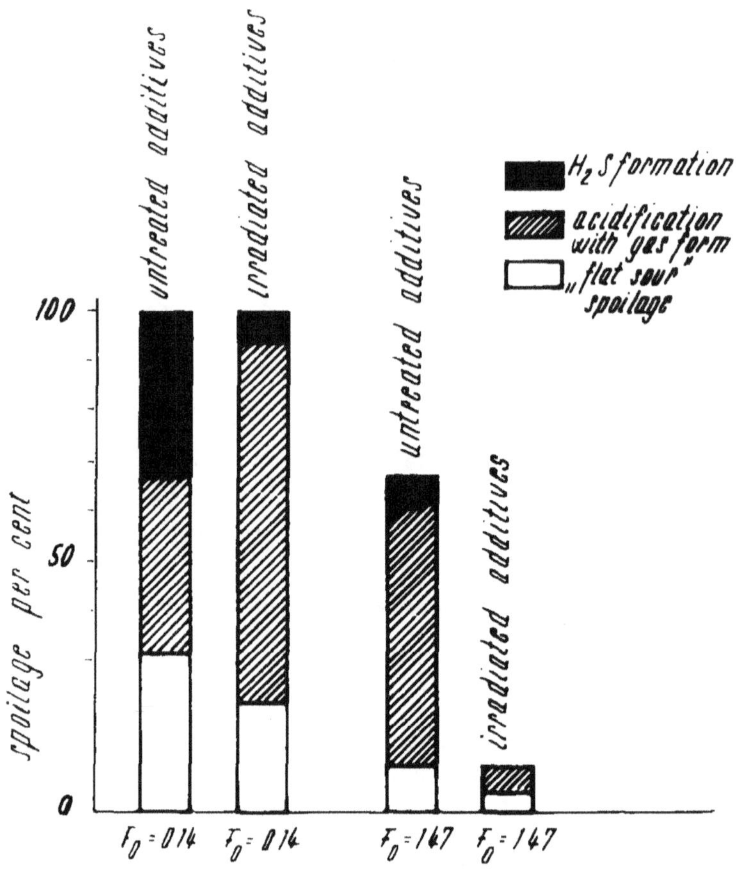

FIGURE 9. The distribution of symptoms of spoilage of canned pork liver paste after 80 days' storage at 30°C as a function of heat treatment and decontamination of the additives. (From Farkas, J., *Aspects of the Introduction of Food Irradiation in Developing Countries*, IAEA, Vienna, 1973, 43. With permission.

4. The dose requirement to achieve practical decontamination is a medium one.
5. The retail prices of spices are relatively high so the cost of radiation would amount to only a small fraction of the cost of the final product.
6. Because of compactness, high value, and transportability of packaged spices, their radiation treatment could be done in centralized facilities and/or on a service base, e.g., in existing food irradiation pilot plants.
7. The demand for decontaminated spices is increasing and, therefore, achieving a good microbiological quality would justify a higher price.

G. Wholesomeness Testing of Irradiated Spices

In the wholesomeness testing of spices, we face the difficulty that feeding tests with hot and pungent materials containing high concentrations of biologically active compounds like those found in many of the important spices are extremely difficult or even impracticable to perform, particularly at the high levels required in the protocols for irradiated staple foods. There are exceptions to this rule, e.g., the powder of sweet capsaicin-free paprika. This product was consumed by albino rats at a 35% level of the total diet for months without adverse effects in the Central Food Research Institute, Budapest, Hungary. A subacute wholesomeness test was performed without any difficulty with radicidated (1.5 Mrad) and/or heat-treated (autoclaved) paprika powder at the same 35% level in the diet.[131,132]

On the other hand, it should be taken into consideration that spices are minor ingredients of our food and consumed in small quantities. For such products, the 25 to 35% dietary level during feeding tests is unrealistically high.

Spices have been accorded a high priority in the research program of the International Project in the Field of Food Irradiation. As a consequence, a wholesomeness testing program on irradiated spices has been embarked upon in Hungary under contract to the International Project.[133,134] Preliminary feeding studies in rats fed a diet containing a spice mixture at levels ranging from 2 to 25% have now been completed. Apart from the lower growth rate caused by the lower food consumption by rats of feed containing high levels of spice, the data show no evidence of adverse effects from feeding of spices whether irradiated (1.5 Mrad), unirradiated, or treated with EO even up to a 25% dietary level. Nevertheless, it was felt that in the course of a long-term study, feeding high levels of spice would produce underweight animals and this might lead to further complications. The new research program includes short-term mutagenicity testing and the identification of radiation-induced chemical changes in individual spices. The possible mutagenic effect of untreated paprika, heat treated (at 121°C with wet steam for 5 min), and irradiated (5-Mrad gamma radiation), respectively, was already studied using the Ames test as well as the host-mediated assay. None of these experiments has indicated induction of or increase in reverse mutations due to irradiated or unirradiated paprika.[135]

H. Legislation and Commercialization of Irradiated Spices

Legislation for the radiation treatment of spices is only in its initial stages. In the U.S.S.R. radiation disinfestation of dried food concentrates for unlimited consumption was approved in 1966. The dose level permitted is 70 krad. In the Netherlands an approval was granted in 1970 for experimental batches of irradiated spices and condiments. The permitted dose range is 800 to 1000 krad.[136] Similar approval for experimental batches of mixed seasonings and ingredients was given by the Hungarian authorities in 1972 and 1976.

To promote commercialization, 10 tons of spices were radiation treated in 1976 by the Dutch Pilot Plant for Food Irradiation in Wageningen.[138]

Within the framework of the Dutch-Hungarian Food Irradiation Research Cooperation Program, the MONIMPEX Trade Company of Hungary sent about 100 kg of radiation-treated paprika powder to Holland for evaluation during 1975 and 1976.

For test marketing in Hungary in 1977, 18,000 cans of hash were produced using radiation-decontaminated mixed dry ingredients.[137] The mixture contained wheat flour, powdered garlic, sodium caseinate, powdered onions, and ground paprika.

V. SUMMARY

Radurization and radicidation of spices are technically feasible and economically promising processes. They offer the best solution to the problem of decontaminating natural spices.

Dose requirements are moderate. A radiation dose of 0.4 to 0.5 Mrad proved sufficient to reduce the viable cell count to the level required by the meat and canning industries. No sensory changes of spices can be detected at these radiation doses. Recontamination can be prevented since the product can be irradiated in its final packaging. The demand for spices of good microbiological quality is ever increasing and the industry could therefore absorb the higher cost of the irradiation treatment. With special reference to the role played by many developing countries in producing and marketing spices, radiation decontamination of spices has particular importance in those countries for which the export of spices is a valuable contribution to much needed convertible currency earnings.

REFERENCES

1. **Clark, W. R. E.**, Modern trends in the application of spice, *Food Manuf.*, 45, 53, 1970.
2. International Trade Centre UNCTA/GATT, *Market for Spices in North America, Western Europe and Japan*, Geneva, 1970.
3. **Christensen, C. M., Fanse, H. A., Nelson, G. H.**, Bates, F., and Mirocha, C. J., Microflora of black and red pepper, *Appl. Microbiol.*, 15, 622, 1967.
4. **Hadlok, R.**, Schimmelpilzkontamination von Fleischerzeugnissen durch naturbelassene Gewürze, *Fleischwirtschaft*, 49, 1601, 1969.
5. **Jensen, L. B., Wood, I. H., and Jansen, C. E.**, Swelling in canned chopped hams, *Ind. Eng. Chem.*, 26, 1118, 1934.
6. **Hobbing, E.**, Die bakteriologische Bedeutung der pflanzlichen Gewürze in der Wurst, *Tierärztl. Rundsch.*, 41, 252, 1935.
7. **Hille, J.**, Der Keimgehalt der Gewürze Majoran, weisser Pfeffer, Nelken und seine Bedeutung in der Wurstwarenfabrikation, Veterinary Medicine Dissertation, Free University of Berlin, 1937.
8. **Yesair, J. and Williams, M. H.**, Spice contamination and its control, *Food Res.*, 7, 118, 1942.
9. **Hüttman, H.**, Untersuchung über den Keimgehalt der Deutschland üblichen Gewurze und seine Beziehung zur Entstehung von Fehlfabrikaten, Veterinary Medicine Dissertation, Hannover, 1954.
10. **Coretti, K.**, Der Keimgehalt von Gewürzen, *Fleischwirtschaft*, 7, 305, 1955.
11. **Mossell, D. A. A.**, The importance of the bacteriological condition of ingredients used as minor components in some canned meat products, *Ann. Inst. Pasteur Lille*, 7, 171, 1955.
12. **Schönberg, F.**, Die Bedeutung keimfreier bzw. Keimarmer Naturgewürze für die Erzielung einwandfreier Rohwurst, *Fleischwirtschaft*, 49, 461, 1955.
13. **Walz, E.**, Ein eindruckvolles Beispiel von Verkeimung (Bazillen- und Schimmelpilzbefall) in der Fleischwirtschaft verwendeter Gewürze, *Arch. Lebensmittelhyg.*, 7, 138, 1956.
14. **Coretti, K.**, Kaltentkeimung von Gewürzen mit Äthylenoxyd, *Fleischwirtschaft*, 9, 183, 1957.
15. **Pohja, M. S.**, Vergleichende Untersuchungen über den Mikrobengehalt fester und flüssiger Gewürze, *Fleischwirtschaft*, 9, 547, 1957.
16. **Kammer, J.**, Über die besonderen Eigenschaften der aeroben Bazillen in Gewürzen unter besonderer Berücksichtigung der Eiweisszersetzung, Veterinary Medicine Dissertation, Hannover, 1961.
17. **Eschmann, K. H.**, Gewürze — eine Quelle bakteriologischer Infektionen, *Alimenta*, 4, 83, 1965.
18. **Glas, A.**, *Praktisches Handbuch der Lebensmittel*, Bayerischer Landwirtschaftsverlag GmbH., Munich, 1965.
19. **Proctor, B. E.**, Goldblith, S. A., and Fram, H., Effect of supervoltage cathode rays on bacterial flora of spices and other dry food materials, *Food Res.*, 15, 490, 1950.
20. **Kopplow, E.**, Die hygienische Bedeutung des Keimgehaltes von Gewürzen für die Wurst — und Fleischwarenherstellung, *Z. Hyg.*, 135, 396, 1952.
21. **Lerke, P. A. and Farber, L.**, Effect of electron beam irradiation on the microbial content of spices and teas, *Food Technol.*, 14, 266, 1960.
22. **Török, G. and Farkas, J.**, Kisérletek fűszerpaprikaőrlemények ionizalo sugárzásos csiraszámcsökkentésére, (Investigations into the Reduction of Viable cell Count in Paprika Powder by Ionizing Radiation) (in Hungarian), *Commun. Cent. Food Res. Inst., Budapest*, 3, 1, 1961.
23. **Elter, B. and Scharner, E.**, Über den Keimgehalt von Gewürzen und seine Bedeutung für die Fleischwarenproduktion, *Fleisch*, 22, 219, 1968.
24. **Szabó, P.**, Kisérletek a füszerpaprika csiraszámának csökkentésere (Experiments for reduction of microbial cell count in paprika) (in Hungarian), *Konzerv-Paprikaip.*, 128, 1968.
25. **Gerhardt, U.**, Entkeimung von Gewürzen, *Gordian*, 9, 427, 1969.
26. Bulgarian Meat Research Institute, Obezzarazhivanie specij prednaznachennüh dlya proizvodstva myasnüh konservov, posredstvom oblutcheniya ioniziruyuschimi lutcheniyami (Sterilization of spices intended for use in the production of canned meat products, by means of ionizing radiations), Manuscript, 1970.
27. **Hill, M. H.**, Seasoning in meat products, *Process Biochem.*, 6(12), 27, 1971.
28. **Krishnaswamy, M. A., Patel, J. D., and Parthasarathy, N.**, Enumeration of microorganisms in spices and spice mixtures, *J. Food Sci. Technol.*, 8(4), 191, 1971.
29. **Farkas, J., Beczner, J., Kiss, I., and Incze, K.**, Cell Count Reduction in Seasoning, Particularly in Ground Paprika, by Radiation Treatment, 3rd Progress Rep. to International Atomic Energy Agency, Contract No. 931/R1/RB, January 1, 1972 to June 30, 1972, Central Food Research Institute, Budapest, 1972.
30. **Tjaberg, T. B., Underdal, B., and Lunde, G.**, The effect of ionizing radiation on the microbial content and the volatile constituents of spices, *J. Appl. Bacteriol.*, 35, 473, 1972.
31. **Delkinova, S. M. and Dupuy, P.**, Sanitation of onion powder by irradiation and heating, *Acta Aliment. Acad. Sci. Hung.*, 2, 181, 1973.

32. **Farkas, J. and El-Nawawy, A. S.**, Effect of gamma radiation upon the viable cell count and some other quality characteristics of dried onions, *Acta Aliment. Acad. Sci. Hung.*, 2, 137, 1973.
33. **Farkas, J., Beczner, J., and Incze, K.**, Feasibility of irradiation of spices with special reference to paprika, in *Radiation Preservation of Food*, International Atomic Energy Agency (IAEA), Vienna, 1973, 389.
34. **Julseth, R. M. and Deibel, R. H.**, Microbial profile of selected spices and herbs at import, *J. Milk Food Technol.*, 37, 414, 1974.
35. **Vajdi, M.**, Comparative Effects of Ethylene Oxide, Gamma Irradiation and Microwave Treatments on the Control of Microorganisms in Selected Spices, M. S. Thesis. University of Manitoba, Winnepeg, 1970.
36. **Krishnaswamy, M. A., Patel, J. D., Mair, K. K. S., and Muthu, M.**, Microbioogical quality of certain spices, *Indian Spices*, 11(1—2), 6, 1974.
37. **Inal, T., Keskin, S., Tolgay, Z., and Zezcan, I.**, Sterilization of spices by means of gamma rays, *Fleischwirtschaft*, 55, 675, 1975.
38. **Powers, E. M., Lawyer, R., and Masuoka, Y.**, Microbiology of processed spices, *J. Milk Food Technol.*, 38, 683, 1975.
39. **Flannigan, B. and Hui, S. C.**, The occurrence of aflatoxin-producing strains of *Aspergillus flavus* in the mould floras of ground spices, *J. Appl. Bacteriol.*, 41, 411, 1976.
40. **Palumbo, S. A., Rivenburgh, A. I., Smith, J. L., and Kissinger, J. C.**, Identification of *Bacillus subtilis* from sausage products and spices, *J. Appl. Bacteriol.*, 38, 99, 1975.
41. **Inal, T.**, Untersuchungen über die Bakterienflora der Gewürze unter besonderer Berücksichtigung der aeroben sporenbilder, *Veteriner Fakültesi clergisi*, 12(1—2), 82, 1965.
42. **Kim, H. U. and Goepfert, J. M.**, Occurrence of *Bacillus cereus* in selected dry food products, *J. Milk Food Technol.*, 34(1), 12, 1971.
43. **Powers, E. M., Latt, T. G., and Brown, T.**, Incidence and levels of *Bacillus cereus* in processed spices, *J. Milk Food Technol.*, 39(10), 668, 1976.
44. **Volkova, R. S.**, *Bacillus cereus* contamination of foods and environment at institutional feeding points, *Gig. Sanit.*, 36(2), 108, 1970.
45. **Smith, L. D. S.**, in *Microbiological Quality of Foods*, Slanetz, L. W., Chichester, C. O., Gaufin, A. R., and Ordal, Z. J., Academic Press, New York, 1963, 77.
46. **Strong, D. H., Canada, J. C., and Griffiths, B. B.**, Incidence of *Clostridium perfringens* in American foods, *Appl. Microbiol.*, 11, 42, 1963.
47. **Krishnaswamy, M. A., Nair, K. K. S., Patel, J. D., and Parthasarathy, N.**, Preliminary observations on the survival of *Salmonella* in curry, samber, coriander and red-chilly powders, *J. Food Sci. Technol.*, 12, 195, 1975.
48. **Wilson, C. R. and Andrews, W. H.**, Sulfite compounds as neutralizers of spice toxicity for *Salmonella*, *J. Milk Food Technol.*, 39, 464, 1976.
49. **Bodrossy, L.**, A husipari füszerek csiratartalmának csökkentése (Reduction of microbial contamination of spices used in the meat industry) (in Hungarian), *Husipar*, 9, 160, 1960.
50. World Health Organization, Salmonella surveillance, *Weekly Epidemiological Record*, No. 42, World Health Organization, Geneva, 1974, 351.
51. **Horie, Y., Yamazaki, M., Miyaki, K., and Udagawa, S.**, On the fungal contents of spices, *J. Food Hyg. Soc. Jpn.*, 12, 516, 1971.
52. **Moreno-Martinez, E. and Christensen, C. M.**, Fungus flora of black and white pepper (*Piper nigrum* L.), *Rev. Latinoam. Microbiol.*, 15, 19, 1973.
53. **Pal, N. and Kundu, A. K.**, Studies on *Aspergillus* spp. from Indian spices in relation to aflatoxin production, *Sci. Cult.*, 30, 252, 1972.
54. **Shank, R. C., Wogan, G. N., and Gibson, J. F.**, Dietary aflatoxins and human liver cancer. I. Toxigenic moulds in foods and foodstuffs of tropical South-East Asia, *Food Cosmet. Toxicol.*, 10(1), 51, 1972.
55. **Christensen, C. M.**, Pure spices — how pure?, *Am. Soc. Microbiol. (ASM) News.*, 38(4), 165, 1972.
56. **Scott, P. M. and Kennedy, B. P. C.**, Analysis and survey of ground black, white and capsicum peppers for aflatoxins, *J. Assoc. Off. Agric. Chem.*, 56, 1452, 1973.
57. **Scott, P. M. and Kennedy, B. P. C.**, The analysis of spices and herbs for aflatoxins, *Can. Inst. Food Technol. J.*, 8, 124, 1975.
58. **Suzuki, J. I., Dainius, B., and Kilbuck, J. H.**, A modified method for aflatoxin determination in spices, *J. Food Sci.*, 38, 949, 1973.
59. **Jensen, L. B.**, *Microbiology of Meats*, 2nd ed., Garrard Press, Champaign, Ill., 1945, 345.
60. **Gisske, W.**, Keimfreie Gewürzpräparate für die Herstellung von Fleischgerichten und Wurstwaren, *Fleischwirtschaft*, 6, 280, 1957.
61. **Kelch, F.**, Die Bedeutung der Pflanzlichen Gewurze fur die Wurstfabrikation, *Fleischwirtschaft*, 10, 215, 1958.

62. **Schönberg, F.**, Über die Bedeutung bakterienhaltiger Gewürze für das Verderben von Wurstwaren und zur Keimverminderung in Gewürzen durch UV Entkeimungslampen, *Fleischwirtschaft*, 1, 132, 1952.
63. **Schönberg, F.**, Zum Wert der Untersuchung von Gewürzen auf proteolytische Keime mit Hilfe des Calcium-Casein-Pepton Agars und gleichzeitig zur Prüfung der Frage des Einflusses der Temperatur auf das Wachstum der in Gewürzen vorkommenden Bacillen, *Arch. Lebensmittelhyg.*, 10, 281, 1959.
64. **Schönberg, F.**, Zur werwendung von Gewürzessenzen aus Naturgewürzen für die Herstellung von Fleischwaren, *Fleischwirtschaft*, 14, 272, 1962.
65. **Heath, H. B.**, Bacteria free spices in food processing, *Food Process. Packag.*, 4, 144, 1964.
66. **Bartels, H. and Hadlok, R.**, Organoleptische und bakteriologische Untersuchungen von Gewürzextrakten aus Naturgewürzen, *Fleischwirtschaft*, 46, 234, 1966.
67. **Gola, J., Singer, V., and Lukesova, M.**, The utilisation of liquid concentrates of spice, especially of liquid concentrates of garlic and onion at the production of meat products on mechanized and continuous lines in meat industry, IXth Conf. Eur. Meat Res. Workers, Budapest, September 4—11, 1963.
68. **Pekhov, A. V., Katyuzhanskaya, A. N., and Dyobankova, N. F.**, Ispol'zovanie ekstraktov pryanuh rastenii v proizvodstve konservov (The use of seasoning extracts in the canning industry) (in Russian), *Konservn. i Ovoshchesush. Prom.*, 7, 12, 1969.
69. **Schurm, H.**, Seasoning hard sausages with spice extracts, *Food Trade Rev.*, 41(3), 31, 1969.
70. **Wyler, O. D.**, Hilfs- und Zusatzstoffe und ihr Einfluss auf Technologie und Qualitat von Fleischwaren, *Die Fleischwirtschaft*, 52, 727, 1972.
71. **Mihályi, V. and Zukál, E.**, Az izprofil-analizis használata fűszerek és fűsserkivonatok izbirálatára (Using the flavor profile analysis in the examination of spices and spice extracts) (in Hungarian), *Husipar*, 21, 127, 1972.
72. **Körmendy, L.**, Különböző fűszerfélék és fűszerolajok vizsgálata. I. Bors és borskivonat összehasonlitása (Examination of domestic and foreign spices and spice oils. I. Comparing pepper and pepper extracts) (in Hungarian), *Husipar*, 22, 123, 1973.
73. **James, L. H.**, Reducing the microbial content of spices, *Food Ind.*, 23, 428, 1938.
74. **Hall, L. A.**, Sterilized spices, new factor in food quality control, *Food Ind.*, 426, 464, 1938.
75. **Hall, L. A.**, Sterilization of spices by treatment with ethylene oxide, *Food Packer*, 32(12), 26, 1961.
76. **Pappas, H. I. and Hall, L. A.**, The control of thermophilic bacteria, *Food Technol.*, 6, 456, 1952.
77. American Meat Institute, Committee on Textbooks, *Sausage and Ready-to-Serve Meats*, Rev. ed., Institute of Meat Packing, University of Chicago, Illinois, 1953.
78. **Makó, I., Mėhes, G., and Takács, J.**, Fűszerek csirátlanitása etilénoxiddal (Disinfection of spices by ethylene oxide) (in Hungarian), *Magyar Allatorv. Lapja*, 14, 118, 1959.
79. **Makó, I., Mėhes, G., and Takács, J.**, Husipari fűszerek csirátlanitása (Sterilization of spices for the meat industry) (in Hungarian), *Husipar*, 8, 23, 1959.
80. **Mayr, G. E. and Suhr, H.**, Preservation and sterilization of pure and mixed spices, Conf. Proc. Spices, Tropical Products Institute, April 1972, 201.
81. **Coretti, K. and Inal, T.**, Rückstandprobleme bei der Kaltentkeimung von Gewürzen mit T-gas (Äthylenoxid), *Fleischwirtschaft*, 5, 599, 1969.
82. **Hadlok, R. and Toure, B.**, Mykologische und bakteriologische Untersuchungen entkeimter Gewürze, *Arch. Lebensmittelhyg.*, 24(1), 20, 1973.
83. **Murányi, L., Nagy, L., and Halász, K.**, Fűszerpaprika csirátlanitásának lehetősége etilénoxiddal (Possibility of sterilization of paprika by ethylene oxide) (in Hungarian), 3rd Natl. Meet. Food Hyg., Nagykőrös, May 3 — 7, 1971.
84. **Lerche, L.**, Vorkommen und Unschädlichmachung von Salmonellabakterien in Trockeneiweiss, *Arch. Lebensmittelhyg.*, 8, 267, 1957.
85. **Kelch, H. and Coretti, K.**, Über die Anwendungsmöglichkeit von Äthylenoxyd in bakteriologischen Instituten, *Fleischwirtschaft*, 11, 833, 1959.
86. **Gottschalk, H. M.**, A review on spices — present status of decontamination techniques such as gamma irradiation, *Food Irradiat. Inf.*, 7, 7, 1977.
87. **Scudamore, K. A. and Heuser, S. G.**, Ethylene oxide and its persistant reaction products in wheat flour and other commodities. Residues from fumigation or sterilization and effect of processing, *Pest. Sci.*, 2(2), 80, 1971.
88. **Kröller, E.**, Untersuchungen zur Begasung von Lebensmitteln mit Äthylenoxyd und zu dessen Restmengen-Bestimmung, *Dtsche Lebensm.-Rundsch.*, 102, 227, 1966.
89. **Wesley, F., Rourke, B., and Darbishire, O.**, The formation of persistent toxic chlorohydrins in foodstuffs by fumigation with ethylene oxide and with propylene oxide, *J. Food Sci.*, 30, 1037, 1965.
90. **Reith, J. F.**, Vorming van schadelijke produkten bij het steriliseren van levensmiddelen met ethyleenoxide en propyleenoxide, *Voeding*, 28, 531, 1967.

91. **Fischbach, H.**, The food additives amendment and safety of U.S. food, in *The Safety of Foods*, Ayres, J. C., Blood, F. R., Chichester, C. O., Graham, H. D., McCutcheon, R. S., Powers, J. J., Schweigert, B. S., Stevens, A. D., and Zweig, G., AVI Publishing, Westport, Conn., 1968, 44.
92. **Ragelis, E. P., Fischer, B. S., Klimeck, B. A., and Johnson, C.**, Isolation and determination of chlorohydrins in foods fumigated with ethylene oxide or with propylene oxide, *J. Assoc. Off. Agric. Chem.*, 51, 709, 1968.
93. **Stijve, T., Kalsbach, R., and Eyring, G.**, Determination and occurrence of ethylene chlorohydrin residues in foodstuffs fumigated with ethylene oxide, *Trav. Chim. Aliment. Hyg.*, 67, 403, 1976.
94. **Strekalova, E. E.**, Toward the question of the mutagenic action of ethylene oxide on mammals, *Toksikol. Nov. Prom. Khim. Veshchestv.*, 12, 72, 1971.
95. **Embree, J. W., Lyon, J. P., and Hine, C. H.**, The mutagenic potential of ethylene oxide using the dominant-lethal assay in rats, *Toxicol. Appl. Pharmacol.*, 40, 261, 1977.
96. **WHO-IARC**, Cadmium, nickel, some epoxides, miscellaneous industrial chemicals, and general considerations on volatile anesthetics, *WHO-IARC Monographs on Evaluation of Carcinogenic Risk of Chemicals to Man*, Vol. 11, International Agency for Research on Cancer, Lyon, 1976.
97. **Kucerová, M. and Zhurkov, V. S.**, Population cytogenetic study of mutagenic effect of epichlorohydrin, *Mutat. Res.*, 46, 227, 1977.
98. **Scharf, M. M.**, Sterilization of Spices by In Situ Salt Formation, U.S. Patent 3,316,100 Ser. No. 455,327, April 25, 1967.
99. **Incze, K.**, Uj módszerek fűszerek csirátlanitására (New methods in sterilization of spices) (in Hungarian), in Colloq. Cent. Food Res. Inst. and Food Sci. Comm. Hung. Acad. Sci., Budapest, May 1970.
100. **Thiessen, F.**, Effect of heat on spices, *Fleischerei*, 22(11), 15, 1970.
101. **Thiessen, F.**, Behaviour of natural spices and spice extracts when subjected to heat. II. The effect of heat on natural spices and spice extracts in sausage mixtures, *Fleischwirtschaft*, 50(6), 813, 1970.
102. **Thiessen, F. and Scheide, J.**, Heat sensibility of natural spices and spice essences. I. The effect of heating on natural spices and spice essences in model emulsions, *Fleischwirtschaft*, 50(3), 317, 1970.
103. **Vajdi, M. and Pereira, N. N.**, Comparative effects of ethylene oxide, gamma irradiation and microwave treatment on selected spices, *J. Food Sci.*, 38(5), 893, 1973.
104. **Coretti, K.**, Gewürzentkeimung, eine hygienische Notwendigkeit. Zur Entkeimung durch UV-Bestrahlung, *Fleischwirtschaft*, 7, 386, 1955.
105. **Bergman, C. and Pradella, L.**, Ein Beitrag zur Frage des Keimgehaltes der Gewürze und seiner Abtötung mit der Asta Analysen Quarzlampe, *Monatsh. Veterinaermed.*, 10, 301, 1955.
106. **Dunn, C. G., Campbell, W. L., Fram, H., and Hutchins, A.**, Biological and photochemical effects of high energy electrostatically produced roentgen rays and cathode rays, *J. Appl. Phys.*, 19, 605, 1948.
107. **Hall, R. L.**, Effect of Ionizing Radiation on Spices, Rep. I-VIII, U.S. Army Quartermaster Corps, Contract 7-84-01-002, N-28, McCormick Company, Inc., Hunt Valley, Md., 1953 to 1958.
108. **Robinson, R. F., Overbeck, R. C., and Porter, F. E.**, X-ray sterilization of spices, *Coffee Tea Ind. Flavor Field*, 77, 61, 1954.
109. **Rigler, J.**, Paper presented at Zweites Symposium der Internationalen Vereinigung von Tierärtzlicher Lebensmittelhygieniker (I.A.V.F.H.); from **Kelch, F.**, *Fleischwirtschaft*, 12, 563, 1960.
110. **Maurer, H. J.**, Elektronenbeschleuniger und Sterilization, GDCh-Hauptversammlung, in *Angew. Chem.*, 69, 736, 1957.
111. **Hansen, P.-I.E.**, Radiation treatment of meat products and animal by products, in *Food irradiat.*, International Atomic Energy Agency, Vienna, 1966, 411.
112. **Bachman, S. and Gieszczynska, J.**, Studies on some microbiological and chemical aspects of irradiated spices, in *Aspects of the Introduction of Food Irradiation in Developing Countries*, IAEA, Vienna, 1973, 33.
113. **Szabad, J.**, Az etilénoxidos kezelés és a gamma sugárzás csiracsökkentő hatásának összehasonlító vizsgálata (Comparative Investigation of Cell-Count Reducing Effects of Ethylene Oxide Treatment and Gamma Radiation)(in Hungarian), M.S. Thesis, Universtity of Horticulture, Budapest, 1977.
114. **Van Kooij, J. G.**, Neuere Erkentnisse über die Bestrahlung von Früchten und Gemüsen, in *Nahrungsmittelbestrahlung*, Somogyi, J. C., Ed., Forster Verlag, Zürich, 1960, 44.
115. **Farkas, J.**, Cell Count Reduction of Seasoning, Particulaly in Ground Paprika, by Radiation Treatment, 1st Prog. Rep. to the IAEA, Contract No. 931/RB, Central Food Research Institute, Budapest, Hungary, October 1, 1970 to June 30, 1971.
116. **Farkas, J., Török, G., and Horváth, Zs.**, Tárolási kisérletek sugárzással pasztőrözött fűszerpaprikaőrleménnyel (Storage experiments with powdered paprika pasteurized by irradiation) (in Hungarian), *Commun. Cent. Food Res. Inst., Budapest*, 2, 19, 1962.
117. **Farkas, J.**, Response of bacterial spores to combinations of heat, irradiation and some chemical factors controlling the spore state, in Res. Coord. Meet. Microbiol. Aspects Food Irradiat., FAO/IAEA, Vienna, May 1970.

118. **Morgan, B. H. and Reed, J. M.**, Resistance of bacterial spores to gamma irradiation, *Food Res.*, 19, 357, 1954.
119. **Roberts, T. A., Ditchett, M., and Ingram, M.**, The effect of sodium chloride on radiation resistance and recovery of irradiated anaerobic spores, *J. Appl. Bacteriol.*, 28, 336, 1965.
120. **Roberts, T. A.**, Symposium on bacterial spores: recovering spores damaged by heat, ionizing radiations or ethylene oxide, *J. Appl. Bacteriol.*, 33, 74, 1970.
121. *Begriffbestimmungen für Gewürze, Ersatzgewürze und daraus hergestellte Erzeugnisse*, Schriftenreihe des Bundes für Lebensmittelrecht und Lebensmittelkunde, Heft 51, Hamburg-Berlin-Düsseldorf, B. Behr's Verlag, 1964.
122. **Farkas, J.**, Radurization and radicidation of spices, in *Aspects of the Introduction of Food Irradiation in Developing Countries*, IAEA, Vienna, 1973, 43.
123. **Beczner, J., Farkas, J., Watterich, A., Buda, B., and Kiss, I.**, Study into the identification of irradiated ground paprika, in *Proc. Int. Colloq. Identificat. Irradiated Foodstuffs, Karlsruhe*, October 1973, Commission of the European Communities, Directorate-General Scientific and Technical Information and Information Management, 1974, 255.
124. **Beczner, J. and Farkas, J.**, Tárolási kísérletek besugárzással csiraszegényitett füszerpaprika-örleménnyel (Storage experiments with ground seasoning paprika treated by irradiation for reduction of cell count) (in Hungarian,) *Kísérletügyi Közl., E.*, 67(1—2), 3, 1974.
125. **Farkas, J., Kiss, I., and Andrássy, É.**, The survival and recovery of irradiated bacterial spores affected by population density and some external factors, in *Radiosterilization of Medical Products*, IAEA, Vienna, 1967, 343.
126. **Duncan, C. L.**, Arrest of growth from spores in semipreserved foods, *J. Appl. Bacteriol.*, 33, 60, 1970.
127. **Silliker, J. H.**, Total counts as indexes of food quality, in *Microbiological Quality of Foods*, Slanetz, L. W., Chichester, C. O., Gaufin, A. R., and Ordal, Z. J., Eds., Academic Press., New York, 1963, 102.
128. **Crossley, E. L.**, Bacteriology of meat and fish pastes including new method of detection of certain anaerobic bacteria, *J. Hyg.*, 38, 205, 1938; as cited in **Baumgartner, J. G.**, *Canned Foods, An Introduction to their Microbiology*, J. & A. Churchill, London, 1949.
129. **Saint-Lebe, L. and Berger, G.**, La conservation par irradiation des produits alimentaires pulverulents. Intérêt et possibilitiés de la méthode et application au cas de l'amidon de mais, in *Peaceful Uses of Atomic Energy*, Vol. 12, United Nations, New York and Vienna, 1972.
130. **Dutova, W. N., Kardashev, A. V., and Goftarsh, M. M,**. The use of gamma irradiation in cold sterilization of salt and spice mixtures, *All Union Res. Mar. Fish. Oceanogr., Moscow*, 73, 69, 1970; (English Transl. Fisheries Research Board, Canada).
131. **Barna, J.**, Toxicity Test of Irradiated and Heat Treated Paprika (in Hungarian), Res. Rep., Central Food Research Institute, Budapest, 1973.
132. **Farkas, J.**, Hungary — progress in food irradiation, *Food Irradiat., Inf.*, 4, 11, 1975.
133. **Barna, J.**, Wholesomeness of an irradiated spice mixture, *Food Irradiat. Inf.*, Suppl. 4, 48, 1975.
134. **Anon.**, Spices — a new approach, *Food Irradiat. Inf.*, 7, 67, 1977.
135. **Farkas, J., Andrássy, É., and Incze, K.**, Evaluation of possible mutagenicity of irradiated spices, *Acta Aliment.*, 10, 129, 1981.
136. **Ulmann, R. M.**, Introducing irradiated foods to the producer and consumer, in *Peaceful Uses of Atomic Energy*, Vol. 12, United Nations, New York and IAEA, Vienna, 1972.
137. **Farkas, J.**, Progress in food irradiation in Hungary, *Food Irradiat. Inf.*, (9), 45, 1978.
138. **Anon.**, Application of Atomic Energy in Agriculture, Annu. Rep., 1976, Association Euratom — ITAL, Wageningen, The Netherlands, 139.

Chapter 5

DELAY IN POSTHARVEST RIPENING AND SENESCENCE OF FRUITS

Ernest K. Akamine and James H. Moy

TABLE OF CONTENTS

I. INTRODUCTION

A major contributing factor to the shortage of crop foods in the world is losses due to spoilage in the production, transportation, and storage of fresh commodities, and approximately one third of the produce harvested is discarded due to the spoilage resulting from improper handling. According to Salunkhe,[1] this spoilage amounts to more than a billion dollars which are annually lost to growers and handlers of fruits and vegetables.

Extensive research has been conducted in the use of irradiation to extend the shelf life of fresh commodities in postharvest handling. Several review publications concerning this area of research have appeared such as those of Massey[2] (dealing with irradiation of foods including fruits), Maxie and Abdel-Kader[3] (the physiology of fruits), Pablo et al.[4] (irradiation studies on tropical and subtropical fruits and vegetables), and Moy et al.[5] (irradiation research in Hawaii).

The title of this chapter needs clarification. A delay in postharvest ripening can occur only in a climacteric fruit, such as a green banana, which ripens normally after harvest. It cannot occur in a nonclimacteric fruit such as an orange, because upon harvest, it does not undergo normal physiological and biochemical processes associated with quality improvement as the fruit ripens. Senescence or physiological breakdown, on the other hand, occurs in both climacteric and nonclimacteric fruits as they age in postharvest storage, and its onset may be delayed. The term "fruit" in the title is used in the botanical sense and not in the horticultural sense which separates fruit vegetables such as tomatoes from the general category of fruits such as pears. With the exception of one investigation in which ^{137}Cs was used as the source of gamma rays, ^{60}Co was the source used in all the investigations reported in the literature reviewed here, and the term "irradiation" or "radiation" as used here refers to exposure to gamma rays.

In many instances, postharvest decay is a major obstacle to retention or extension of shelf life in storage and decay will normally hasten the ripening and senescing processes. Therefore, any discussion of attempts to delay physiological ripening and/or senescence must also take into consideration the effects of decay. Furthermore, before a beneficial delay in ripening and/or senescence by irradiation can be recognized, the effects of the treatment on the physiological, biochemical, nutritional, and organoleptic aspects of the fresh commodity must be determined, that is, the tolerance of the commodity to irradiation must be considered. In the following section on individual fruits, the effects of irradiation are evaluated in terms of the basic requirements for the beneficial use of irradiation for shelf-life extension by employing the criteria in Table 1.

II. FRUITS

A. Pears (*Pyrus communis* L.)

Bramlage and Covey[6] reported that Bartlett pears irradiated at 100 to 400 krad were delayed in ripening by several days, but upon ripening, they developed mealy pulp texture and failed to develop the characteristic ripe flavor of the variety. Somogyi and Romani[7] found that textural changes induced by irradiation doses of 300, 600, and 900 krad corresponded with a decrease in protopectin content and an increase in pectin and pectate fractions of the fruit accompanied by an increase in pectin methylesterase activity. Since gamma irradiation produces ozone from O_2, it was thought possible to protect pears from ozone injury by removing O_2 from the fruit (approximately 20% of internal gas is O_2) before irradiation. Therefore, Bussel and Maxie[8] removed the internal O_2 by flushing the fruit with N_2 and then immediately irradiated the fruit in air. The method was found to be impractical because of the rapid reentry of O_2 into the fruit when it was returned to air in the irradiation chamber. Perhaps irradiation in N_2 could have improved the tolerance of the fruit.

Table 1
BASIC REQUIREMENTS AND CRITERIA FOR EVALUATING RADIATION EFFECTS FOR DETERMINING THE FEASIBILITY OF RADIATION TECHNOLOGY FOR EXTENDING THE SHELF LIFE OF FRUITS

Basic requirements	Criteria for evaluating radiation effects
High tolerance to radiation dose sufficient to:	Fruit surface (color changes, injury)
Effectively control decay	Pathological effects (control of pathogens)
Effectively delay ripening and senescence	Physiological effects (respiration, ethylene production, rate of ripening, rate of senescence)
	Organoleptic effects (taste, aroma, texture, color)
	Nutritional effects (soluble solids, sugars, pH, total acids, vitamins, proteins, carotenoids)
	Biochemical effects (enzymes, metabolism, cytotoxicity)
Economic feasibility	Comparison with other methods
Quarantine use (if applicable)	Disinfestation (insect control)

According to Maxie et al.,[9] doses of irradiation up to 200 krad did not interfere with normal ripening of Bartlett pears and doses of 300 and 400 krad allowed the fruit to soften eventually, but the pulp was poor in color, mealy in texture, and insipid in flavor. Pears subjected to 300 or 400 krad did not ripen even when held in flowing atmospheres containing 1000 ppm ethylene (C_2H_4) for 8 days at 20°C. Ripening was inhibited by irradiation only when applied to preclimacteric fruit. Tolerance to irradiation was related to C_2H_4 production and respiratory behavior. Immediately after irradiation, there was a transitory burst of C_2H_4 which subsequently declined. In fruits subjected to inhibitory doses (300 and 400 krad), the decline continued, but in those subjected to noninhibitory doses (100 and 200 krad), C_2H_4 production increased about the same time this gas increased in the untreated fruit. As in C_2H_4 production, there was a transitory increase in respiration immediately after irradiation which was followed by a decline. In fruits exposed to noninhibitory doses, normal climacteric rise took place and the fruits ripened normally, but in fruits exposed to inhibitory doses, after the initial rise, respiration remained erratically at high levels and the fruits never ripened. Maxie et al.[9] concluded that the adverse effects of gamma rays in Bartlett pears involve both an inhibition of C_2H_4 production and a decreased sensitivity of the fruit to the ripening action of the gas. Romani[10] observed that anaerobic pretreatment and anaerobic irradiation did not appreciably change the postirradiation respiratory pattern of pears and concluded that this suggested some direct radiation damage rather than ozone damage.

According to the above results, 100- to 300-krad doses delayed the ripening of pears, but the organoleptic scores of the pulp were decreased. In any event, a delay in ripening occurred only when irradiation was applied to preclimacteric fruits.

B. Peaches (*Prunus persica* L.)

Bramlage and Covey[6] reported that doses of 125 to 300 krad softened peaches during irradiation and the irradiated fruit subsequently developed more red color but less flavor than untreated fruit. Brown rot, but not Rhizopus rot, was controlled by irradiation and water on the surface of the fruit during irradiation caused spotting of the skin. Maxie et al.[11] showed that irradiation doses of 100 to 600 krad initiate the climacteric and ripening sequence by inducing the preclimacteric fruits to produce stimulatory amounts of C_2H_4. Irradiated fruits produced more C_2H_4 and ripened 4 to 5 days earlier than untreated fruits. Irradiated fruits were also redder (more anthocyanin). These researchers speculated that since the loss of texture made the fruits very susceptible to mechanical injury, irradiation may not be feasible as a protective technology.

Dhaliwal and Salunkhe[12] presumably using eating-stage Elberta and Gem peaches, found that a dose range of 100 to 300 krad in conjunction with packaging in polyethylene bags increased the refrigerated life of the fruits by 15 to 20 days. They claimed that irradiation controlled the spoilage fungi and the increased concentration of CO_2 and decreased concentration of O_2 in the package caused by the initial burst of respiration in the irradiated fruit modified the storage atmosphere and possibly contributed to the extension of shelf life. Somogyi and Romani[7] reported that textural changes of peaches exposed to 300- to 900-krad doses were accompanied by a decrease in protopectin content and an increase in pectin and pectate fractions due to an increase in pectin methylesterase activity after irradiation.

According to the above findings on the effects of irradiation on peaches, the general effects were accelerated ripening and softening of the fruit. However, the refrigerated storage life of eating stage irradiated fruits could be extended if packaged in polyethylene bags.

C. Nectarines (*P. persica nectarina* L.).

Irradiation at doses of 200 to 400 krad controlled brown rot in nectarines but it adversely affected the flavor of the fruit without affecting the texture or color according to the findings of Bramlage and Covey.[6] Maxie et al.[11] reported that nectarines irradiated at 100 to 600 krad were acceptable in flavor and appearance, but poor in aroma and texture. They predicted that loss of firmness made the fruits so susceptible to mechanical injury that irradiation as a protective technology may be precluded. Irradiated fruits produced more C_2H_4 and ripened 4 to 5 days earlier than untreated preclimacteric fruits; they were also redder than the controls.

Data from the above sources indicate some contradiction as to the effects of irradiation on the organoleptic evaluations of nectarines. However, the accelerated ripening caused by irradiation precludes the possibility of using this treatment for delaying ripening.

D. Plums (*P. cerasifera* L.)

Bramlage and Covey[6] found that a dose range of 125 to 300 krad inhibited the blue coloring of plums and 200 krad and above softened the fruits.

E. Strawberries (*Fragaria chiloensis* L.)

Lovell and Flick[13] reported improvement of the keeping quality of strawberries by irradiating them at 150 krad and storing at 4.4 or 15.6°C. Irradiation at this dose controlled molding with a slight softening of the berries. Higher doses caused the development of soft spots on the fruit surface. Irradiation did not affect the concentrations of reducing or nonreducing sugars, but it significantly decreased the titratable acidity levels, and this accounted for the increase in sweetness in the irradiated fruits.

Pierson et al.[14] considered 200 krad as the optimum dose for strawberries stored at 4.4°C. Partially green strawberries were irradiated at 100 to 300 krad and stored at 4.4°C for 15 days, then at ambient temperatures for ripening by Salunkhe.[1] All treatments were unacceptable in taste quality due to lack of flavor development, which was probably the result of improper or abnormal ripening. The refrigeration life of ripe strawberries was increased by irradiation, but organoleptic quality was progressively decreased as the storage period increased.

According to Cooper and Salunkhe[15] a dose of 200 krad extended the shelf life, but the use of certain fungicides in conjunction with irradiation did not improve the shelf life any further. Salunkhe et al.[16] irradiated strawberries at 93 and 465 krad at two dose rates (9.3 and 93 krad/hr) and found that the rate of degradation of ascorbic acid was more rapid at the higher dose rate with the lower absorbed dose. With the higher absorbed dose, however, the disruption had already occurred and therefore there was no apparent difference in ascorbic acid content between fruits treated at the two dose rates.

Strawberries irradiated at 100 to 200 krad were stored under simulated air and railway shipping conditions by Bramlage and Covey.[6] They found that irradiation decreased storage decay and extended shelf life by several days. Maxie et al.[17] irradiated Shasta strawberries at 200 krad and subjected them to simulated and actual transit (railway) tests. They reported that irradiation significantly reduced the amount of decayed fruits without apparent ill effects, provided the berries were cooled immediately after harvest and kept cool, although the principal pathogen *Botrytis cinerea* Pers. was not completely inactivated. Strawberries irradiated at 200 krad were shipped on a commercial airline from California to New York by Guerrero et al.[18] Although irradiation was beneficial in controlling decay, they stated that if airlines can improve their service by providing adequate facilities to keep the fruits cool in transit and handling, the benefit from irradiation may not be so apparent.

In general, laboratory studies indicated a dose of approximately 200 krad to be the optimum for the extension of shelf life of refrigerated, ripe strawberries, and this benefit was accrued from decay control in storage. These results were verified in simulated and actual shipments of irradiated fruits.

F. Sweet Cherries (*P. avium* L.)

Salunkhe[1] increased the refrigerated life of ripe sweet cherries with irradiation doses of 200 to 400 krad which delayed the rate of senescence. For storage at 4.4°C, the optimum dose for delaying senescence was found to be 300 krad by Pierson et al.[14] Cooper and Salunkhe[15] also extended the shelf life of four varieties of sweet cherries irradiated at 300 krad and stored at 4.4°C and 80% relative humidity, and it was further increased when the irradiated cherries were treated with some fungicide preparations and packaged in a plastic film.

From the above observations, it is clear that the refrigerated life of sweet cherries can be extended by irradiation at 300 krad and that a further extension of shelf life can be obtained when irradiation is combined with fungicidal treatment and packaging.

G. Apples (*Pyrus malus* L.)

Apples irradiated at doses up to 150 krad and stored at 0 and 5°C for 3 to 6 months in air or modified atmospheres by Bramlage and Covey[6] were softer, more shriveled, and had less flavor than unirradiated fruits.

H. Apricots (*Prunes armeniaca* L.)

The optimum dose of irradiation to extend the storage life of firm, ripe apricots without loss of quality was found to be 200 krad by Salunkhe[1] who also found that treatment with 400 krad induced a breakdown of tissues which then became vulnerable to penetration of decay organisms and resulted in a high incidence of mold growth.

I. Figs (*Ficus carica* L.)

Bramlage and Covey[6] reduced the decay incidence and thus significantly extended by several days the shelf life of figs irradiated at 200 to 400 krad and stored at 3.3 to 21.1°C for various periods.

J. Avocados (*Persea americana* Mill.)

In tests conducted on Fuerte avocados by Bramlage and Covey[6] a dose of only 10 krad caused browning of vascular strands. Doses of 20 krad and above caused severe flesh discoloration and skin browning. Some 60 varieties of avocados including many seedlings in Hawaii were irradiated at a dose of 33 krad by Akamine and Goo,[19] and all were severely injured with symptoms similar to those described by Bramlage and Covey.[6] Akamine and Goo[19] used several supplementary treatments (cooling and heating fruit during irradiation, intermittent irradiation, use of chemical radiation protectors, elimination of ozone during irradiation, and variable postirradiation temperature and relative humidity storage) in attempts to improve the tolerance of the fruits to irradiation, but to no avail.

The nature of apparent lack of tolerance to irradiation by a seedling variety of avocado was investigated physiologically by Akamine and Goo[20] who determined, on the basis of respiration studies, that the fruits can tolerate only doses up to 2.5 krad which also accelerated ripening. The magnitude of increase in respiration immediately after irradiation was inversely related to the degree of tolerance by five fruit species in the preclimacteric stage when irradiated. Thus, avocados, the least tolerant species, exhibited the greatest increase in respiration and papayas (*Carica papaya* L.), the most tolerant species, produced the smallest increase in respiration. When doses up to 2.5 krad were administered to preclimacteric avocados, after the initial increase after irradiation, the respiration declined approximately to the level of the unirradiated fruit and then rose very rapidly and attained the respiratory peak in 3 days at 21.1°C as compared with 6 days for the untreated fruit. When nontolerant doses were applied to avocados, the initial increase in respiration continued to increase very rapidly to a peak and then decreased as rapidly as it increased and injured the fruit. Young[21] also found that there was an initial burst of respiration after irradiation in Fuerte avocados irradiated at 5 or 10 krad and this was followed by a second burst of respiration which coincided with the normal climacteric rise of an untreated fruit. It peaked, however, a few days later with the result that ripening was delayed. The C_2H_4 production pattern coincided with the respiratory pattern.

The above findings indicate that avocados are in general very sensitive to irradiation. In the Fuerte avocado, approximately 10 krad is the maximum dose this variety can tolerate and apparently ripening is delayed by this dose. In another variety, even a dose of 2.5 krad accelerates ripening.

K. Table Grapes (*Vitis vinifera* L.)

Salunkhe[1] found no adverse effects of irradiation at doses as high as 500 krad 3 months after Thompson Seedless grapes were irradiated, but Bramlage and Covey[6] found that irriadiation up to 300 krad was less effective than SO_2 for the preservation of Emperor and Thompson varieties. Maxie et al.[22] in concurrence with Bramlage and Covey[6] concluded that loss in firmness and lack of decay control in fresh Emperor and Tokay grapes irradiated at 100 to 300 krad make irradiation impractical as a substitute for the standard SO_2 treatment.

L. Pineapples (*Ananas comosus* L. Merr.)

Upadhya and Brewbaker[23] found that irradiation at doses of 30 to 50 krad delayed the degreening of the pineapple fruit in storage at 7.2°C, and irradiated fruit had lower "sweetness" rating based on total soluble solids analysis than untreated fruit. These findings were interpreted as evidence of delayed ripening by these investigators. Unfortunately, this interpretation is invalid because the pineapple is a nonclimacteric fruit which upon harvest does not undergo any physiological or biochemical changes to improve the fruit quality in storage and the lower total soluble solids content of the irradiated fruit was probably a symptom of radiation injury. Therefore, it follows that the delay in degreening noticed in irradiated fruit was probably a manifestation of a delay in senescence. However, Upadhya and Brewbaker[23] did not hold the irradiated pineapples long enough to actually determine the 4 to 7 days extension of shelf life claimed by them.

Wenkam and Moy[24] conducted assays on moisture, total carotenoids, and reduced and total ascorbic acids on irradiated (33 to 200 krad) pineapples stored at 7.2 and 21.1°C for various periods and found no significant differences in nutrients between irradiated and unirradiated fruits. Doses over 150 krad slightly increased the total soluble solids content after 6 days of storage, but there were no consistent changes in pH resulting from irradiation according to Upadhya et al.[25] who also noticed the loss of chlorophyll in crown leaves irradiated at 100 to 500 krad. Upadhya et al.[26] compared summer with winter fruits and found that irradiation at doses of 30 to 50 krad did not produce any significant differences among treatments or between seasons in certain biochemical constituents in fruits stored at 7.2 to 12.8°C. No significant differences were detected by taste panels in organoleptic attributes between pineapples irradiated at 25 to 200 krad and unirradiated fruits as reported by Moy et al.[27] Pineapples irradiated by Upadhya et al.[28] at doses of 30, 50, 100, and 500 krad and tested immediately after irradiation produced cytotoxic substances which significantly decreased mitotic frequency in onion root tips and increased the frequency of chromosome breaks. After 8 days of storage at 18°C, whereas fruits irradiated at 100 and 500 krad still exhibited cytotoxic effects, those irradiated at 30 and 50 krad did not. These investigators concluded that the nature of the cytotoxic substances was not known and that further studies to explore the nature of the antimitotic substances and animal feeding experiments were needed to assess the safety of irradiated pineapples for human consumption.

It is seen from the above investigations that, in general, pineapples can tolerate irradiation doses up to about 50 krad without the loss of organoleptic attributes or certain biochemical and nutritional constituents. However, the problem of cytotoxicity must be resolved.

M. Bananas (*Musa sapientum* L.)

Mathur[29] concluded that the optimum irradiation doses for shelf-life extension of Cavendish bananas was 20 krad, and when stored at 19°C (65 to 85% relative humidity), irradiated bananas required 45 days to ripen without C_2H_4 treatment as against 25 days of the unirradiated fruits. Irradiated fruits were no different from control fruits in color, texture, or taste.

Ferguson and Yates[30] found ozone injury (skin splitting and discoloration) in Gros Michel bananas irradiated in a closed chamber. Those irradiated at the optimum dose range of 25 to 50 krad in a ventilated chamber were delayed in ripening about 2 days in a ripening room, but at the ripe stage (conversion of starches to sugars progressed normally), they compared favorably with untreated bananas in sensory evaluations.

According to Maxie et al.,[31] only preclimacteric bananas were delayed in ripening. For the Gros Michel and Valery varieties, Maxie et al.[31] observed that 35 to 50 krad inhibited ripening without harming the fruit quality, apparently by reducing the sensitivity of the fruit to the ripening action of C_2H_4; to reach a comparable stage of ripeness, irradiated bananas needed 12- to 24-hr longer treatment with 1000 ppm C_2H_4 than did unirradiated bananas. The nature of the reduced sensitivity is not known. After the onset of the respiratory climacteric, the maximum tolerable dose of gamma radiation increased with increasing degree of ripening. Maxie et al.[31] concluded that although accelerated C_2H_4 production and ripening induced by physical injury may be partially offset by irradiation, the added care in temperature control to prevent the intensification of stress effects such as chilling imposed on the fruit and longer C_2H_4 treatment time required to ripen the fruit, might make irradiation impractical for commercial use. Moy et al.[32] irradiated Brazilian bananas of different stages of color development at 25 to 100 krad and subjected them to sensory evaluations at the edible ripe stage and found no significant effects of irradiation on pulp taste, aroma, color, or texture. There was no shelf-life extension with irradiation even in the "mature green" bananas, because they were irradiated 3 to 4 days after storage at 21.1°C, during which time the respiratory rise had apparently begun and the fruits were no longer in the preclimacteric stage.

In general, it appears from the above findings that delayed ripening can be attained in preclimacteric bananas with irradiation at the dose range of 20 to 50 krad without affecting the eating quality of the fruit at the ripe stage.

N. Olives (*Olea europaea* L.)

Fresh green olives irradiated at 100 to 400 krad and stored at 10°C for 2 to 4 weeks by Bramlage and Covey[6] were all damaged with external and internal discoloration and softening.

O. Lychees (*Litchi chinensis* Sonn.)

Upadhya and Brewbaker[33] investigated the effects of irradiation on total soluble solids, acidity, total phenolic substances, reducing sugars, and total proteins on lychees. They found that, in general, fruits irradiated at 30, 50, or 100 krad and stored at 15.6 to 21.1°C for 6 days were not significantly affected in these biochemical parameters. Six varieties of lychees irradiated at 50 and 100 krad and stored at 7.2°C for 1 to 3 weeks by Moy et al.[34] were not injured according to organoleptic evaluations of aroma, flavor, texture, and color of the pulp, and irradiation appeared to decrease the incidence of decay.

Akamine and Goo[35] conducted extensive experiments on the effects of gamma irradiation on the shelf life of freshly harvested, partially ripe, and ripe fruits of five lychee varieties (Brewster, Hak Ip, Heung Lai, Kwai Mi, and Groff). Fruits irradiated at 25 to 400 krad were packaged in polyethylene bags (to prevent desiccation and resulting darkening of fruit pericarp) and stored at 1.7, 7.2, 12.8, 22.2, and 26.1 to 31.1°C. In addition to organoleptic evaluations on flavor, aroma, and texture, observations were made on degree of pericarp darkening, percentage of decayed and molded fruits, percentage of weight loss, percentage of total soluble solids, pH, respiration, and C_2H_4 production. Based on these parameters, it was established that lychees are nonclimacteric, the tolerable level of irradiation is limited by the pericarp darkening at 1.7°C and decay and molding at 7.2°C and above, and that the optimum irradiation dose/storage temperature combination is 25 krad/7.2°C for shelf-life maintenance but not shelf-life extension. Out-of-state shipments of lychees from Hawaii are allowed after a disinfestation treatment with an ethylene dibromide fumigant which kills fruit flies, and the optimum storage temperature for both fumigated and untreated lychees is 1.7°C. A comparison of shelf life between fruits irradiated at 25 krad and stored at 7.2°C and those fumigated and stored at 1.7°C resulted in a shelf life of 4 weeks for the irradiated and 5 weeks for the fumigated lot according to Akamine and Goo.[35] These investigations indicate that delay in senescence is not accomplished by irradiation.

P. Oranges (*Citrus sinensis* L.)

Although irradiation doses of 100 to 300 krad caused severe peel injury to Florida oranges, only the 300-krad-treated fruits were distinguished by tasters as being different from untreated fruits in organoleptic evaluations according to Grierson and Dennison.[36] These authors also reported that there was no selective killing of *Penicillium* as against *Diplodia*. Irradiation at doses up to 200 krad increased rind pitting in shipping tests conducted by Bramlage and Covey[6] who also found that the effectiveness of irradiation as a decay control measure depended on the degree of initial infection at the time of irradiation.

Guerreo et al.[37] reported that, based on acceptance of organoleptic evaluations by tasters, 200 krad was probably the optimum dose for the irradiation of Washington Navel oranges. Juice from fruits irradiated at this dose and stored at −1.7 to 0°C had higher levels of soluble solids content, higher pH, lower titratable acidity, and slightly lower (not significant) ascorbic acid content than juice from unirradiated fruits. This dose controlled decay caused by *P. italicum* Wehmer. However, although rind pitting and browning were more evident in unwaxed irradiated than in waxed irradiated fruits, more scuff marks appeared on irradiated than on unirradiated fruits, and crushing resistance of the pulp was reduced by irradiation. Guerreo et al.[37] concluded that it was unlikely that gamma irradiation would ever be a feasible technology for oranges in California where handling procedures in the field and packing house are so well developed that losses are not severe, but it may be possible to use 200 krad to protect oranges from decay in countries where refrigeration facilities are lacking or inadequate.

Romani[10] found that respiration of oranges increased immediately after irradiation, and it continued to increase with prolonged exposure. According to the data of Romani et al.[38] a 200-krad treatment reduced the ascorbic acid content by 4% and a 400-krad treatment by 15% immediately after irradiation, but 24 hr after irradiation, there was a 100% recovery of ascorbic acid in fruits irradiated at the lower dose and 88% recovery in those irradiated at the higher dose. There was recovery for doses above 400 krad also, but in decreasing amounts with increasing doses, indicating the degree of irreversible damage.

Kahan and Monselise[39] irradiated Valencia-type oranges which had been inoculated with *P. digitatum* spores and stored them at 25 to 27°C. They claimed that doses of 45 to 175 krad increased shelf life linearly with increasing dose by controlling decay, and the average shelf-life extension within this dose range was 4 to 5 days. They also found no change in external or internal appearance in uninoculated oranges irradiated at 100 to 360 krad. Dennison et al.[40] determined the pectinesterase activity of irradiated Valencia oranges and found that the increase in the activity of this enzyme due to irradiation at doses of 100 krad and above could be minimized with prolonged storage of 14 days.

The nature of irradiation injury to the peel of oranges was thoroughly investigated by Monselise and Kahan,[41] Riov et al.,[42] Belli-Donini et al.,[43] and Riov.[44] Monselise and Kahan[41] found the following effects of irradiation on Shamouti oranges at doses above 56 krad and up to 280 krad: (1) peel pitting was greater at the stem end than at the blossom end, (2) there was a very high reduction in ascorbic acid content in the flavedo with only slight decrease in the juice, (3) peroxidase activity increased and catalase activity decreased in the flavedo, (4) reducing sugars in the flavedo decreased, and (5) there was no change in the total soluble solids or citric acid of the juice.

Immediately after irradiating oranges at 200 krad, Riov et al.[42] found a marked increase in phenylalanine ammonia-lyase (PAL) activity, which then rapidly decreased to control levels in fruits which did not show visible injury symptoms, but remained relatively high for several days in fruits which showed damage in the peel later. The enhanced PAL activity was accompanied by an accumulation of phenolics. In three varieties of irradiated oranges investigated by Belli-Donini et al.,[43] terpene compounds (main components of oil glands in citrus fruits) increased with time and irradiation dose, but there was varietal difference in the magnitude of accumulation of the compounds. They hypothesized that irradiation facilitated the diffusion of terpene compounds in the exocarp cells and this damaged the skin tissues of oranges. Riov[44] using histochemical procedures demonstrated the accumulation of phenolic compounds only in damaged flavedo cells immediately after irradiation, and suggested that phenolic compounds caused the death of cells and subsequent peel necrosis (pitting) in irradiated oranges.

In summary, although there are some discrepancies in the data presented above, a dose of approximately 200 krad seems to be the optimum for irradiating oranges. This dose apparently causes some peel injury, but it controls storage decay with minimum effects on sensory attributes or concentrations of soluble solids and acids including ascorbic acid. However, because the texture is affected adversely by irradiation, the softened fruit is more susceptible to scuffing and crushing injuries.

Q. Lemons (*C. limonia* Osbeck)

Beraha et al.[45] reported that lemons of unknown provenance were damaged by irradiation doses of 250 to 300 krad. Maxie et al.[46] reported severe injuries to Eureka lemons irradiated at 100 krad and above. Injury symptoms as described by Maxie et al.[46] included formation of cavities along segment walls of the pulp, dramatic losses in ascorbic and citric acids, early death of the button followed by invasion of *Alternaria* rot, and atypical upsurge of respiration immediately after irradiation followed by a decline. A similar rise in respiration was reported by Romani[10] who suggested that radiation products extended to the fruit did not mediate the rise. Auxin and kinetin treatments applied after irradiation did not protect the buttons from injury. Maxie et al.[46] concluded that irradiation is of doubtful value as a supplement to cool storage for lemons. In lemons irradiated at 400 krad by Romani et al.,[38] the loss in ascorbic acid of 13% immediately after irradiation was accompanied by a nearly equivalent increase in dehydroascorbic acid. There was a further increase in dehydroascorbic acid and 91% of the ascorbic acid was recovered 24 hr after irradiation.

Maxie et al.[47] investigating the respiratory behavior and C_2H_4 production of green Eureka lemons subjected to irradiation doses of 25 to 100 krad, found that irradiation induced the typical respiratory behavior of a true climacteric fruit, and this was accompanied by degreening of the fruit. Internal C_2H_4 also increased with irradiation. They concluded that irradiation-induced C_2H_4 production caused the degreening of the fruit. Since irradiation was injurious to lemons at doses as low as 50 krad, they surmised that the irradiation-induced C_2H_4 production could have been an injury response. In summary, Maxie et al.[47] stated that until biochemical events such as degradation of starch or other reserve carbohydrates, increases in soluble carbohydrates, solubilization of pectic materials, and softening of tissues were investigated, they were assuming that the respiratory behavior in irradiated lemon fruit was not that of a true (autogenous) climacteric fruit. Shikhande and Kaewubon[48] using controlled atmospheres improved the tolerance of lemons to irradiation. High CO_2 during irradiation suppressed aerobic postirradiation C_2H_4 production and respiration and CO_2 in various combinations with O_2 after irradiation in air suppressed C_2H_4 production, minimized chlorphyll and ascorbic acid loss, and delayed physiological disorders in storage.

From the above, it may be surmised that because of the extreme sensitivity of the fruit to irradiation, practical use of this treatment to preserve lemons seems nil.

R. Grapefruit (*C. grandis* Osbeck)

Marsh grapefruit irradiated at 100 to 300 krad by Grierson and Dennison[36] developed severe peel injury; however, only fruit irradiated at 300 krad were distinguished by taste panelists.

According to Riov et al.,[49] whereas early fruit tolerated a 200-krad dose, fully ripe grapefruit tolerated only 100 krad. When fruits of the two stages of development were irradiated at their respective tolerable doses, C_2H_4 production and PAL activity patterns were similar, i.e., immediately after irradiation there was an upsurge in C_2H_4 production and PAL activity which decreased to control levels, and no rind pitting ensued.

The peel of grapefruit has been the subject of much research in attempts to elucidate the causes of injury by irradiation. Riov et al.[42] demonstrated that in fruits not injured by irradiation, after the rise in PAL activity immediately after irradiation, it decreased to levels of untreated fruit, but in fruits showing peel injury later, PAL activity did not decrease after the initial increase, and accumulation of phenolics accompanied enhanced PAL activity. Riov and Goren[50] found a linear relationship between increase in doses and inhibition of protein synthesis in the peel. A dose of 400 krad almost completely inhibited protein synthesis which was not recovered 5 days after irradiation (no repair event). Two heretofore unknown compounds found in the peel of gamma-irradiated grapefruit were isolated and identified as scopoletin (6-methoxy-7-hydroxy-coumarin) and scopolin (7-glucoside of scopoletin) by Riov et al.,[51] and the content of these compounds and radiation damage (peel pitting) were positively correlated with doses up to 300 krad. The authors suggested that the accumulation of phenolic compounds in the peel after irradiation contributed to the development of radiation injury. Riov,[44] after detecting considerable accumulation of phenolic compounds in cells damage by irradiation and none in undamaged cells, suggested that these compounds caused cell death and eventual necrosis (pitting) of the grapefruit rind.

The results of these investigations indicate the extreme sensitivity of the peel of grapefruit to irradiation, and this limits the application of irradiation for preserving this fruit.

S. Tangerines (*C. deliciosa* Tenore)

Dancy tangerines were found to be very susceptible to irradiation injury by Akamine and Wong.[52] Doses as low as 20 krad induced bronzing of the fruit surface, impairment of flavor and aroma, and increase in decay incidence in refrigerated fruits.

T. Mangoes (*Mangifera indica* L.)

Whereas Mather[29] in India claimed that the optimum irradiation dose for Alphonso mangoes was 75 krad for extension of shelf life, Dharkar et al.[53,54] found 25 krad to be the optimum. Mathur[29] packaged the irradiated fruits in perforated polyethylene bags and stored them at 15°C (85 to 90% relative humidity), thus obviating the necessity of ripening after storage. Under these conditions, the shelf life of mature green mangoes was extended nearly two-fold with no impairment of organoleptic qualities. He also suggested that a combination technique involving low-dose irradiation coupled with mild refrigeration may offer a commercial possibility.

Dharkar et al.[53] achieved an extension of shelf life of about 6 days at 25 to 30°C after irradiating Alphonso mangoes at 25 krad in air, N_2, or CO_2. Irradiation in N_2 atmosphere was of some advantage in minimizing changes in organoleptic quality, ascorbic acid, and cartenoids, and doses of 50 krad and above increased spoilage as evidenced by increased spotting and darkening of fruit surface. When unirradiated, mature, unripe Alphonso mangoes were skin-coated with an emulsion made of an acetylated monoglyceride preparation, they were physiologically damaged presumably due to an excessive inhibition of respiration according to Dharkar et al.,[54] but this was rectified by a spurt in respiratory activity when skin-coated fruits were irradiated at 25 krad in air or N_2. These workers concluded that, judging by organoleptic evaluations, analysis for fruit constituents, skin color, and pressure tests, the storage life of Alphonso mangoes could be increased by combining skin-coating with irradiation in air or N_2.

Hatton et al.[55] in Florida found that Irwin mangoes irradiated at 10 to 25 krad delayed ripening 14 days and 50 krad delayed 17 days, but palatability was preserved only at 10 krad and below. In Sensation mangoes, although 25 krad delayed ripening 3 days (lower doses were ineffective), palatability was impaired at this dose. Though anthracnose control by irradiation was erratic and inconclusive in these two cultivars, it seemed to be dose dependent.

Dennison and Ahmed[56] irradiated Florida mature green Kent mangoes at 100 to 300 krad and stored them at a ripening temperature of 20°C (92% relative humidity). After 0, 4, and 8 days of storage, they were analyzed for chemical constituents. Results showed that in the unirradiated fruits, soluble solids and sucrose increased, and alcohol insoluble solids, titratable acidity, reducing sugars, and firmness decreased more rapidly than those in irradiated fruits during ripening, indicating that irradiation induced a delay in ripening. Dennison and Ahmed[56] also found higher levels of water-soluble and lower versene-insoluble pectic fractions in irradiated than in unirradiated fruits which indicated a higher pectin esterase activity in irradiated fruit.

Upadhya et al.[5] in Hawaii stated that there were no significant changes in soluble solids content in mangoes irradiated at 75 to 100 krad, but this dose reduced fruit weight loss in storage as compared with untreated fruit. Total protein content was decreased at 300 krad, and adverse changes in appearance and color of the rind and pulp occurred only at doses above 250 krad. Later Upadhya and Brewbaker[58] reported that mature green Haden and Pirie mangoes tolerated only up to 150 krad, and tolerance was limited by surface scalding (darkening) and pulp off-flavor development. In a subsequent report, Upadhya et al.[59] further reduced the tolerant level of Pirie mangoes to 100 krad and recommended that fruits irradiated at 75 to 100 krad be stored at 7.2 to 12.8°C (90% relative humidity) to extend their shelf life by 1 to 2 weeks. This speculated shelf life was based only on lack of significant biochemical changes in the pulp up to about 2 weeks of refrigerated storage and not on an actual determination of shelf life under simulated retail conditions.

Moy et al.[60] evaluated organoleptically Hawaii-grown Haden and Golden Glow mangoes irradiated at 33 to 150 krad and found no effects on taste qualities. They also found that fruits irradiated at 50 krad were no different from fruits fumigated with ethylene dibromide in a shipment of fruits from Hawaii to California. Analyses for β-carotene in mangoes immediately after irradiation at 100 to 500 krad by Hilker and Wong[61] indicated no significant changes in the retention of this vitamin precursor. Wenkam and Moy[62] assayed various nutritional constituents, moisture, total carotenoids, reduced and total ascorbic acids, and reducing sugars (before and after conversion), in two varieties of mangoes (Golden Glow and Haden) irradiated at 33 to 150 krad and stored at 7.2 and 21.2°C for various periods, but found no significant differences in nutrient levels between irradiated and unirradiated fruits. According to Kojima and Buddenhagen[63] storage diseases and their severity in mangoes were reduced by irradiation, but reduction was significant only at doses above 240 krad.

The respiratory behavior of irradiated mangoes was compared with that of other irradiated climacteric fruits (avocado, sour sop, yellow passion fruit, and papaya) by Akamine and Goo[20] in Hawaii. Based on this comparison, the mango was judged moderate in degree of tolerance to irradiation. Akamine and Goo[64] conducted numerous experiments on irradiation of Haden mangoes in which the effects of treatment on physical appearance, organoleptic attributes, and physiological and biochemical aspects were determined (Plate 2).* They found that doses of 25 to 100 krad did not damage the pulp of mature green to three-quarter-ripe fruits when stored at room temperature (22.8 to 25.6°C) or at 12.8°C initially, then at room temperature, but doses above 100 krad injured the pulp in that ripening was impaired and off-aromas developed. The fruit surface of refrigerated and unrefrigerated mature green fruits was scalded (darkened) by doses of 25 krad and above. The degree of scaling in irradiated one-quarter- to three-quarter-ripe fruits was influenced by irradiation dose and storage temperature; thus, in unrefrigerated one-quarter- to three-quarter-ripe fruits, doses up to 100 krad produced no scald, but in refrigerated fruits, scaling occurred in one-

* Plate 2 appears after page 144.

quarter-ripe fruit irradiated at 75 krad and above; in one-half-ripe fruits, scalding occurred at 100 krad; and in three-quarter-ripe fruits, no scalding occurred at all doses (25 to 100 krad). Thus it is clear that surface scalding, rather than internal injury, limited the tolerance level to irradiation, and in general only one-quarter-ripe and riper fruits tolerated doses of 25 to 100 krad. An irradiation dose of 25 krad had no effect on pH, percent total soluble solids, percent titratable acids, flavor, aroma, weight loss or rate of ripening or senescence of one-quarter- and one-half-ripe fruits. At this dose, the shelf life was not affected as additionally evidenced by a comparison of respiration data of irradiated fruits with control fruits.

From the above, it is clear that there is considerable discrepancy on the reported effects of gamma irradiation on mangoes. Some of the causes of the discrepancy are varietal differences, varying pre and postirradiation treatments used by different researchers, and observations on different parameters of irradiation effects by the various workers. Nevertheless, it seems that in general, the maximum dose that mangoes in India can tolerate without any detectable injury is approximately 75 krad, those of Florida 10 krad, and those from Hawaii 100 krad. At these dose ranges, there is no significant storage decay control, but shelf life is extended through delayed ripening. For mangoes grown in Hawaii, surface scalding limits the use of irradiation to only one-quarter- to one-half-ripe fruits when irradiated at 25 to 100 krad.

U. Papayas (*Carica papaya* L.)

From 1964 to 1969, extensive research was conducted in Hawaii on the effects of gamma irradiation on fruits and vegetables including papayas.* This research was supported largely with grants from the U.S. Atomic Energy Commission and because nearly all fresh commodities exported from Hawaii must be disinfested against fruit flies and other insects, the direction of the research effort was focused on the effects of irradiation on insect mortality as well as preservation of shelf life of the commodities. Effects of irradiation on insects will be discussed later in the section on feasibility of irradiation technology as a method to preserve fresh commodities.

In initial experiments with papaya irradiation, it was discovered by Akamine and Wong[65] that continuous aeration of the chamber with humidified air during irradiation to prevent accumulation of ozone and desiccation of fruit did not prevent scalding (darkening) of the surface of fruits irradiated at doses of above 100 krad. With or without aeration, doses of 100 krad and lower did not scald the fruits. Thus in subsequent treatments, papayas were subjected to gamma rays in a sealed container without aeration.

The approximate upper limits of irradiation doses for overall fruit tolerance of papayas of various stages of ripeness stored at room temperature were found by Akamine and Wong[65,66] to be as follows for the usual quality factors: for surface scald, 100 krad; surface color development, 200 krad; flavor and aroma, 400 krad; and tissue breakdown, 500 krad. Thus the maximum dose papayas can tolerate is limited by surface scalding and this limit is approximately 100 krad. At the edible ripe stage (full yellow color on surface), irradiated fruits were firmer than unirradiated ones (an indication of delayed ripening), and maximum firmness was attained at 75 and 100 krad. However, after prolonged storage beyond the edible ripe stage, fruits irradiated at the higher dose were softer. Therefore, 75 krad was considered the optimum dose for retention of firmness. The correlation between retention of firmness and rate of senescence (overripening) is clearly seen in Table 2 which compares irradiated with unirradiated fruits.

* The results of the extensive work on papayas are reported in considerable detail not only because of the commercial importance of this fruit but because the work done covers the whole range of research parameters, tests, and market potentials of the irradiated product.

Akamine and Wong[65,66] also found that doses of 75 to 100 krad had only a slight control of storage decay and that damaging doses of 600 krad and higher were required for significant decay control. When irradiation was combined with a hot water treatment currently used commercially to control decay in ethylene dibromide fumigated papayas for export, decay was controlled and senescence delayed as seen in Table 3. Hot water alone slightly softened the fruit, but loss in firmness was regained upon subsequent irradiation, and the fruit attained the firmness of another fruit subjected to irradiation only. When the shelf life of papayas subjected to the hot water plus ethylene dibromide fumigation treatment was compared with that of fruits subjected to the hot water plus 75-krad treatment, the retention of firmness in the irradiated fruit resulted in delay in senescence to the extent that the salable life of the irradiated fruits was approximately 3 days greater than that of the fumigated fruits (Plate 3).*

Papayas grown in Hawaii are shipped out of the state by air and marine transportation. The effects of irradiation on shelf life of fruits stored under conditions simulating shipping by air were investigated by Akamine and Goo[67] who, since air transportation does not entail refrigeration, stored the treated fruits at ambient conditions (22.8 to 26.7°C, 60 to 78% relative humidity). The results of a typical experiment (Table 4) showed that the hot water plus irradiation treatments significantly improved the shelf life, with the 75- and 100-krad combination being more effective than the 50-krad combination. Because the hot water plus 100-krad treatment caused some scalding of the fruit surface, Akamine and Goo[67] again concluded that the hot water plus 75-krad treatment is the optimum treatment for shelf-life extension for unrefrigerated fruit.

Using 12.8°C, which was found to be the optimum temperature for storage of refrigerated partially ripe fruits by Akamine,[68] Akamine and Wong[69] conducted preliminary experiments simulating marine shipment of irradiated papayas stored at 12.8°C. It was soon discovered that this temperature caused scalding of papayas subjected to the hot water plus 75-krad treatment, that 15.6°C eliminated scalding, and that higher temperatures increased the rate of ripening and senescence. That 15.6°C was the optimum temperature for refrigerating irradiated papayas is seen in Table 5. Data in the table indicate that after the initial storage period of 7 days at this temperature, the shelf life of the hot water plus irradiation treatment at room temperature was 5 days longer than that of the control, 3 days greater than hot water only, and 6 days greater than irradiation only. After 14 and 21 days of refrigeration, the control and irradiated fruits were 100% unmarketable, but hot water plus irradiation-treated fruits had 3 or 4 more days of salable life than fruits treated only with hot water.

In additional work, Akamine and Goo[70] again confirmed that for unrefrigerated papayas, the optimum irradiation treatment for extension of shelf life is a combination of hot water and 75 krad. The authors reconfirmed that for refrigerated fruits, the optimum storage temperature is 15.6°C for this hot water-irradiation treatment which simultaneously controls storage decay, delays ripening and senescence, and extends the shelf life of papayas. Akamine and Goo[70] compared the shelf life of papayas treated with the currently used hot water plus fumigation treatment for export papayas and stored at 12.8°C with that of fruits treated with hot water plus 75 krad and stored at 15.6°C. They concluded that the extension of shelf life over the control fruits was 2 days for the fumigated and 5 days for the irradiated papayas.[70]

The effects of controlled atmosphere (CA) storage on the shelf life of papayas subjected to the optimum irradiation treatment for refrigerated fruits (hot water plus 75 krad plus 15.6°C) were investigated by Akamine and Goo.[71] CA storage further extended the shelf life by 2 days, i.e., hot-water-treated, irradiated fruits stored in mod-

* Plate 3 appears after page 144.

Table 2
RELATION BETWEEN RETENTION OF FIRMNESS AND RATE OF SENESCENCE IN IRRADIATED AND UNIRRADIATED PAPAYAS AT ROOM TEMPERATURE (23—27°C)

Treatment (krad)	Edible ripe		2 Days later		3 Days later		4 Days later		6 Days later		Decrease in pressure (%) (6 days)
	Pressure (lb)	Overrripe fruits (%)	Pressure (lb)	Overripe fruits (%)	Pressure (lb)	Overripe fruits (%)	Pressure (lb)	Overripe fruits (%)	Pressure (lb)	Overripe fruits (%)	
0	23.4	0	19.2	20	18.0	100	14.4	—	10.3	—	56.0
75	25.3	0	25.7	0	22.1	0	22.6	0	19.0	20	24.9

From Akamine, E. K. and Wong, R. T. F., *Hawaii Farm Sci*, 15, 4, 1966. With permission.

Table 3
EFFECT OF HOT WATER ON CONTROLLING DECAY AND DELAYING SENESCENCE IN IRRADIATED PAPAYAS AT ROOM TEMPERATURE (23—27°C)

Treatment	Decayed fruits (%) Edible ripe stage	1 Day later	2 Days later	3 Days later	Overripe fruits on last day (%)
Control	0	0	40	60	100[a]
Hot water[b] only	0	0	0	0	100[a]
75 krad only	0	0	20	40	0
Hot water + 75 krad	0	0	0	0	0

[a] 100% Overripe 1 day after edible ripe stage.
[b] 48.9°C, 20 min followed by cooling in running tap water, 20 min.

From Akamine, E. K. and Wong, R. T. F., *Hawaii Farm Sci.*, 15, 4, 1966. With permission.

Table 4
EFFECT OF IRRADIATION ON SHELF LIFE OF PARTIALLY RIPE PAPAYAS AT ROOM TEMPERATURE (23—27°C)

Treatment	Average salable days (± SE)	Increase over control (days)
Control	5.8 ± 0.3	—
Hot water + 50 krad	8.9 ± 0.3	3.1 ± 0.2
Hot water + 75 krad	9.6 ± 0.2	3.8 ± 0.2
Hot water + 100 krad[a]	9.3 ± 0.2	3.5 ± 0.2

[a] Fruit surface slightly scalded.

From Akamine, E. K. and Goo, T., Hawaiian Food Irradiation Program, Rep. 1968 to 1969, Contract AT(26-1)-374 and Contract No. AT(04-3)-235, Proj. Agreement No. 5, Division of Isotopes Development, U.S. Atomic Energy Commission, Washington, D.C., 1970, 34.

Table 5
EFFECT OF IRRADIATION ON EXTENSION OF SHELF LIFE OF REFRIGERATED PAPAYAS

Treatment	Total number days for 50% of fruits to be unmarketable: 7 Days at 15.6°C + number days at room temp[a]	14 Days at 15.6°C + number days at room temp[a]	21 Days at 15.6°C + number days at room temp[a]
None	10	100% Unmarketable	100% Unmarketable
Hot water only	12	16	21
75 krad only	9	100% Unmarketable	100% Unmarketable
Hot water + 75 krad	15	20	24

[a] 23—27°C.

From Akamine, E. K., and Wong, R., Dosimetry, Tolerance, and Shelf Life Extension Related to Disinfestation of Fruits and Vegetables by Gamma Irradiation, Rep. 1965 to 1966, Contract No. AT(04-3)-235, Proj. Agreement No. 5, Division of Isotopes Development, U.S. Atomic Energy Commission, Washington, D.C., 1966, 17.

PLATE 1. The beneficial effect of ionizing radiation in delaying spoilage of strawberries during refrigerated storage.

PLATE 2. Effect of irradiation on Haden mangoes of two different stages of maturity when treated. Top row: mature green fruits. Bottom row: one-quarter-ripe fruits. (A) 0 krad; (B) 25 krad; (C) 50 krad; (D) 100 krad. Note scalding (darkened areas) of surface on all irradiated mature green fruits and no surface effects on one-quarter-ripe fruits. Photographed 4 days after treatments and storage at 23 to 27°C.

PLATE 3. Effect of 75-krad irradiation and hot water treatment (48.9°C, 20 min) on decay control in one-quarter-ripe papayas when treated. (A) Untreated; (B) irradiation only; (C) hot water only; (D) hot water plus irradiation. Note decay control only in hot water and hot water plus irradiation treatments. Photographed 9 days after treatments and storage at 23 to 27°C.

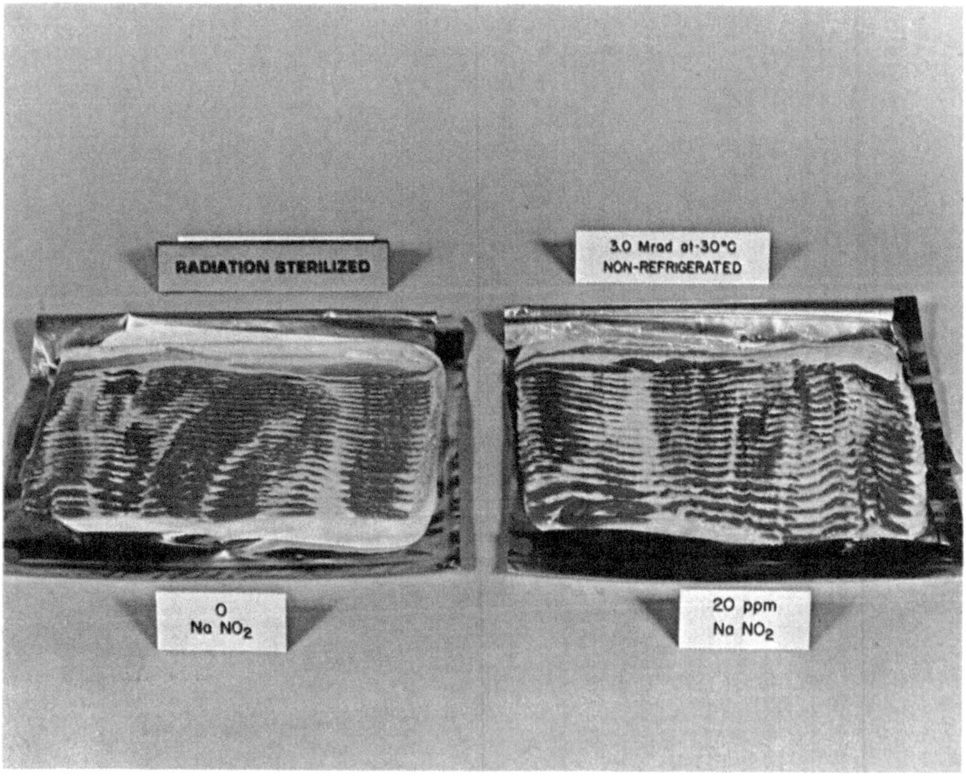

PLATE 4. Appearance of radappertized (30 kGy at −30°C) bacon with no nitrite and only 20 ppm sodium nitrite upon opening sealed packages following 6 months storage without refrigeration.

Table 6
EFFECT OF CA STORAGE ON SHELF LIFE OF HOT-WATER-TREATED, IRRADIATED PAPAYAS

Treatment	Initial storage atmosphere	Shelf life (days) at room temp[a] after initial storage at 15.6°C for	
		6 Days	12 Days
Hot water only	Air	3	0
Hot water + 75 krad	Air	6	3
Hot water + 75 krad	CA	8	5

[a] 23—27°C.

Akamine, E. K. and Goo, T., Dosimetry, Tolerance, and Shelf Life Extension Related to Disinfestation of Fruits and Vegetables by Gamma Irradiation, Rep. 1967 to 1968, Contract No. AT(04-3)-235, Proj. Agreement No. 5, Division of Isotopes Development, U.S. Atomic Energy Commission, Washington, D.C., 1969, 63.

ified atmospheres consisting of 2 to 4% O_2 and 98 to 96% N_2 at 15.6°C for 6 to 12 days, then removed from CA and stored at room temperature had 2 more marketable days than comparable fruits stored in air (Table 6).

An increase in respiratory activity immediately after treatment is typical of irradiated fruits. Akamine and Goo[20] found that the papaya was most tolerant to irradiation among fruit species investigated, and this was correlated with the least increase in respiratory rise immediately after irradiation. Respiration and C_2H_4 production were used by Akamine and Goo[72,73] to elucidate the effects of irradiation on ripening in refrigerated and unrefrigerated papayas. These authors showed that the optimum hot water plus 75-krad treatment delayed the attainment of the climacteric peak (ripe stage), thus delaying ripening. However, this delay occurred only in papayas with less than 25% surface color (yellow) which confirmed previous findings by Akamine et al.[65-67,69] According to Akamine and Goo[72,73] in preclimacteric fruit not injured (no scalding) by irradiation, C_2H_4 production preceded respiratory rise by 1 to 2 days, but in injured fruit, the two physiological processes coincided.

Effects of irradiation on papayas grown and treated in Hawaii, Taiwan, and Venezuela were investigated by Moy et al.[74] Because of differences such as variety, fruit size, ambient storage temperatures, and adequacy of equipment in the three countries, the accumulated data are not completely valid for comparison among geographical areas; nevertheless, some general assumptions could be made: papayas from the three areas tolerated 100 krad without any impairment of chemical or nutritional qualities of the pulp, and their shelf life was extended by the hot water plus 75-krad treatment. Buddenhagen and Kojima[75] working with papayas inoculated with several pathogens (*Ascochyta*, *Phytophthora*, *Colletotrichum*, *Botryodiplodia* and *Rhizopus*) found that irradiation at doses of 75 to 100 krad did not significantly influence the susceptibility of the fruit tissues to these pathogens.

Organoleptic evaluations of papayas shipped from Hawaii to California by air and marine transportation by Moy et al.[76] indicated that there were no significant differences in sensory attributes or total soluble solids content between irradiated (hot water plus 75 krad) and fumigated (hot water plus ethylene dibromide) fruits, except that irradiated fruits were firmer than fumigated ones and developed a characteristic "flowery aroma" after prolonged refrigerated storage. However, this odor was not judged objectionable by taste panelists. Moy and Hsia[77] investigated the influence of radiation dose, degree of ripeness, and postirradiation storage temperatures on the development of the characteristic aroma. They concluded the less ripe fruit, when irradiated, was more likely to develop the aroma than riper fruit and that doses of 33 to 100 krad and storage at 12.8, 21.1, and 26.7°C had little influence on the degree of aroma development. Treated papayas were air-shipped from Hawaii to California and back to Hawaii for sensory evaluations by Moy et al.[78] who concluded that quality of the irradiated fruit was superior to that of the fumigated or vapor heat-treated fruit.

Hilker[79] and Hilker and Young[80] found no appreciable change in ascorbic acid content in papayas irradiated at 25 to 300 krad, but Wenkam and Moy[62] noted increased reduced ascorbic acid content in irradiated fruits. Irradiation did not affect protein or pectin content according to the data of Hilker and Young[80] nor cryptoxanthin content as reported by Hilker and Moy.[62] Hilker and Young,[80] and Young and Lin[81] found increased deesterification of pectin leading to an increase in pectic acid content with increases in irradiation dose immediately after irradiation. Although this increase was accompanied by a decrease in fruit firmness, these workers noticed that the plant enzymes restored the pectin esters to normal amounts 3 to 5 days after irradiation as the fruit metabolically repaired the damage if the dosage was not excessive. These observations are probably related to the increase in respiration immediately after irradiation followed by a decrease to the control fruit level and subsequent delayed attainment of the climacteric peak as observed by Akamine and Goo.[20,72,73] and the retention of firmness after postirradiation storage as observed by Akamine and Wong[65,66,69] and Akamine and Goo.[70]

Although Upadhya et al.[82] found significant increases in peroxidases in papayas irradiated at 400 to 500 krad. Moy et al.[83] found no significant changes in peroxidase activity caused by 75- to 85.5-krad irradiation in the peel, placenta, or pulp of the fruit. Upadhya et al.[82] also found no appreciable reduction in total proteins, total soluble solids, or reducing sugars in fruits irradiated up to 500 krad, but phenolic compounds were slightly reduced and proteolytic activity of the enzyme, papain, was reduced approximately 40%. In papayas irradiated at a dose of 100 krad, membrane instability increased more rapidly as the fruit ripened than in control papayas according to the findings of Buddenhagen and Takata.[84] These investigators also stated that this reduction in membrane stability could have profound effects on certain cellular enzyme activities which take place in the ripening process and senescence. This phenomenon may possibly be related to the delay in ripening and senescence of irradiated papayas.

In his survey of market potential for irradiated papayas in certain cities in California, Spielman[85] found that the public demand for irradiated papayas was the same as untreated fruits, provided irradiation is approved by proper governmental agencies. After further market surveys, Spielmann[86] concluded that gamma irradiation would favorably affect market demand, because it would improve the texture consistency, reduce spoilage, and permit extended mass displays.

In order to satisfy, in part, the requirements for clearance of irradiation as a technology for treating papayas, plans were formulated and implemented for larger simulated and actual shipping tests in Hawaii. McClish and Dollar[87] studied the feasibility of alternative packaging methods for irradiated papayas destined for shipping. Mature green irradiated papayas (hot water plus 25 to 250 krad) were stored by Dollar et al.[88] at 12.8°C for 7 days, then at 21.1°C for 9 days to simulate the conditions for marine shipping from Hawaii to the West Coast, U.S. and for retail handling. At the end of the simulated retail storage condition, they found no significant changes caused by irradiation in the analyses of the following biochemical parameters of the edible pulp: dry matter, total and reduced ascorbic acid, carotenoids, pH, titratable acidity, total soluble solids, and total reducing sugars. However, irradiated fruits were firmer. Dollar et al.[88] did not observe fruit surface effects of the treatments and storage temperatures.

Dollar et al.[89] formulated a simulated air and marine shipping study protocol for comparing the effects of irradiation, fumigation, and vapor heat (the latter two are current approved quarantine procedures for export fruit) on the physical, physiological, biochemical, and organoleptic parameters. Simulated shipping studies conducted by Dollar et al.[90] according to procedures of Dollar et al.[89] for marine shipment showed that shelf life of irradiated papayas (25 to 150 krad) was in general better or equal to that of currently used procedures. There were no differences among the treatments in color development, reduced ascorbic acid, carotenoid values, or flavor, but irradiated fruits were firmer and scored higher in aroma than vapor heat-treated fruits.

With a modified protocol of Dollar et al.,[89] an actual shipping test was conducted by sea to California by Dollar et al.[91] Part of the shipment was left at the Davis campus of the University of California for evaluation and the remainder was shipped back to Hawaii via air for evaluation at the University of Hawaii. In general, the shelf life of the irradiated papayas was better than that of the fumigated or vapor heat-treated fruits with no significant differences in weight loss, total soluble solids, dry matter, reduced ascorbic acid, pH, titratable acids, or flavor. The irradiated fruits had better aroma than the vapor heat-treated ones, and 75-krad-treated fruits scored higher than other treatments in color development. In shelf-life extension, there was no difference between 25- and 75-krad doses according to the data of Dollar et al.[91] This was probably due to the fact that the temperature of the hot water used in combination with irradiation was only about 46°C which is below the recommended 48.9°C for maximum decay control.

Again modifying the original protocol of Dollar et al.[89] slightly, Dollar et al.[92] conducted an actual air shipment of papayas from Hawaii to the West Coast, U.S. and returned it to Hawaii (5000 mi round trip) for evaluation. They reported that irradiated papayas were superior to fumigated or vapor heat-treated ones.

The above results of extensive research on the preservation of fresh papayas in Hawaii clearly indicate the possible use of irradiation as a means of delaying the processes of ripening and senescence to extend the shelf life. The feasibility of radiation technology for papayas will be discussed further in another section.

V. Cantaloupes (*Cucumis melo* L., var. *reticulatus* Naud.)

Bramlage and Lipton[93] found that irradiation did not consistently affect decay incidence, but it softened the fruit. Softening of fruit by irradiation was also reported by Ravetto et al.[94] who found that up to 400-krad dose, irradiation had minimal effects on eating quality, total soluble solids, or flesh color. Respiration rate was increased but C_2H_4 production was suppressed by 100- to 300-krad doses which also reduced the ascorbic acid content. The doses above 400 krad predisposed cantaloupes to develop off-aromas and atypical flavors and to increase storage decay. Ravetto et al.[94] concluded that irradiation at doses tolerated by cantaloupes appeared not to be beneficial for the preservation of this fruit. However, no long-term storage studies were conducted to determine the effects of irradiation on actual shelf life.

W. Honeydew Melons (*C. melo* L., var. *inodorus*)

Ravetto et al.[95] determined that irradiation doses up to 400 krad did not adversely affect texture, skin color, flesh color, or total soluble solids content of honeydew melons, but doses above 400 krad severely softened the fruit and predisposed it to bruising. Up to 800-krad dose, no differences in flavor were detected by tasters, but in fruits irradiated at doses above 400 krad, the flesh was dry and mealy. Until long-term storage studies are conducted, Ravetto et al.[95] concluded that the possible postharvest benefits to be gained from irradiation will not be clear.

X. Tomatoes (*Lycopersicon esculentum* Mill.)

Salunkhe et al.[16] reported that rate of ripening in green and pink tomatoes irradiated at about 186 krad was delayed (delayed softening and pigment development) and shelf life was extended. Green tomatoes irradiated by Salunkhe[1] at doses of 100 to 300 krad did not ripen evenly and their red color development was also impaired. Bramlage and Lipton[93] reported that storage decay of some lots of tomatoes they studied was reduced by irradiation, but the treatment caused mottling (uneven surface red pigmentation) and softening of fruit. Irradiated tomatoes were subjected to simulated railway transit conditions by Ravetto et al.[96] They reported that an irradiation dose of 300 to 400 krad did not significantly increase the incidence or severity of transit injury provided the tomatoes were held securely in place within the commercial package.

Mathur[29] considered a dose of 75 krad optimum for tomatoes held in storage at 9 to 10°C and 85 to 90% relative humidity. A dose of 100 krad adversely affected ripening although decay was controlled. Susceptibility to decay increased with irradiation dose, but Mathur[29] was able to control the decay with a postirradiation treatment with 500 ppm methyl ester of indolyl-3-acetic acid which also reduced losses due to transpiration and respiration. The storage life of irradiated tomatoes treated with the hormone and stored at 9 to 10°C (lowest ripening temperatures) was 80 days as against 47 days of control fruits.

Abdel-Kader et al.[97-99] conducted physiological investigations on effects of irradiation on tomato fruits. In their series of experiments on respiration and C_2H_4 production, Abdel-Kader et al.[97] found that gamma irradiation induced an immediate stimulation of respiration with its magnitude and duration of response increasing with dose. Whereas doses less than 500 krad stimulated C_2H_4 production, 600 krad or more depressed it. Stimulation in C_2H_4 production did not accompany increase in ripening rate. The inhibition of color development by irradiation was less with advanced fruit maturity at time of treatment. The color development in mature green fruits irradiated at 500 krad or more was very abnormal and the fruits never attained table ripeness. The color development in pink fruits subjected to 600 krad or more failed to reach table ripeness.

Shelf life of two varieties of tomatoes ("Early Pak No. 7" and "Ace") irradiated at 250 to 300 krad and stored at 12 to 15°C was extended 4 to 12 days due to decay control, provided the fruits were of the pink or riper stage when irradiated according to the findings of Abdel-Kader et al.[98] Doses higher than 300 krad caused physiological breakdown resulting in more decay and lower doses were ineffective for decay control. Because of their adverse sensitivity to irradiation, there was no shelf-life extension in irradiated mature green or breaker stage tomatoes. Abdel-Kader et al.[98] cautioned that the importance of storage and transit temperatures for tomatoes could not be overemphasized and that irradiation could not be used to substitute for or modify the temperature requirements of the fruit. They concluded with a sobering remark to the effect that loss of texture and development of off-flavor, although this did not preclude acceptance by the taste panelists, might limit the application of irradiation technology on pink and riper fruits, despite the control of decay. According to the data of Abdel-Kader et al.[99] irradiation reduced the ascorbic acid content in greater proportion in pink and table-ripe than in mature green tomatoes. However, the final ascorbic acid concentration was higher in the irradiated pink and table-ripe fruits than in irradiated mature green fruit subsequently ripened, because of higher initial concentration of the vitamin in the riper fruits. Irradiation reduced total acidity and increased pH as compared with control fruits; it did not affect soluble solids content, but it reduced the firmness, especially immediately after irradiation.

The physiological nature of radiation damage and recovery was investigated by Lee et al.[100,101] and McGlasson and Lee.[102] In tomatoes of four stages of development irradiated at 200 and 400 krad, Lee et al.[100] recorded increases in respiration and C_2H_4 production immediately after irradiation, but 3 to 6 days after irradiation, they reported that these physiological processes decreased to the levels of preclimacteric control fruits. There was pronounced softening immediately after irradiation, but less mature fruits regained practically all of their original firmness within 4 days. These observations suggested that immature (comparable to "mature green stage" used by other investigators) tomatoes could recover from or repair damage caused by irradiation doses up to 400 krad and could thus pass through normal climacteric rise in respiration, but C_2H_4 evolution and pigment production associated with the climacteric were abnormal. Lee et al.[100] speculated that it was likely that ionizing radiation damaged some other components of the ripening process, possibly those involved in the action of C_2H_4 for ripening.

The increase in respiration activity immediately after irradiation was further investigated by Lee et al.[101] They studied the effects of treatment on the metabolism of ^{14}C-labeled organic acids in the tricarboxylic acid (TCA) cycle by infiltrating disks of pericarp tissue of tomatoes of various stages of development with the labeled respiratory intermediates. Metabolism of the organic acids was increased, but it was not affected by the difference in developmental stage of the fruit. This led Lee et al.[101] to conclude that the increase in respiration and metabolism of the organic acids in the TCA cycle could be explained by an increase in mitochondrial membrane permeability or a partial loss of respiratory control, and the apparent recovery from the damage to these processes is reflected in the restoration of the normal function of the mitochondria. They further stated that the independence of respiration and metabolism of the respiratory substrates from the physiological age of the fruit suggested that the delay in ripening of older irradiated fruit was a result of partially irreversible damage to some system controlling the ripening process and was not directly related with respiratory metabolism. This system could involve the synthesis of proteins (enzymes).

This possibility was investigated and verified by McGlasson and Lee[102] who found that the trichloroacetic acid soluble N/Total N ratio in the pericarp tissue of fruit increased immediately after irradiation. The ratio of immature fruit returned almost to the level of the control fruit 2 days after irradiation, but the ratio of the nearly mature fruit remained unchanged for 4 days. These observations were related to those of previous investigations of Lee et al.[100,101] on the dependence on age for the ability of tomatoes to recover from irradiation effects.

From the above, it is clearly seen that the effects of irradation on tolerance, as measured in terms of surface (pigment development and firmness) and internal (color, flavor, aroma, texture, ascorbic acid, and soluble solids) effects, on delay in ripening and/or senescence (shelf life extension), and on disease control in tomatoes, depend on irradiation dose and fruit maturity. The nature of delayed ripening has been elucidated by physiological investigations involving respiration, metabolism of respiratory substrates, and C_2H_4 evolution.

Y. Cucumbers (*C. sativus* L.)

In cucumbers irradiated by Bramlage and Lipton,[93] decay incidence was not consistently affected, but the fruit was softened and its fresh green appearance was impaired. Morris et al[103] also reported deterioration (loss of chlorophyll and firmness) of cucumbers irradiated at doses above 50 krad which reduced the shelf life.

Z. Summer Squash (*Cucurbita pepo* L.)

Bramlage and Lipton[93] reported that irradiation softened summer squash and impaired its appearance with no consistent control of decay.

AA. Bell Peppers (*Capsicum annuum* L.)

Irradiation caused softening, yellowing, and calyx discoloration of bell peppers irradiated by Bramlage and Lipton.[93]

III. FEASIBILITY OF IRRADIATION TECHNOLOGY

A. Shelf-Life Extension

Aside from the factors involving physical handling, the shelf life or marketable life of fresh commodities is determined by physiological and pathological factors. The physiological factor consists of rate of ripening and onset of senescence, and the pathological factor involves storage diseases. For shelf-life extension of a climacteric fruit, the rate of ripening and/or senescence must be delayed without damaging the fruit, and for a nonclimacteric fruit, the rate of senescence must be delayed without injuring the fruit. Furthermore, if present on the fruits, decay organisms will decrease the shelf life, so they must be controlled. Maxie and Abdel-Kader[3] listed a number of suggested criteria to determine whether irradiation can be used as a practical technology: (1) it must reduce losses due to decay or physiological disorders, (2) it must not induce major losses in nutritional or organoleptic qualities, and (3) it must not induce excessive changes in texture or appearance. Maxie and Sommer,[104] in reviewing the major chemical changes induced by irradiation at doses tolerated by fruits and vegetables, stated that the most limiting phenomenon in the use of irradiation to extend shelf life was the textural changes (softening) which resulted from radiation-induced breakage of cell wall constituents (pectin, cellulose, and hemicellulose macromolecules).

Based on the above criteria, the effects of irradiation, as gleaned from the available literature (admittedly very limited for some fruits), were evaluated for each of the 27 fruits discussed in this chapter. Depending on their responses to irradiation, these fruits are categorized in several groups in Figure 1 as follows:

Lack of tolerance to irradiation — In this group are fruits that are, in general, too sensitive to low-dose irradiation and are thus affected with various external and internal injuries. They include pears, avocados, lemons, grapefruit, oranges, tangerines, cucumbers, summer squash, bell peppers, olives, plums, apples, table grapes, and cantaloupes.

Accelerated ripening — Irradiation accelerates the ripening of peaches and nectarines.

Only tolerance to irradiation — Pineapples, lychees, and possibly honeydew melons are tolerant to low-dose irradiation, but their ripening or senescence is not affected.

Delayed ripening — The rate of ripening is delayed in bananas, mangoes, and papayas by irradiation.

Delayed senescence — Senescence is delayed in sweet cherries, apricots, and papayas.

Decay control — Storage decay is controlled in tomatoes, strawberries, and figs.

Only in fruits in which irradiation results in a delay in ripening and/or senescence or decay control without injury can there be an extension of shelf life. It follows then that only those eight fruits in the above categories 4, 5, and 6 are probably benefited by irradiation, and these include bananas, mangoes, papayas, sweet cherries, apricots, tomatoes, strawberries, and figs (Figure 1). In 1971, Maxie et al.[105] summarized their 15 years of research on irradiation in the laboratory and simulated and actual shipping experiments. They concluded that irradiation was not commercially feasible for the preservation of fresh fruits and vegetables because it produced no beneficial effects at doses tolerated by the commodities concerned and because it was economically unfavorable as compared with existing competing processes of preservation. They presented a selected list of 20 commodities with factors limiting the use of irradiation on each

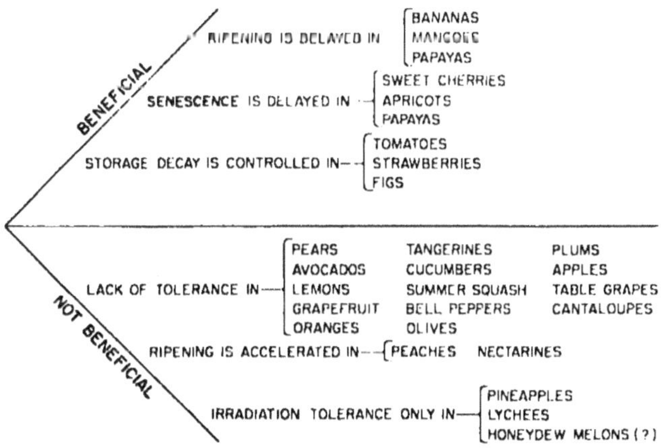

FIGURE 1. The response of 27 fruits to irradiation.

of them. Of the above eight fruits probably benefiting from irradiation according to the literature, bananas, papayas, apricots, tomatoes, and strawberries are listed in the compilation of Maxie et al.[105] According to Maxie et al.,[105] economics limit the commercial application of irradiation of bananas, papayas, and strawberries, the economic factors being available, cheaper, equally effective or more effective alternative methods for bananas and strawberries and inadequate acreage for papayas. For apricots and tomatoes, the chief factor limiting the application of irradiation on them is tissue softening, so stated these investigators.

It is of interest to note that Maxie and Abdel-Kader[3] once enthusiastically wrote that for developing countries where there is a lack of adequate facilities and techniques, irradiation could be the technology to preserve fresh commodities, but later Maxie et al.[105] lost their enthusiasm with a statement to the effect that the hope that irradiation could be a substitute for refrigeration in emerging countries must be dispelled.

The pessimistic conclusions by Maxie et al.[105] in 1971 are not in accord with recent reports on irradiated foods.[106-110] A 1979 compilation of irradiated foods legally approved for human consumption[106] (see Table 1, Chapter 1, Volume 1) includes radurized apricots, cherries, raspberries, grapes, tomatoes, and peaches (Bulgaria); radurized strawberries (World Health Organization, Hungary, the Netherlands, and South Africa), and ripening-delayed avocados (South Africa). After the above-listed approvals were granted, South Africa conducted 3-month market tests in two cities using 4 tons each of irradiated strawberries, papayas, and mangoes,[107] and 50 tons of mangoes of which 9000 kg were shipped in February 1977 via Southampton, England, and then sold, with Dutch approval, in the Netherlands.[108,109] In Hungary, the Budapest Vegetable and Fruit Marketing Cooperative Enterprise conducted in 1973 a field market trial of radurized strawberries.[110]

B. Disinfestation

A discussion on the use of irradiation on fresh fruits is not complete without mentioning an additional possible use of this technology, and that is, its use as a method to destroy insect pests. Because of the prevalence of fruit flies in Hawaii, all fruits, except pineapples, destined for shipment to areas in the U.S. and foreign countries where fruit flies can survive, must be disinfested, i.e., they must be treated to destroy the insects. Currently, fumigation with ethylene dibromide is the standard procedure for this purpose, but vapor heat is another approved procedure for treating fruits destined for export. Export pineapples are exempt from quarantine regulations, because although they are attacked by fruit flies, the insects do not emerge from the fruit. Because it was demonstrated by Balock et al.,[111] Balock et al.,[112] and Steiner[113] that low doses of gamma rays (approximately 25 krad) could be used to disinfest fresh commodities in Hawaii, irradiation was considered as a possible quarantine alternative treatment for currently used methods.

Since papayas and lychees can tolerate doses of 25 krad and above, irradiation can be used as an alternative treatment for these export fruits. Mangoes are currently not allowed to be exported because of a lack of method to kill the seed weevil, although fruit flies on this fruit are eliminated by present methods. Irradiation can also be used to disinfest mangoes for the export trade, because they can tolerate doses above those required to destroy fruit flies and seed weevil. Because avocados are injured at doses far below those required for disinfestation, irradiation cannot be used for this fruit.

There are no fruits approved for treatment with gamma rays to date. As a step toward clearance and approval of irradiation (26 krad) as a disinfestation technology for export papayas, long-term animal feeding studies and short-term genetic susceptibility studies were conducted on irradiated papayas, and these studies demonstrated no differences attributable to radiation at or twice the proposed dose range. However, the U.S. Department of Energy and the U.S. Department of Agriculture have not formally filed a petition for certification of irradiation for export papayas. Clearance for papayas would probably lead to a similar approval for Hawaiian mangoes and lychees for export. Papayas are considered to be the best candidate for irradiation treatment. If irradiation is approved for this fruit, papayas represent the epitome of what benefits can be bestowed on any fresh commodity by this procedure. At a dose of 26 krad, the viability and reproductive ability of insects are controlled without affecting quality of the export fruit and in combination with the hot water treatment, storage decay is controlled and shelf life is extended. If slightly higher doses (up to about 75 krad) are allowed, a further extension of shelf life is achieved. This extension, in contrast to other fruits, such as irradiated strawberries for which refrigeration is required, is obtained in refrigerated as well as unrefrigerated papayas, thus making the technology applicable to shipments by sea as well as air.

C. Economics

Economic aspects of irradiation are not considered here. However, suffice it to say that economic considerations will decide whether the technology is economically feasible or not after the benefits of irradiation are weighed against the cost of the technology.

IV. SUMMARY

Literature perusal was employed to determine the effects of gamma irradiation on tolerance and shelf life of 27 botanical fruits. Tolerance was determined in terms of external and internal visible and/or measurable injury symptoms and effects on sensory evaluations and biochemical parameters of the edible pulp. Shelf life was determined in terms of physiological and biochemical changes affecting the rate of respiration and senescence. Since decay incidence plays an important role in the life of fresh commodities, the effect of irradiation on decay organisms was also considered.

According to the available literature on the 27 fruits, apparently only 8 (bananas, mangoes, papayas, sweet cherries, apricots, tomatoes, strawberries, and figs) may benefit from irradiation in that their shelf life is extended through a delay in ripening and/or senescence or control of storage decay without sustaining any injury. Of the eight fruits, papayas seem to benefit the most by irradiation, because the treatment not only extends the shelf life, but also sterilizes insects. It can thus be used as a quarantine procedure in areas such as Hawaii, where disinfestation is required for export fruit. Economic considerations will, however, determine the commercial feasibility of irradiation as a technology for preserving papayas for export as well as other fruits for which benefits of irradiation seem apparent, provided the treatment is approved by a regulatory agency of the government.

REFERENCES

1. **Salunkhe, D. K.**, Gamma radiation effects on fruits and vegetables, *Econ. Bot.*, 15, 28, 1961.
2. **Massey, L. M., Jr.**, Food irradiation, *U.S. Dep. Agric. Sci. Rev.*, 5, 29, 1967.
3. **Maxie, E. C. and Abdel-Kader, A.**, Food irradiation — physiology of fruits as related to feasibility of the technology, *Adv. Food Res.*, 15, 105, 1967.
4. **Pablo, I. S., Akamine, E. K., and Chachin, K.**, Irradiation, in *Postharvest Physiology, Handling and Utilization of Tropical and Subtropical Fruits and Vegetables*, Pantastico, E. B., Ed., AVI Publishing, Westport, 1975, Chap. 11.
5. **Moy, J. H., Akamine, E. K., Brewbaker, J. L., Buddenhagen, I. W., Ross, E., Spielmann, H., Upadhya, M. D., Wenkam, N., Helber, D., Dollar, A. M., Hanaoka, M., and McGlish, G. A.**, Dosimetry, tolerance and shelf-life extension related to disinfestation of tropical fruits by gamma irradiation, in *Disinfestation of Fruit by Irradiation*, STI/PUB/299, International Atomic Energy Agency (IAEA), Vienna, 1971, 43.
6. **Bramlage, W. J. and Covey, H. M.**, Gamma radiation of fruits to extend market life, Mark. Res. Rep. No. 717, Agricultural Research Service, U.S. Department of Agriculture Washington, D.C., 1975, 1.
7. **Somogyi, L. P. and Romani, R. J.**, Irradiation-induced textural change in fruits and its relation to pectin metabolism, *J. Food Sci.*, 29, 366, 1964.
8. **Bussel, J. and Maxie, E. C.**, Gas exchange in Bartlett pears in relation to gamma irradiation, *Proc. Am. Soc. Hortic. Sci.*, 88, 151, 1966.
9. **Maxie, E. C., Sommer, N. F., Muller, C. J., and Rae, H. L.**, Effect of gamma radiation on the ripening of Bartlett pears, *Plant Physiol.*, 41, 437, 1966.
10. **Romani, R. J.**, Radiation physiology of fruit — respiration during and immediately following kilorad doses of ionizing radiation, *Radiat. Bot.*, 4, 299, 1964.
11. **Maxie, E. C., Johnson, C. F., Boyd, C., Rae, H. L., and Sommer, N. F.**, Effect of gamma irradiation on ripening and quality of nectarines and peaches, *Proc. Am. Soc. Hortic. Sci.*, 89, 91, 1966.
12. **Dhaliwal, A. S. and Salunkhe, D. K.**, Ionizing radiation and packaging effects on respiratory behavior, fungal growth, and storage-life of peaches, *Prunus persica, Radiat. Bot.*, 3, 75, 1963.
13. **Lovell, R. T. and Flick, G. J.**, Irradiation of Gulf Coast area strawberries, *Food Technol. (Chicago)*, 20, 99, 1966.
14. **Pierson, N. W., Wilcox, E. B., and Salunkhe, D. K.**, Radiation preservation of fruits, *Utah Farm Home Sci.*, 23, 14, 1962.
15. **Cooper, G. M. and Salunkhe, D. K.**, Effect of gamma-radiation, chemical, and packaging treatments on refrigerated life of strawberries and sweet cherries, *Food Technol. (Chicago)*, 17, 123, 1963.
16. **Salunkhe, D. K., Gerber, R. K., and Pollard, L. H.**, Physiological and chemical effects of gamma radiation on certain fruits, vegetables, and their products, *Proc. Am. Soc. Hortic. Sci.*, 74, 123, 1959.
17. **Maxie, E. C., Sommer, N. F., and Rae, H. L.**, Effect of gamma irradiation on Shasta strawberries under marketing conditions, *Isot. Radiat. Technol.*, 2, 50, 1964.
18. **Guerreo, F. P., Ravetto, D., and Maxie, E. C.**, Air test shipment of irradiated strawberries, in Radiation Technology in Conjunction with Postharvest Procedures as a Means of Extending the Shelf Life of Fruits and Vegetables, Rep. 1966 to 1967, Contract No. AT (11-1)-34, Proj. Agreement No. 80, Division of Isotopes Development, U.S. Atomic Energy Commission, Washington, D.C., 1967, 34.
19. **Akamine, E. K. and Goo, T.**, The effect of irradiation on avocados, in Dosimetry, Tolerance, and Shelf Life Extension Related to Disinfestation of Fruits and Vegetables by Gamma Irradiation, Rep. 1966 to 1967, Contract No. AT(04-3)-235, Proj. Agreement No. 5, Division of Isotopes Development, U.S. Atomic Energy Commission, Washington, D.C., 1967, 115.

20. **Akamine, E. K. and Goo, T.**, Respiration of gamma-irradiated fresh fruits, *J. Food Sci.*, 36, 1074, 1971.
21. **Young, R. E.**, Effect of ionizing radiation on respiration and ethylene production of avocado fruit, *Nature (London)*, 205, 1113, 1965.
22. **Maxie, E. C., Nelson, K. E., and Johnson, C. F.**, Effect of gamma irradiation on table grapes, *Proc. Am. Soc. Hortic. Sci.*, 84, 263, 1964.
23. **Upadhya, M. D. and Brewbaker, J. L.**, Effects of gamma irradiation on the pineapple, *Hawaii Farm Sci.*, 15, 8, 1966.
24. **Wenkam, N. S. and Moy, A. P.**, Nutritional composition of irradiated fruit. II. Pineapple, in Dosimetry, Tolerance, and Shelf Life Extension Related to Disinfestation of Fruits and Vegetables by Gamma Irradiation, Rep. 1967 to 1968, Contract No. AT(04-3)-235, Proj. Agreement No. 5, Division of Isotopes Development, U.S. Atomic Energy Commission, Washington, D.C., 1969, 157.
25. **Upadhya, M. D., Brewbaker, J. L., and Ching, K. W.**, Biochemical changes in gamma-irradiated pineapples, in Dosimetry, Tolerance, and Shelf Life Extension Related to Disinfestation of Fruits and Vegetables by Gamma Irradiation, Rep. 1966 to 1967, Contract No. AT(04-3)-235, Proj. Agreement No. 5, Division of Isotopes Development, U.S. Atomic Energy Commission, Washington, D.C., 1967, 85.
26. **Upahhya, M. D., Brewbaker, J. L., Ching, K. W., and Lin, M.**, Comparative study of irradiated summer and winter pineapples under varying storage periods and temperatures, in Dosimetry, Tolerance, and Shelf Life Extension Related to Disinfestation of Fruits and Vegetables by Gamma Irradiation, Rep. 1966 to 1967, Contract No. AT(04-3)-235, Proj. Agreement No. 5, Division of Isotopes Development, U.S. Atomic Energy Commission, Washington, D.C., 1967, 95.
27. **Moy, J. H., Hsia, S., and Sawato, M.**, Organoleptic evaluation of gamma irradiated pineapple, in Dosimetry, Tolerance, and Shelf Life Extension Related to Disinfestation of Fruits and Vegetables by Gamma Irradiation, Rep. 1967 to 1968, Contract No. AT(04-3)-235, Proj. Agreement No. 5, Division of Isotopes Development, U.S. Atomic Energy Commission, Washington, D.C., 1969, 152.
28. **Upadhya, M. D., Makinen, Y., and Brewbaker, J. L.**, Post-irradiation storage recovery of pineapple fruits from induced cytotoxicity following disinfestation doses of gamma irradiation, in Dosimetry, Tolerance, and Shelf Life Extension Related to Disinfestation of Fruits and Vegetables by Gamma Irradiation, Rep. 1966 to 1967, Contract No. AT(04-3)-235, Proj. Agreement No. 5, Division of Isotopes Development, U.S. Atomic Energy Commission, Washington, D.C., 1967, 105.
29. **Mathur, P. B.**, Application of atomic energy in the preservation of fresh fruits and vegetables, *Indian Food Packer*, 22, 1, 1968.
30. **Ferguson, W. E. and Yates, A. R.**, The effects of gamma radiation on bananas, *Food Technol. (Chicago)*, 20, 105, 1966.
31. **Maxie, E. C., Amezquita, R., Hasson, B. M., and Johnson, C. F.**, Effect of gamma irradiation on the ripening of banana fruit, *Proc. Am. Soc. Hortic. Sci.*, 92, 235, 1968.
32. **Moy, J. H., Ross, E., Goo, T., and Hsia, S. T.**, Sensory evaluation of irradiated bananas, in Dosimetry, Tolerance, and Shelf Life Extension Related to Disinfection of Fruits and Vegetables by Gamma Irradiation, Rep. 1966 to 1967, No. AT(04-3)-235, Proj. Agreement No. 5, Division of Isotopes Development, U.S. Atomic Energy Commission, Washington, D.C., 1967. 120.
33. **Upadhya, M. D. and Brewbaker, J. L.**, Biochemical changes during storage of gamma-irradiated lychees, in Dosimetry, Tolerance, and Shelf Life Extension Related to Disinfestation of Fruits and Vegetables by Gamma Irradiation, Rep. 1966 to 1967, Contract No. AT(04-3)-235, Proj. Agreement No. 5, Division of Isotopes Development, U.S. Atomic Energy Commission, Washington, D.C., 1967, 140.
34. **Moy, J. H., Chang, G. K. L., and Hsia, S. T.**, Organoleptic evaluation of gamma irradiated lychee, in Dosimetry, Tolerance, and Shelf Life Extension Related to Disinfestation of Fruits and Vegetables by Gamma Irradiation, Rep. 1967 to 1968, Contract No. AT(04-3)-235, Proj. Agreement 5, Division of Isotopes Development, U.S. Atomic Energy Commission, Washington, D.C., 1969, 163.
35. **Akamine, E. K. and Goo, T.**, Effects of gamma irradiation on shelf life of fresh lychees *(Litchi chinensis Sonn.)*, *Hawaii Agric. Exp. Stn. Res. Bull.*, 169, 1, 1977.
36. **Grierson, W. and Dennison, R. A.**, Irradiation treatment of "Valencia" oranges and "Marsh" grapefruit, *Proc. Fla. State Hortic. Soc.*, 78, 233, 1965.
37. **Guerrero, F. P., Maxie, E. C., Johnson, C. F., Eaks, I. L., and Sommer, N. F.**, Effects of postharvest gamma irradiation on orange fruits, *Proc. Am. Soc. Hortic. Sci.*, 90, 575, 1967.
38. **Romani, R. J., van Kooy, J., Lim, L., and Bowers, B.**, Radiation physiology of fruit — ascorbic acid, sulfhydryl and soluble nitrogen content of irradiated citrus, *Radiat. Bot.*, 3, 363, 1963.
39. **Kahan, R. S. and Monselise, S. P.**, Extension of storage life of citrus fruit by irradiation, *Food Technol. (Chicago)*, 19, 122, 1965.
40. **Dennison, R. A., Ahmed, E. M., and Martin, F. G.**, Pectinesterase activity in irradiated "Valencia" oranges, *Proc. Am. Soc. Hortic. Sci.*, 91, 163, 1967.

41. **Monselise, S. P. and Kahan, R. S.**, Changes in composition and in enzymatic activities of flavedo and juice of Shamouti oranges following gamma radiation, *Radiat. Bot.*, 6, 265, 1966.
42. **Riov, J., Monselise, S. P., and Kahan, R. S.**, Effect of gamma radiation on phenylalanine ammonialyase activity and accumulation of phenolic compounds in citrus fruit peel, *Radiat. Bot.*, 8, 463, 1968.
43. **Belli-Donini, M. L., Baraldi, D., and Taggi, R.**, Relationship between peel damage and the accumulation of terpene compounds in irradiated oranges, *Radiat. Bot.*, 14, 1, 1974.
44. **Riov, J.**, Histochemical evidence for the relationship between peel damage and the accumulation of phenolic compounds in gamma-irradiated citrus fruit, *Radiat. Bot.*, 15, 257, 1975.
45. **Beraha, L., Ramsey, G. B., Smith, M. A., and Wright, W. R.**, Factors influencing the use of gamma radiation for control of decay of lemons and oranges, *Phytopathology*, 49, 91, 1959.
46. **Maxie, E. C., Eaks, I. L., and Sommer, N. F.**, Some physiological efects of gamma irradiation on lemon fruit, *Radiat. Bot.*, 4, 405, 1964.
47. **Maxie, E. C., Eaks, I. L., Sommer, N. F., Rae, H. L., and El-Batal, S.**, Effect of gamma radiation on rate of ethylene and carbon dioxide evolution by lemon fruit, *Plant Physiol.*, 40, 407, 1965.
48. **Shikhande, A. J. and Kaewubon, N.**, Effects of controlled atmosphere storage on irradiated lemon fruits, *Radiat. Bot.*, 14, 315, 1974.
49. **Riov, J., Monselise, S. P., and Kahan, R. S.**, Radiation damage to grapefruit in relation to ethylene production and phenylalanine ammonia-lyase activity, *Radiat. Bot.*, 10, 281, 1970.
50. **Riov, J. and Goren, R.**, Effects of gamma radiation on protein synthesis in peel of mature grapefruit, *Radiat. Bot.*, 10, 155, 1970.
51. **Riov, J., Gordon, R., Monselise, S. P., and Kahan, R. S.**, Effect of gamma radiation on the synthesis of scopoletin and scopolin in grapefruit peel in relation to radiation damage, *Radiat. Res.*, 45, 326, 1971.
52. **Akamine, E. K. and Wong, R.**, Tolerance of tangerines to irradiation, in Dosimetry, Tolerance, and Shelf Life Extension Related to Disinfestation of Fruits and Vegetables by Gamma Irradiation, Rep. 1965 to 1966, Contract No. AT(04-3)-235, Proj. Agreement No. 5, Division of Isotopes Development, U.S. Atomic Energy Commission, Washington, D. C., 1966, 94.
53. **Dharkar, S. D., Savagaon, K. A., Srirangarajan, A. N., and Sreenivasan, A.**, Irradiation of mangoes. I. Radiation-induced delay in ripening of Alphonso mangoes, *J. Food Sci.*, 31, 863, 1966.
54. **Dharkar, S. D., Savagaon, K. A., Srirangarajan, A. N., and Sreenivasan, A.**, Irradiation of mangoes. II. Radiation effects on skin-coated Alphonso mangoes, *J. Food Sci.*, 31, 870, 1966.
55. **Hatton, T. T., Jr., Beraha, L., and Wright, W. R.**, Preliminary trials of gamma radiation on mature green Irwin and Sensation mangoes, *Proc. 21st Annu. Meet. Fla. Mango Forum*, 1961, 15.
56. **Dennison, R. A. and Ahmed, E. M.**, Irradiation effects on the ripening of Kent mangoes, *J. Food Sci.*, 32, 702, 1967.
57. **Upadhya, M. D., Young, R. L., and Brewbaker, J. L.**, Post-irradiation storage changes in mangoes following gamma irradiation, in Dosimetry, Tolerance, and Shelf Life Extension Related to Disinfestation of Fruits and Vegetables by Gamma Irradiation, Rep. 1965 to 1966, Contract No. AT(04-3)-235, Proj. Agreement No. 5, Division of Isotopes Development, U.S. Atomic Energy Commission, Washington, D.C., 1966, 53.
58. **Upadhya, M. D. and Brewbaker, J. L.**, Irradiation of mangoes for control of the mango seed weevil, *Hawaii Farm Sci.*, 15, 6, 1966.
59. **Upadhya, M. D., Ching, K. W., and Brewbaker, J. L.**, Post-irradiation storage changes in irradiated Pirie mangoes, in Dosimetry, Tolerance, and Shelf Life Extension Related to Disinfestation of Fruits and Vegetables by Irradiation, Rep. 1966 to 1967, Contract No. AT(04-3)-235, Proj. Agreement No. 5, Division of Isotopes Development, U.S. Atomic Energy Commission, Washington, D.C., 1967, 70.
60. **Moy, J. H., Hsia, S. T., and Sawato, M.**, Organoleptic evaluation of gamma irradiated mango, in Dosimetry, Tolerance, and Shelf Life Extension Related to Disinfestation of Fruits and Vegetables by Gamma Irradiation, Rep. 1967 to 1968, Contract No. AT(04-3)-235, Proj. Agreement No. 5, Division of Isotopes Development, U.S. Atomic Energy Commission, Washington, D.C., 1969, 136.
61. **Hilker, D. M. and Wong, E.**, Effects of gamma-irradiation on carotenoid retention in papayas and mangoes, in Dosimetry, Tolerance, and Shelf Life Extension Related to Disinfestation of Fruits and Vegetables by Gamma Irradiation, Rep. 1966 to 1967, Contract No. AT(04-3)-235, Proj. Agreement No. 5, Division of Isotopes Development, U.S. Atomic Energy Commission, Washington, D.C., 1967, 66.
62. **Wenkam, N. S. and Moy, A. P.**, Nutritional composition of irradiated fruit. I. Mango and papaya, in Dosimetry, Tolerance, and Shelf Life Extension Related to Disinfestation of Fruits and Vegetables by Gamma Irradiation, Rep. 1967 to 1968, Contract No. AT(04-3)-235, Proj. Agreement No. 5, Division of Isotopes Development, U.S. Atomic Energy Commission, Washington, D.C., 1969, 126.

63. **Kojima, E. S. and Buddenhagen, I. W.**, Studies on gamma-irradiation in relation to postharvest diseases of papayas and mango, in Dosimetry, Tolerance, and Shelf Life Extension Related to Disinfestation of Fruits and Vegetables by Gamma Irradiation, Rep. 1966 to 1967, Contract No. AT(04-3)-235, Proj. Agreement No. 5, Division of Isotopes Development, U.S. Atomic Energy Commission, Washington, D.C., 1967, 54.
64. **Akamine, E. K. and Goo, T.**, Effects of ionizing irradiation on Haden mangoes, *Hawaii Agric. Exp. Stn. Res. Rep.* 205, 1979, 1.
65. **Akamine, E. K. and Wong, R.**, Shelf life extension of papayas with gamma irradiation, in Dosimetry, Tolerance, and Shelf Life Extension Related to Disinfestation of Fruits and Vegetables by Gamma Irradiation, Rep. 1964 to 1965, Contract No. AT(04-3)-235, Proj. Agreement No. 5, Division of Isotopes Development, U.S. Atomic Energy Commission, Washington, D.C., 1965, 55.
66. **Akamine, E. K. and Wong, R. T. F.**, Extending the shelf life of papayas with gamma irradiation, *Hawaii Farm Sci.*, 15, 4, 1966.
67. **Akamine, E. K. and Goo, T.**, Effect of gamma irradiation on shelf life extension of fresh papayas, in Hawaii Food Irradiation Program, Rep. 1968 to 1969, Contract AT(26-1)-374 and Contract No. AT(04-3)-235, Proj. Agreement No. 5, Division of Isotopes Development, U.S. Atomic Energy Commission, Washington, D.C., 1970, 34.
68. **Akamine, E. K.**, Temperature effects in fresh papayas processed for shipment, *Hawaii Agric. Exp. Stn. Bull.*, 122, 1, 1960.
69. **Akamine, E. K. and Wong, R.**, Refrigerated storage of irradiated papayas, in Dosimetry, Tolerance, and Shelf Life Extension Related to Disinfestation of Fruits and Vegetables by Gamma Irradiation, Rep. 1965 to 1966, Contract No. AT(04-3)-235, Proj. Agreement No. 5, Division of Isotopes Development, U.S. Atomic Energy Commission, Washington, D.C., 1966, 17.
70. **Akamine, E. K. and Goo, T.**, Effects of gamma irradiation on shelf life extension of fresh papayas (*Carica papaya* L. var. Solo), *Hawaii Agric. Exp. Stn. Res. Bull.*, 165, 1, 1977.
71. **Akamine, E. K. and Goo, T.**, Controlled atmosphere storage of irradiated papayas (*Carica papaya* var. Solo), in Dosimetry, Tolerance, and Shelf Life Extension Related to Disinfestation of Fruits and Vegetables by Gamma Irradiation Rep. 1967 to 1968, Contract No. AT(04-3)-235, Proj. Agreement No. 5, Division of Isotopes Development, U.S. Atomic Energy Commission, Washington, D.C., 1969, 63.
72. **Akamine, E. K. and Goo, T.**, Respiration and ethylene production in irradiated papayas, in Hawaii Food Irradiation Program, Rep. 1968 to 1969, Contract AT(26-1)-374 and Contract No. AT(04-3)-235, Proj. Agreement No. 5, Division of Isotopes Development, U.S. Atomic Energy Commission, Washington, D.C., 1970, 54.
73. **Akamine, E. K. and Goo, T.**, Respiration, ethylene production, and shelf life extension in irradiated papaya fruit after storage under stimulated shipping conditions, *Hawaii Agric. Exp. Stn. Tech. Bull.*, 93, 1, 1977.
74. **Moy, J. H., Akamine, E. K., Wenkam, N., Dollar, A. M., Hanaoka, M., Kao, H. Y., Lin, W. L., and Revetti, L. M.**, Tolerance, quality and shelf life of gamma-irradiated papaya grown in Hawaii, Taiwan and Venezuela, *Radiation Preservation of Food*, STI/PUB/317, IAEA, Vienna, 1973, 375.
75. **Buddenhagen, I. W. and Kojima, E. S.**, Biology of postharvest disease fungi of papaya in relation to disinfestation of fruits and vegetables by gamma irradiation, in Dosimetry, Tolerance, and Shelf Life Extension Related to Disinfestation of Fruits and Vegetables by Gamma Irradiation, Rep. 1965 to 1966, Contract No. AT(04-3)-235, Proj. Agreement No. 5, Division of Isotopes Development, U.S. Atomic Energy Commission, Washington, D.C., 1966, 30.
76. **Moy, J. H., Ross, E., Hsia, S. T., and Sawato, M.**, Quality evaluation of gamma irradiated papaya in shipping studies, in Dosimetry, Tolerance, and Shelf Life Extension Related to Disinfestation of Fruits and Vegetables by Gamma Irradiation, Rep. 1967 to 1968, Contract No. AT(04-3)-235, Proj. Agreement No. 5, Division of Isotopes Development, U.S. Atomic Energy Commission, Washington, D.C., 1969, 3.
77. **Moy, J. H. and Hsia, S. T.**, The influence of radiation dose, degree of ripeness, and post-irradiation storage temperatures on the development of flowery aroma in gamma irradiated Solo papaya, in Dosimetry, Tolerance, and Shelf Life Extension Related to Disinfestation of Fruits and Vegetables by Gamma Irradiation, Rep. 1967 to 1968, Contract No. AT(04-3)-235, Proj. Agreement No. 5, Division of Isotopes Development, U.S. Atomic Energy Commission, Washington, D.C., 1969, 112.
78. **Moy, J. H., Ross, E., Hsia, S. T., and Sawato, M.**, Quality evaluation of gamma irradiated papaya in shipping studies, *Food Irradiat.*, 11, 14, 1971.
79. **Hilker, D. M.**, Effect of gamma radiation on ascorbic acid content in papaya, in Dosimetry, Tolerance, and Shelf Life Extension Related to Disinfestation of Fruits and Vegetables by Gamma Irradiation, Rep. 1964 to 1965, Contract No. AT(04-3)-235, Proj. Agreement No. 5, Division of Isotopes Development, U.S. Atomic Energy Commission, Washington, D.C., 1965, 52.
80. **Hilker, D. M. and Young, R. L.**, The effect of ionizing radiation on some nutritional and biochemical properties of papaya, *Hawaii Farm Sci.*, 15, 9, 1966.

81. **Young, R. L. and Lin, M.**, Radiation effects on protein and pectin in papaya, in Dosimetry, Tolerance, and Shelf Life Extension Related to Disinfestation of Fruits and Vegetables by Gamma Irradiation, Rep. 1964 to 1965, Contract No. AT(04-3)-235, Proj. Agreement No. 5, Division of Isotopes Development, U.S. Atomic Energy Commission, Washington, D.C., 1965, 7.
82. **Upadhya, M. D., Brewbaker, J. L., and Ching, K. W.**, Biochemical changes during storage of gamma irradiated papaya, in Dosimetry, Tolerance, and Shelf Life Extension Related to Disinfestation of Fruits and Vegetables by Gamma Irradiation, Rep. 1966 to 1967, Contract No. AT (04-3)-235, Proj. Agreement No. 5, Division of Isotopes Development, U.S. Atomic Energy Commission, Washington, D.C., 1967, 3.
83. **Moy, J. H., Dollar, A. M., and Hsia, S. T.**, Peroxidase acitivity in papaya (*Carica papaya* L. var. Solo), in Hawaii Food Irradiation Program, Rep. 1968 to 1969, Contract AT(26-1)-374 and Contract No. AT(04-3)-235, Proj. Agreement No. 5, Division of Isotopes Development, U.S. Atomic Energy Commission, Washington, D.C., 1970, 19.
84. **Buddenhagen, I. W. and Takata, G. S.**, Ultrastructural studies on irradiated and unirradiated papaya fruit tissue, in Dosimetry, Tolerance, and Shelf Life Extension Related to Disinfestation of Fruits and Vegetables by Gamma Irradiation, Rep. 1967 to 1968. Contract AT(04-3)-235, Proj. Agreement No. 5, Division of Isotopes Development, U.S. Atomic Energy Commission, Washington, D.C., 1969, 115.
85. **Spielmann, H.**, Demand analysis and market development for fresh and potentially gamma irradiated papaya on U.S. mainland markets, *Hawaii Agric. Exp. Stn. Agric. Econ. Bull.*, 29, 1, 1968.
86. **Spielmann, H.**, Demand characteristics for fresh and potentially gamma irradiated papaya on selected U.S. mainland markets, in Dosimetry, Tolerance, and Shelf Life Extension Related to Disinfestation of Fruits and Vegetables by Gamma Irradiation, Rep. 1967 to 1968, Contract No. AT(04-3)-235, Proj. Agreement No. 5, Division of Isotopes Development, U.S. Atomic Energy Commission, Washington, D.C., 1969, 9.
87. **McClish, G. A. and Dollar, A. M.**, Papaya package development, in Hawaii Food Irradiation Program, Rep. 1968 to 1969, Contract AT(26-1)-374 and Contract No. AT(04-3)-235, Proj. Agreement No. 5, Division of Isotopes Development, U.S. Atomic Energy Commission, Washington, D.C., 1970, 71.
88. **Dollar, A. M., Hanaoka, M., Moy, J. H., Cinnamon, A. D., Hammill, E., Helber, D., Hsia, S. T., and Wenkam, N.**, Physiological, chemical, and physical changes during ripening of papaya, in Hawaii Food Irradiation Program, Rep. 1968 to 1969, Contract AT(26-1)-374 and Contract No. AT(04-3)-235, Proj. Agreement No. 5, Division of Isotopes Development, U.S. Atomic Energy Commission, Washington, D.C., 1970, 85.
89. **Dollar, A. M., Hanaoka, M., McClish, G. A., Ross, E., and Moy, J. H.**, Simulated shipping study protocol, in Hawaii Food Irradiation Program, Contract AT(26-1)-374 and Contract No. AT-(04-3)-235, Proj. Agreement No. 5, Division of Isotopes Development, U.S. Atomic Energy Commission, Washington, D.C., 1970, 129.
90. **Dollar, A. M., Hanaoka, M., McClish, G., Moy, J. H., Chan, K. C., Cinnamon, A. D., Helber, D., Hsia, S. T., and Jaw, R. C.**, Simulated shipping studies, in Hawaii Food Irradiation Program, Rep. 1968 to 1969, Contract AT(26-1)-374 and Contract No. AT(04-3)-235, Proj. Agreement No. 5, Division of Isotopes Development, U.S. Atomic Energy Commission, Washington, D.C., 1970, 155.
91. **Dollar, A. M., Hanaoka, M., McClish, G., Moy, J. H., Chan, K. C., Co, H. Y. H., Cinnamon, A. D., and Hsia, S. T.**, Large scale shipping studies — papaya, in Hawaii Food Irradiation Program, Rep. 1968 to 1969, Contract AT(26-1)-374 and Contract No. AT(04-3)-235, Proj. Agreement No. 5, Division of Isotopes Development, U.S. Atomic Energy Commission, Washington, D.C., 1970, 193.
92. **Dollar, A. M., Hanaoka, M., McClish, G. A., and Moy, J. H.**, Semicommercial scale studies on irradiated papaya, *Disinfestation of Fruit by Irradiation*, STI/PUB/299, IAEA, Vienna, 1971, 137.
93. **Bramlage, W. J. and Lipton, W. J.**, Gamma radiation of vegetables to extend market life, Mark. Res. Rep. No. 703, Agricultural Research Service, U.S. Department of Agriculture, Washington, D.C., 1965.
94. **Ravetto, D., Morris, L. L., Johnson, C. F., and Maxie, E. C.**, Effect of gamma irradiation on postharvest behavior of the cantaloupe, in Radiation Technology in Conjunction with Postharvest Procedures as a Means of Extending the Shelf Life of Fruits and Vegetables, Rep. 1966 to 1967, Contract No. AT(11-1)-34, Proj. Agreement No. 80, Division of Isotopes Development, U.S. Atomic Energy Commission, Washington, D.C., 1967, 65.
95. **Ravetto, D., Morris, L. L., and Maxie, E. C.**, Effects of gamma irradiation on postharvest behavior of honeydew melons, in Radiation Technology in Conjunction with Postharvest Procedures as a Means of Extending the Shelf Life of Fruits and Vegetables, Rep. 1966 to 1967, Contract No. AT(11-1)-34, Proj. Agreement No. 80, Division of Isotopes Development, U.S. Atomic Energy Commission, Washington, D.C., 1967. 92.

96. **Ravetto, D., Morris, L. L., and Maxie, E. C.**, Effect of gamma irradiation on the susceptibility of tomato fruits to transit injury, in Radiation Technology in Conjunction with Postharvest Procedures as a Means of Extending the Shelf Life of Fruits and Vegetables, Rep. 1966 to 1967, Contract No. AT(11-1)-34, Proj. Agreement No. 80, Division of Isotopes Development, U.S. Atomic Energy Commission, Washington, D.C., 1967, 104.
97. **Abdel-Kader, A. S., Morris, L. L., and Maxie, E. C.**, Physiological studies of gamma irradiated tomato fruits. I. Effect on respiratory rate, ethylene production, and ripening, *Proc. Am. Soc. Hortic. Sci.*, 92, 553, 1968.
98. **Abdel-Kader, A. S., Morris, L. L., and Maxie, E. C.**, Physiological studies of gamma-irradiated tomato fruits. II. Effects on deterioration and shelf life, *Proc. Am. Soc. Hortic. Sci.*, 93, 831, 1968.
99. **Abdel-Kader, A. S., Morris, L. L., and Maxie, E. C.**, Physiological studies of gamma-irradiated tomato fruits. III. Effects on ascorbic acid content, acidity, and texture, *Proc. Am. Soc. Hortic. Sci.*, 93, 843, 1968.
100. **Lee, T. H., McGlasson, W. B., and Edwards, R. A.**, Effect of gamma radiation on tomato fruit picked at four stages of development, *Radiat. Bot.*, 8, 259, 1968.
101. **Lee, T. H., McGlasson, W. B., and Edwards, R. A.**, Physiology of disks of irradiated tomato fruit. II. Incorporation of [^{14}C] labelled respiratory intermediates, *Radiat. Bot.*, 11, 137, 1971.
102. **McGlasson, W. B. and Lee, T. H.**, Damage and repair of protein in gamma irradiated tomato fruit, *Radiat. Bot.*, 11, 239, 1971.
103. **Morris, L. L., Herner, R., and Abdel-Kader, A. S.**, Studies of gamma irradiation effects on storage life of cucumbers, in Radiation Technology in Conjunction with Postharvest Procedures as a Means of Extending the Shelf Life of Fruits and Vegetables, Rep. 1963 to 1964, Contract No. AT(11-1)-34, Proj. Agreement No. 80, Division of Isotopes Development, U.S. Atomic Energy Commission, Washington, D.C., 1964, 65.
104. **Maxie, E. C. and Sommer, N. F.**, Changes in some chemical constituents in irradiated fruits and vegetables, *Preservation of Fruit and Vegetables by Radiation*, IAEA, Vienna, 1968, 39.
105. **Maxie, E. C., Sommer, N. F., and Mitchell, F. G.**, Infeasibility of irradiating fresh fruits and vegetables, *HortScience*, 6, 202, 1971.
106. **Anon.**, List of clearances. General survey of irradiated food products cleared for human consumption in different countries, *Food Irradiat. Newsl.*, 3(2), 3, 1979; see also Chapter 1, Volume I of this book.
107. **Anon.**, Six further foods additionally cleared in South Africa, *Food Irradiat. Inf.*, 9, 144, 1978.
108. **van der Linde, H. J.**, South Africa Petitions and Clearances, *Food Irradiat. Inf.*, 9, 66, 1978.
109. **Anon.**, Irradiated mangoes shipped from South Africa to Europe, *Food Irradiat. Inf.*, 8, 67, 1978.
110. **Farkas, J.**, Hungary, Economic feasibility, market trials, consumer acceptance, *Food Irradiat. Inf.*, 6, 27, 1978.
111. **Balock, J. W., Burditt, A. K., Jr., and Christenson, L. D.**, Effects of gamma radiation on various stages of three fruit fly species, *J. Econ. Entomol.*, 56, 42, 1963.
112. **Balock, J. W., Burditt, A. K., Jr., Seo, S. T., and Akamine, E. K.**, Gamma irradiation as a quarantine treatment for Hawaiian fruit flies, *J. Econ. Entomol.*, 59, 202, 1966.
113. **Steiner, L. F.**, Gamma irradiation for disinfestation of export fruits and vegetables, *Hawaii Farm Sci.*, 15, 11, 1966.

Chapter 6

SPROUT INHIBITION IN TUBERS AND BULBS

Akira Matsuyama and Keiji Umeda

TABLE OF CONTENTS

I. INTRODUCTION

One of the most promising applications of ionizing radiations to food preservation is sprout inhibition in potato tubers and onion bulbs because of its obvious technological feasibility at the level of low doses. At present, irradiated potatoes are the only commercial product in the field of food irradiation throughout the world.[1] In the early stages of radiation research, various investigators had already suggested the possible use of ionizing radiation to stimulate plant growth and seed germination at very low doses.[2] In 1936, it was shown by Metlitsky[3]* that vegetable sprouting can be inhibited with X-rays. However, attention to this radiation effect had not been paid until radiation facilities became available industrially.

In a series of studies carried out at Brookhaven National Laboratory to observe reactions of potato tubers and growing potato plants to various doses and kinds of ionizing radiation, Sparrow and Christensen[4] reported in 1950 that a small dose of 4.5 krad of X-rays markedly inhibited the germination of seed potatoes. Several experiments in this series also indicated that the storage quality of irradiated tubers might be considerably better than that of the unirradiated. They proceeded to test the validity of these observations and showed in 1954 that sprouting of white potato tubers (the Katahdin variety) can be inhibited greatly at 5 kR and completely at 20 kR, and that these dosages extend the storage life of the tubers.[5] It should be noted that Sparrow and Schairer[6] suggested a dose range from 8 to 20 kR to be commercially useful for sprout inhibition and weight-loss reduction.

Also, in 1954 Brownell and co-workers[7] found that gamma irradiation reduced the sprouting of the White Pearl and Yellow Dry (Bermuda) onions, although the irradiation effect was not complete in a dose range of 6.5 to 39 krad. Dallyn et al.[8] also indicated that the excellent control of sprouting in onions was achieved by lower doses of gamma rays (about 4 to 8 kR). These initial studies on the inhibitory effect of irradiation on sprouting in tubers and bulbs created considerable interest in the possible use of this unique method for sprout control to extend the storage life of these commodities. So far, many investigations have confirmed the effectiveness of low-dose irradiation for sprout control under different treatment conditions and in various areas of the world. The possibility of the large-scale application of this treatment has long been considered.[9]

* In deference to the authors' wishes, the spelling in this chapter is Metlitsky. The spelling more commonly used in the U.S. is Metlitskii.

Potatoes and onions are consumed in quantity in many countries. Potatoes supply important nutrients — carbohydrates, ascorbic acid, and vitamin B. Onions, garlic, and other similar root crops meet the particular preference of some population groups because of their characteristic flavor and taste. Since these root crops are seasonal, a successful technology to reduce storage loss has been anticipated. This will achieve year-round availability of these commodities, a great advantage for their use at home and in processing plants. In most countries, there are three main reasons for storage losses in these root crops, i.e., (1) sprout growth, (2) shriveling due to moisture losses, and (3) disease or storage rots.[10-12]

As mentioned above, sprouting can be controlled by low-dose irradiation (Figures 1 and 2). Since sprouted tubers lose water more readily than unsprouted ones, the prevention of sprout growth contributes to the reduction of moisture losses. In hot countries, there is a tuber moth problem in addition to sprouting. Both problems can be controlled by low-dose irradiation.[13] Thus, the irradiation treatments of tubers and bulbs facilitate the reduction of losses during storage. On the other hand, the irradiation treatment tends to induce some unfavorable side effects, e.g., enhancement of storage rot and discoloration. Rot incidence is associated with the decrease in the resistance of tubers to phytopathogenic microorganisms caused by the irradiation treatment. Discoloration developed in irradiated potatoes is known to occur through several different processes. The reducing sugar problem in potato storage at low temperature is an obstacle to processing, but it has been shown that this problem can be overcome by reconditioning at higher temperature.

In many countries, extensive studies have been devoted to establishing the most appropriate technological conditions for the irradiation treatment under different agronomic, climatic, and varietal conditions. At present, the irradiation technology of sprout inhibition in potato tubers and onion bulbs has been well established. In some countries, research and development programs have proceeded to the next step, i.e., pilot-plant scale studies, as has begun in the U.S.S.R., and consumer acceptance tests by volunteers or through marketing trials. Although the world's first commercial irradiation plant, constructed in Canada, has been shut down after the treatment of 8 million lb of potatoes, a full commercial-scale potato irradiator has been in operation in Shihoro, Hokkaido, Japan, since the end of 1973. More regarding this plant will be described later in this chapter.[1,14] Similarly to the irradiation treatments of potatoes and onions, the radiation technology for the effective control of sprouting or rooting with low doses in potatoes and onions can be extended to other root crops such as carrots, shallots, garlic, yams, sweet potatoes, gingers, sugar beets, and chestnuts.

II. FUNDAMENTALS OF RADIATION INHIBITION OF SPROUTING

A. Morphological and Histological Observations

Growing points (meristematic tissues) in buds of freshly harvested potato tubers or onion bulbs are in the dormant state. At the end of the dormancy period, cell division and enlargement in growing points take place.[15] Even during the dormancy period, some growth of buds is observed under the microscope. In response to the growth at growing points, doses required for sprout inhibition progressively increase, although in the case of potato tubers the increase is not remarkable. In freshly harvested potato tubers, each of several tiny dormant buds in an eye had young leaves over its growing point. During the course of the dormancy-period storage, the number of leaves over the growing point increases to some extent resulting in a domeshaped bud. Internal growth of buds in onion bulbs also takes place gradually during storage.

When tubers and bulbs are exposed to radiation, remarkable morphological and histological changes in dormant buds are induced to develop the deformed buds and necrosis at growing points during subsequent storage. For example, in the case of the

FIGURE 1. Katahdin potatoes stored for 5½ months at 55°F after treatment with varying amounts of Cobalt 60 radiation. Top left — "controls" (no radiation); bottom right — optimum radiation treatment. (Courtesy of Atomic Energy of Canada, Ltd.)

FIGURE 2. Autumn spice onions stored for 8 months at 50°F after treatment with various amounts of Cobalt 60 radiation. Left — "controls" (no radiation); center and right — radiation treated. (Courtesy of Atomic Energy of Canada, Ltd.)

Danshaku, Shimabara, and Norin-ichigo varieties of potatoes,[16] necrosis in the apical region of growing points — namely, tunica, corpus, procambium, and other tissues — was observed 1 month after irradiation with a comparatively higher dose of 15 or 30 krad. Cells adjacent to the necrotic area were enlarged and their walls were partially lignified and suberized. Samples irradiated with 7 krad developed necrosis only in one or two cells in the secondary layer of tunica. The extent and area of radiation-induced necrosis in growing points and their adjacent cells varied with irradiation dosages. During postirradiation storage, swelling and elongation of young leaves at growing point developed in the buds of tubers, forming the deformed bud shape specific to irradiated tubers (Figure 3).

After the dormancy period, unirradiated tubers develop normal buds with growing points covered by young scaly leaves, which can grow to the stem with an adventitious root.[17] On the other hand, irradiated tubers develop the deformed buds with necrosis in the growing points; these are surrounded by young swollen and abnormal leaves and can not form the stem and root.[18,19] These deformed buds are doomed to failure with regard to further growth.

With unirradiated onion bulbs of the Senshuki and Sapporoki varieties, the scaly leaves had elongated and increased in number during storage; about 1 month after irradiation with 3 to 15 krad, a few cells were observed to be necrotized in the apical region and tissue of shortened stem. Walls of necrotic cells were partially lignified and suberized.[20] The color of inner buds of irradiated bulbs became brownish depending upon doses leading to death.[21] In some examples, irradiated bulbs sprouted and greenish foliage leaves elongated, but the growth of such sprouts was restricted or abnormal. These sprouts have no necrotic zone in the stem of irradiated bulbs. This may be due to the abnormal growth of foliage leaves already developed at the time of irradiation.[20] Histochemical studies have shown that the growth of sprouts and roots in irradiated bulbs is not due to the normal increase through cell division but to an increase in tissue resulting from cell enlargement.[15]

These morphological and histological observations indicate that the sprout-inhibiting effect of radiation is associated with inhibition of cell division and an aberrant enlargement at growing points, resulting in the formation of deformed buds or browning of inner buds, accompanied by necrosis especially at higher doses. These abnormal forms of buds do not undergo the normal development stages of sprouting.

B. Biochemical Basis

Several possible mechanisms have been described for the radiation-induced sprout inhibition in tubers and bulbs, e.g., the inhibition of nucleic acid and nucleotide synthesis, a derangement in the synthesis of auxins, and chromosomal aberration in cells of the meristematic tissues.

The biochemical mechanisms for the emergence from the dormant state of tubers and bulbs and the sprout inhibition by irradiation have been extensively investigated by Russian scientists.[15,22,23] In these studies, it has been shown that cell division at the time of emergence from the dormant state takes place only after the nucleic acids are accumulated in the growing points to a definite level of content and composition. The contents of both RNA and DNA in the growing points were found to be higher in tubers and bulbs of the short dormancy period and to increase in those root crops with the long period dormancy shortly before their transition into active growth. Furthermore, the composition and the extent of polymerization of nucleic acids appeared essential for the initiation of cell division. The increase in nucleic acid contents is accompanied by changes in the isoelectric point of proteins and nucleoproteins in the growing points of potato tubers. The isoelectric point of the meristematic tissues becomes lower after the emergence from the dormant state, suggesting the occurrence of some depolymerization in nucleic acids.

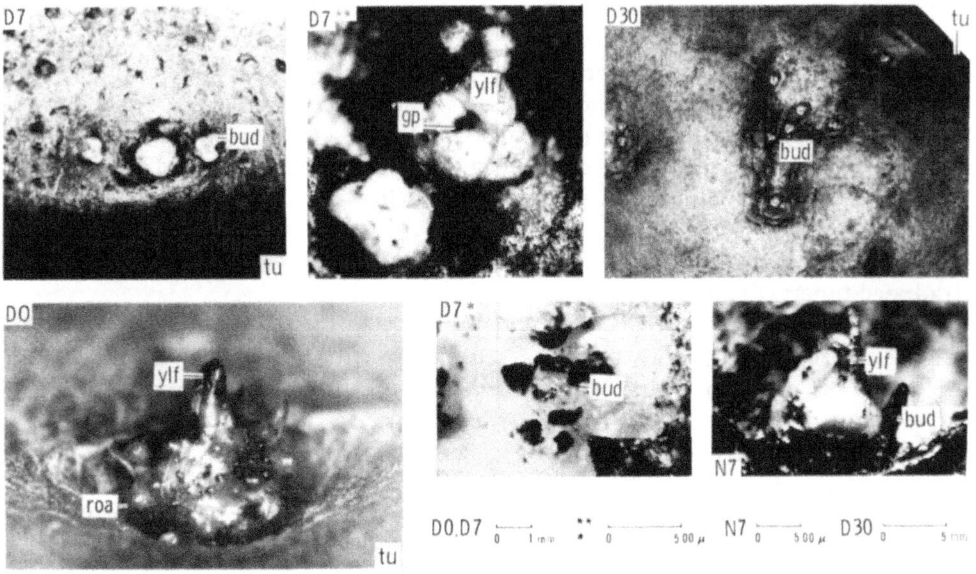

A

FIGURE 3. (A) Sprout buds of potato tubers, "Danshaku" (D) and "Norin ichigo" (N) varieties. Abbreviations: O, unirradiated control; 7 and 30, dormant potato tubers gamma irradiated with 7 and 30 krad, respectively.; bud, deformed bud; gp, growing point; roa, adventitious root; tu, tuber; ylf, young leaf. Each photograph is titled by the combination of variety name (D) or (N) and doses (0), (7), and (30). DO — Normal bud of control tuber. At the end of the resting period the bud was small with only young leaves, but after several days adventitious roots appeared. D7 and D30 — Deformed buds of irradiated tubers. These buds were developed after the period corresponding to the resting period in the control. The growing points of deformed buds were exposed and surrounded by white swollen deformed young leaves which sometimes turned scaly and brown as shown in D7. N7 — Deformed bud of irradiated tuber. The growing point was loosely covered with partially swollen deformed young leaves. The deformed buds of irradiated tuber showed some differences in external appearance, depending upon the variety and time of irradiation. All the growing points caused necrosis and adventitious roots did not appear. No buds grew farther than this state and so these buds could not be observed clearly with irradiated tubers, as shown in D30. (B) Longitudinal sections through the growing point of potato buds, "Shimabara" (S) variety. Abbreviations: O, unirradiated control; 7, 15, and 30, dormant potato tubers gamma irradiated with 7, 15, and 30 krad, respectively; axb, axillary bud; cop, corpus; gp, growing point; lfp, leaf primordium; m, mark; nca, cells adjacent to necrosis; ncr, necrosis; pro, procambium; roa, adventitious root; tun, tunica; vb, vascular bundle; ylf, young leaf. Each photograph is titled in the same way as in Part A. SO — The growing point was covered with young leaves. The tunica was composed of two layers and the corpus was composed of five layers and both were very clear. S7 — Several cells in the second layer of tunica caused necrosis. The bud did not grow farther than the stage of division of leaf primordium. S15 — Necrosis appeared in tunica, corpus, and procambium and their adjacent cells enlarged. S30 — The necrosis spread out more widely than in S15. (The authors thank Dr. H. Shimomura for photographs and legends.)

At the end of the dormancy state there is an accumulation of energy-rich phosphorus compounds such as di- and tri-phosphate nucleotides, especially ATP in the meristematic tissues. These special phosphates play an important role in nucleic acid synthesis and other energy-requiring reactions involved in cell division and enlargement. It is likely that the transition from the dormant state to active growth may begin by the stimulation of nucleotide phosphate synthesis, followed by the accumulation of the quantities and qualities of nucleic acids necessary for sprouting.

Irradiation higher than 5 krad resulted in decreased synthesis of nucleic acids, changes in their composition, and induction of their degradation in the meristematic tissues. The formation of nucleotide di- or tri-phosphates is also suppressed by irradia-

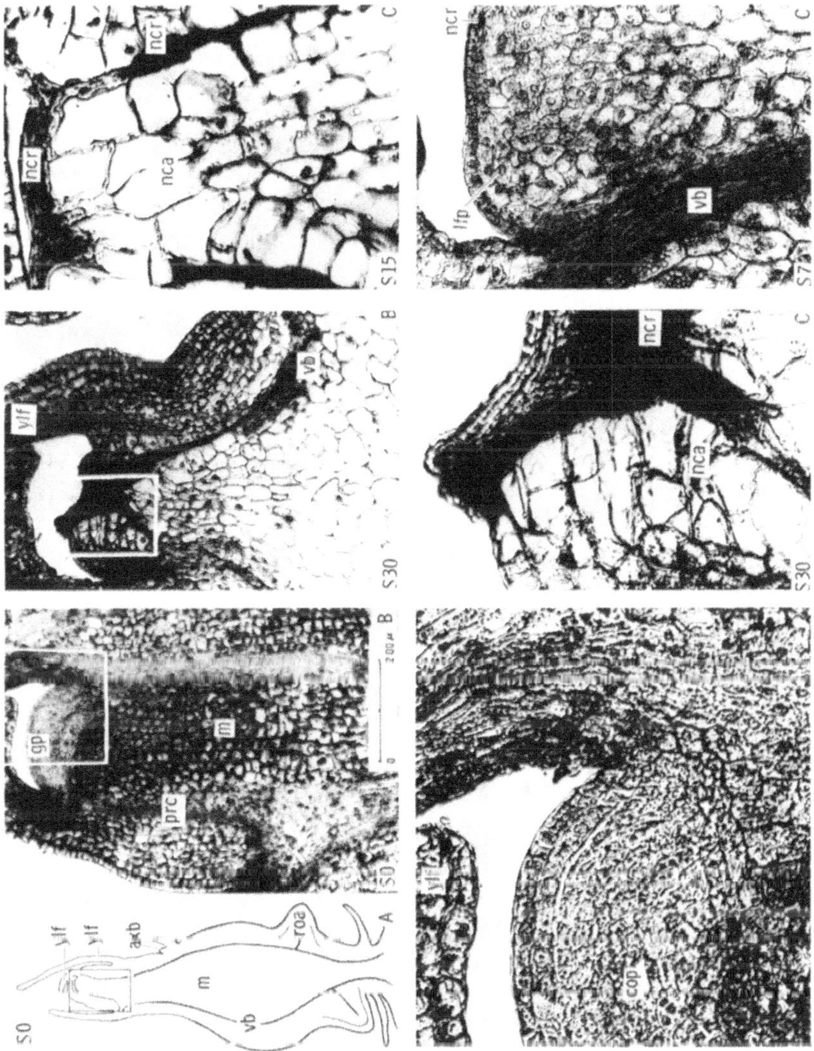
S0
ylf
ylf
axb
m
vb
roa
A
gp
prc
m
B
ylf
vb
S30
B
ncr
nca
ncr
S15
C
ylf
cop
ncr
nca
S30
C
lfp
vb
ncr
C

FIGURE 3B.

tion. Inhibition of ATP synthesis by irradiation disrupts the chain between the respiration and phosphorylation processes. Jaarma[24] has suggested that the phosphorylating mechanism may be disturbed by irradiation, since adenosine triphosphatase is very sensitive to radiation. Subsequently, most energies are dispersed and cannot be used for syntheses of nucleic acids and other substances necessary for active growth. In addition, nucleic acid synthesis is generally very sensitive to radiation. Thus, according to Metlitsky and his co-workers,[15,23] the inhibition of nucleotide and nucleic acid synthesis caused by irradiation in meristematic tissues may be the main basis of sprout inhibition in tubers and bulbs at comparatively low doses.

Next, because of the high sensitivity of the auxin synthesizing system to radiation,[25-27]the mechanism for radiation-induced dormancy has been studied in relation to the metabolism of plant growth regulatory substances. The main action of auxins in promoting the growth processes is the stimulation of the cell enlargement processes. During the dormancy period, auxins are either absent in the growing points or combined in an inactive form. Therefore, the initiation of the growth processes may require an increase in the contents of auxins or their transformation into the active state.

Mathur[28] reported that radiation-induced dormancy of potato tubers was reversed by the postirradiation treatment with 250 and 500 ppm gibberellic acid (GA). After this treatment, new buds were found adjacent to those degenerated by irradiation. Growth regulators such as indole acetic acid (IAA) and GA were found to reverse radiation-induced dormancy in potatoes. IAA was most effective at 20 ppm and within 6 hr of irradiation.[29] At 1000 ppm IAA did not reverse radiation-induced dormancy, but extended the dormancy of tubers by suppressing their susceptibility to decay by irradiation.[30] It is notable that the optimum concentration (20 ppm) for the reversal of the radiation-induced dormancy cannot accelerate sprouting in unirradiated tubers. In contrast, the optimum concentration of GA for the complete reversal of sprout inhibition is much higher than that of IAA. These facts show that the mode of action of IAA in the reversal of sprout inhibition is distinctly different from that of GA. Irradiation with 10 krad completely destroys IAA in potato tubers and interferes with the synthesis of the IAA synthesizing system. The results on the radiation effect on the status of IAA synthesizing system suggest that an exogeneous supply of IAA may trigger the IAA synthesizing system. There are some evidences for the involvement of IAA in the synthesis of the IAA-synthesizing system. Since delayed treatment after 12 hr or a longer period cannot protect the enzyme activity of the IAA synthesizing system from depletion, the genetic process for the synthesis of this enzyme may be permanently repressed shortly after irradiation. Thus, it is likely that synthesis of a key enzyme involved in the IAA formation may be inhibited by irradiation.[13,29] From these results it is plausible that one of the causes for radiation-induced growth inhibition is a derangement in the synthesis of auxins.[29]

Kuzin et al.[31] have proposed the radiotoxin hypothesis which suppresses the essential role of oxidized metabolites (radiotoxins) formed under the influence of radiation to be radiation active on the cells. Irradiation may induce the activation of enzymic oxidation of polyphenols to form orthoquinones which are actively absorbed by cell nuclei to inhibit DNA synthesis and cell division and to induce the chromosomal aberration. They reported the inhibitory effect of extract from irradiated potato tubers on DNA synthesis[31] and their mutagenic action in dominant lethal assay in mice.[32] However, the toxic effect from irradiated tubers has not been reproduced in trace experiments.[33,34,217-219] The chromosomal aberration caused primarily by radiation in cells of meristematic tissue might prevent the normal cell division and development stages of growing points. Thus, it would be one of the possible causes of the radiation-induced dormant state and this view appears to be supported by morphological observations of abnormal buds developed in irradiated tubers and bulbs.

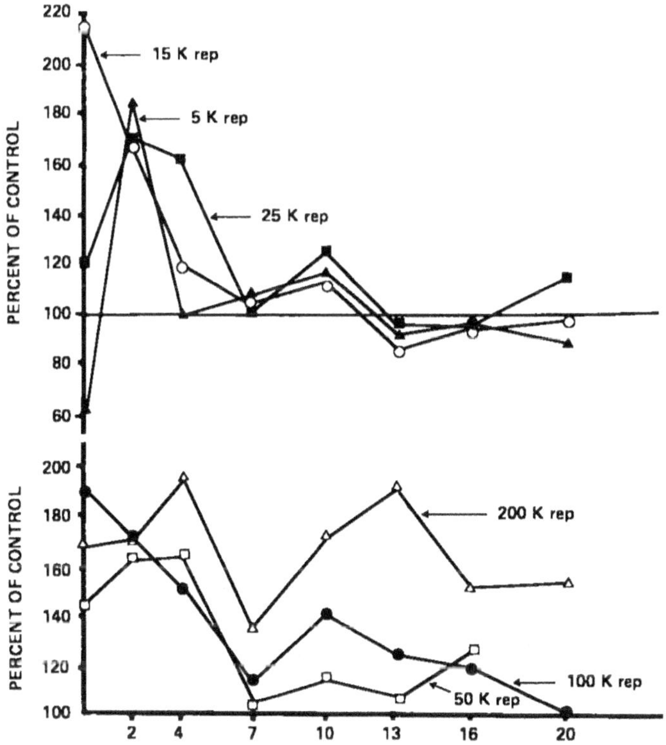

FIGURE 4. Carbon dioxide production by whole tubers (Sebago), presented as percent of the control tubers. Irradiation dosages given in kilorep. (From Gustafson, F. G., Brownell, L. E., and Martens, R. A., *Am. Potato J.*, 34, 177, 1957. With permission.)

III. CHANGES IN METABOLISM AND CHEMICAL CONSTITUENTS BY IRRADIATION

It should be noted that since tubers and bulbs are living tissues and retain their metabolic activities, the effects of irradiation are displayed not only immediately after the treatment but also during the course of postirradiation storage.

A. Respiration

In a dose range of 5 to 15 krep, Gustafson et al.[35] observed an immediate spurt in respiration of potato tubers, both CO_2 production and oxygen uptake, a few hours after irradiation. Following the initial rise, respiration decreases during the first 1 or 2 weeks after irradiation and then increases again (Figure 4). Both activities of CO_2 production and oxygen uptake gradually decrease down to the control level in about 10 weeks, although after the higher doses they remain high.[35] On the other hand, Ussuf and Nair[36] indicated that CO_2 production reaches the maximum at about 24 hr after irradiation without any decrease during the early stage of poststorage. The respiratory quotient was 1 with all doses of 5 to 500 krad.

The increase in respiration for several weeks after irradiation corresponded to the increase in sugar content. Since the process of sugar metabolism in tubers is considered as starch$\rightleftharpoons$sugar$\rightarrow CO_2$, the enhanced respiration may result from the sucrose accumulation.[37] The similar increase in respiration of the whole tubers or tissue slices caused by irradiation has been reported by several authors.[21,22,36,38-42] Rubin and Metlitsky[22] studied respiration of the whole tubers and individual tissues of irradiated potatoes in detail. The respiration of buds decreased proportionally with increasing doses. The decrease in respiration of bud tissues was 10% at 10 kR and 40% at 150 kR as compared with the unirradiated control. On the other hand, the respiration of pulp tissues was somewhat more intensified in the early stage after irradiation and then returns to the control level after a certain period. The respiration of the whole tubers irradiated with 10 kR is stimulated to a certain extent during the first few days after irradiation, but it soon decreases down to the control level. The temperature coefficient of respiration for buds and pulp tissues remained almost unchanged.

Similarly to potato tubers, the respiration of onion bulbs increases soon after irradiation and then begins to decrease after 1 day.[43] By means of the inhibitor experiment, Chachin and Ogata[44] suggested the alternation of the resporatory chain in potatoes may be caused by irradiation.

B. Oxidizing Enzymes and Other Enzyme Activities

Irradiation did not cause any appreciable changes in the activities of oxidizing enzymes such as polyphenoloxidase and cytochromoxidase in crushed bud or pulp tissues.[22] The activity of polyphenoloxidase and a peroxidase in the mitochondria of the buds decreased, as did the cytochromoxidase, particularly in the mitochondria, extracted from them. The increased sensitivity of the oxidizing enzymes in the mitochondria appeared to be due to changes in their structure caused by irradiation. The phospholipid and nucleic acid contents of the tuber pulp were not significantly affected by irradiation, while in the bud tissue irradiation caused an appreciable decrease in the nucleic acid content and a remarkable increase in the phospholipid content, especially in mitochondria. Thus, the radiation-induced suppression of bud-tissue respiration is connected with a slackening of the activity of oxidizing enzymes in the individual structural elements of the cell and especially mitochondria.

The increase in pulp-tissue respiration immediately after irradiation appeared to be due to the increase in starch hydrolysis caused by the radiation and the concomitant increase in the contents of the respiratory materials in tubers.[22] Irradiation enhanced the accumulation of sucrose during storage at 4°C and starch phosphorylase was activated in irradiated tubers stored at 21°C.[45] Since the increase in respiration observed with potatoes irradiated at 10 krad can be due to the mobilization of reducing sugars from storage starch, Ussuf and Nair[36] investigated the starch phosphorylase in irradiated potatoes. They showed more than a 25% increase in starch phosphorylase activity within 2 hr after irradiation which persisted even at 24 hr. Rubin and Metlitsky[22] mentioned that the observed increase in the sugar content of irradiated tubers is due to the considerable increase in the phosphoryl enzyme activity, thus, immediately after irradiation the balance between the processes of the decomposition and synthesis of starch declined toward the side of increased decomposition of the starch for which phosphorylases are responsible. At the same time the phosphoglucomutase activity is intensified. Thus, increased utilization of glucose-1-phosphate is facilitated either via the glycolytic or pentose phosphate pathway.[36]

As a result of irradiation the activity of apyrase (ATP diphosphohydrolase) in tubers increased to about three fold of the control level and decreased 75 to 80% in the activity of hexokinase which is responsible for the process of phosphorylation of sugars. These findings would indirectly indicate the disruption of the respiration-phosphorylation sequence that supplies the energy required for sprouting.[15] With both King Edward and Bintje varieties of potatoes,[46] the activity of β-fructofuranosidase was not observed in freshly harvested control tubers, while it markedly increased after storage of 1 year. In contrast, transferase activity decreased. In irradiated tubers, the invertase activity did not increase, but the transferase activity did increase. The sucrose synthesizing enzyme and the competitive sucrose hydrolyzing enzyme in tubers are respectively influenced in opposite directions by prolonged storage as well as by irradiation.

As mentioned above, irradiation does not affect the activities of oxidizing enzymes including polyphenol-oxidase in both bud and pulp tissues of potato tubers. Ogawa et al.[47] studied changes caused by irradiation in the content of polyphenols and activities of oxidizing enzymes in response to cutting tubers. After irradiation with doses of 5 to 50 krad and the subsequent storage for 10 days or 4 months at room temperature, the content of polyphenols such as chlorogenic acid and isochlorogenic acid increased linearly in the irradiated whole tubers with increasing doses up to 20 krad, while the increase in polyphenol contents induced by cutting tubers into disks was lowered progressively with increasing doses.

Irradiation suppressed the increase in the activities of oxidizing enzymes such as o-diphenol oxidase, peroxidase, and phenylalanine ammonia-lysase (PAL) induced by cutting into tuber disks.[47] The activity of PAL, which is involved in the biosynthetic pathway of polyphenols, was hardly detectable in either irradiated or unirradiated tubers.[47] Since an increase in asparagine and a considerable decrease in phenylalanine were observed within 24 hr after irradiation,[36] the effect of gamma irradiation on PAL activity in potato tubers was investigated.[48] It was found that the irradiation of potato tubers caused the induction of PAL immediately after the treatment and the induced activity was localized in the cortex tissue, mainly concentrated around the bud region with no induction in parenchymatous tissue.[48]

A comparative study of the biochemical activity of potatoes and sweet potatoes showed that the oxygen uptake, polyphenol content, and peroxidase activity were less sensitive to radiation in sweet potatoes than in potatoes.[40] The content of polyphenols and oxidizing enzymes in irradiated tubers was investigated in relation to the browning in raw state.[49-50] Berset and Sandret[51] observed a persistent increase in total polyphenols and chlorogenic acid contents by irradiation and a direct correlation was found between this increase and the after-cooking darkening of tissues (prefried disks).

C. Nucleic Acid and Nucleotide Metabolism

In connection with the biochemical mechanism for sprout inhibition in tubers and bulbs by irradiation, the nucleic acid metabolism in irradiated tubers and bulbs, especially at their growing points, was extensively studied at the Bach Institute of Biochemistry.[15,22,23] As the potato tubers and onion bulbs emerge from the dormant state and proceed to active growth, the synthesis of DNA and RNA in their buds is intensified. If irradiation is applied with sprout-inhibiting doses, the anatomical structure of the buds is drastically changed and the content and distribution of the nucleic acids in the various tissues is altered, i.e., retardation of the synthesis, changes in the composition, and induction of degradation of nucleic acids.

Irradiation does not basically change the nucleic acid content of the parenchyma, but it decreases in the meristem after irradiation. The synthesis of DNA at the growing points and its conversion to RNA are inhibited by irradiation, eventually resulting in the delay of the cell division. However, these changes can be found only some time after irradiation. Immediately after irradiation, a reduction in the basophily of the cytoplasm, i.e., a reduction in the RNA content, can be detected only by cytochemical methods whereas the DNA content remains unchanged. As the storage period after irradiation gets longer, differences in both DNA and RNA contents between the irradiated and unirradiated tubers become increasingly marked. The nucleic acid synthesis continues in the growing points of irradiated tubers at a much lower level than it does in the control. At this stage, retardation of the nucleic acid synthesis can be detected also by the macro-method.

In tuber buds, the nucleic acid contents slightly increased during the first few months after irradiation and then they showed a fairly remarkable decrease after 6 to 7 months of storage. Changes in the nucleic acid occur as well in the pulp of the tubers according to a similar but less intense course. The decrease in the synthesis and contents of nucleic acids caused by irradiation is directly related to the increase in the ribonuclease activity, which is maintained for 4 to 5 months. Irradiation changes the structure of nucleoproteins and their distribution in the growing points, as shown by a decrease in the amount of bases in nucleic acids and of mononucleotides. Irradiation of tubers with 10 krad displaced the isoelectric point of proteins and nucleoproteins in the eye tissues of potato tubers to the acid side. This is an indication of the onset of structural changes in the nucleoproteins, which may be one of the initial effects of radiation.

Chachin et al.[52,53] have obtained similar results with regard to nucleic acid metabolism in the growing points of potato tubers and onion bulbs. The content of sRNA and rRNA in bud disks (buds including cortex tissues) of the unirradiated tubers increased during a period from the dormant to the breaking period; those of tubers irradiated with 5 and 10 krad did not increase. The incorporation rates of ^{32}P into sRNA, rRNA, and DNA in bud disks of irradiated tubers were higher to some extent on the 5th day after irradiation than those of the control. The incorporation rates of ^{32}P into sRNA and mRNA in the 10 krad bud disks decreased between the breaking and sprouting period in contrast to the increase in the control bud disks; those in DNA and rRNA were somewhat higher in irradiated bud disks than in the control.

In inner buds of unirradiated onion bulbs the comparatively high contents of DNA, rRNA, and sRNA were also observed 3 weeks after harvest, at which time some growth of the inner buds took place. On the other hand, the contents of these nucleic acids in inner buds of irradiated onion bulbs decreased 2 months after the treatment to about a half of those of unirradiated bulbs.

D. Carbohydrates

Irradiation of potatoes with sprout-inhibiting doses does not change the carbohydrate content.[54] Varietal differences only were significant for starch content in a dose range up to 15 krep throughout the storage period of 9 months at 40°F (4.5°C) and 68°F (20°C). It appeared that the storage temperature had a much greater effect on total sugars, reducing sugars, and starch contents than the effect of irradiation with doses up to 41 krad.[55]

Although the conversion of reducing sugars to starch took place by reconditioning, there was no significant effect on the starch content of potatoes owing to reconditioning.[10] Many studies showed that irradiation induces the increase in total and reducing sugar contents.[22,36] As mentioned before, the decomposition of starch into sugars is due to the radiation-induced increase in phosphorylase activities. Increase in sugar content is disadvantageous, especially for processing, because of sweetening on cooking and discoloration of crisps. At storage temperatures below 5°C, starch in tubers is converted into sugars whereas at storage temperatures above 10°C, sugars are synthesized to starch. Reconditioning at 15 to 28°C for 1 to 3 weeks before processing decreases reducing sugar content down to the level near the unirradiated control, about 3 to 5 mg/g or less, but the starch content is not affected by reconditioning.[14,56,57]

Immediately after irradiation, potato tubers display a transient increase in the contents of both reducing and nonreducing sugars. Sparrow and Christensen[5] observed a significant degree of sweetness in irradiated potatoes. Burton et al.[58] indicated that with potatoes of the King Edward variety irradiated by 10 krad and stored at 10°C, (Figure 5) the content of nonreducing sugar increased rapidly, up to a maximum of rather more than 1 g/100 g, reached after about 5 days, followed by a rapid decrease down to the control level about 26 days after irradiation. This temporary rise in nonreducing sugar content immediately after irradiation was independent of the date of irradiation, but dependent upon the radiation dose. Irradiation also induced a temporary rise in the content of reducing sugar irrespective of the dose, but the later the date

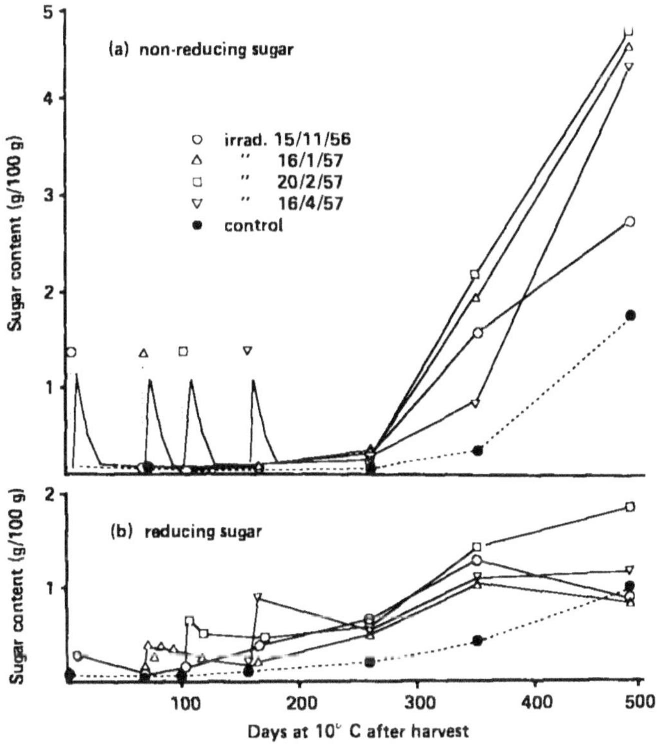

FIGURE 5. Changes in the sugar content of potatoes irradiated with 10,000 rad and stored at 10°C. (From Burton, W. G., Horne, T., and Powell, D. B., *Eur. Potato J.*, 2, 105, 1959. With permission.)

of irradiation the greater was the increase of reducing sugar. On the other hand, unirradiated tubers increased sugar contents after the longer period of storage at 10°C in comparison with irradiated tubers. A higher storage temperature (20°C) resulted in a marked acceleration in the rate of final breakdown of potatoes irradiated at 10 krad.

Cloutier et al.[59] observed with the Katahdin and Netted Gem varieties that contents of reducing sugars and sucrose in irradiated tubers showed an initial rise, followed by a decrease down to the control level after storage of 4.5 months at 40°F (4.5°C), and that the sucrose content in the treated tubers largely increased after 9 months of storage at 68°F (20°C). No significant difference in sugar content was observed between the three doses of 7.9, 10.2, and 14.0 krad. Ogata[21,60] mentioned that the reducing sugar content in tubers increased after irradiation and decreased to that in untreated tubers while there was no change in sugar content of onions by irradiation. Schwimmer et al.[61] observed that the increase in the sucrose content after irradiation continued longer during storage at low temperature while it was rather short at high temperature. With the Bintje variety, the sucrose content increased rapidly during the first few days following irradiation with 8 krad and afterward decreased to the control level over a period of 1 to 3 months after irradiation. After this period, the sugar content was lower in the irradiated tubers than in the unirradiated ones and only minor changes took place in the glucose and fructose contents.[37]

Rubin and Metlitsky[22] also found an increase in the sugar content immediately after irradiation in association with changes in the phosphoryl enzyme activity. Immediately after irradiation, the balance between the degradation and the synthesis of starch, which occur with the help of the phosphorylases, inclines toward the side of increased degradation of the starch. The enhanced decomposition of starch in unirradiated tubers takes place at the end of the dormancy period and at the beginning of sprouting, while it does not occur in irradiated tubers.[15] The α-amylase activity in both cortex and parenchymatous tissues of potato tubers, which were irradiated 1 month or 3 months after harvest, was not significantly affected by irradiation with doses of 10 to 50 krad.[49]

With onions, no significant difference in the total sugar and reducing sugar contents was found between irradiated and unirradiated bulbs in a dose range of 3 to 12 krad[62] or 8 to 16 krep.[63]

E. Amino Acids and Proteins

Some changes were found in the concentration of free amino acids in potato tubers without any alterations in the amino acids consisting of the protein.[64] The result obtained by a French group indicated that the rate of change in amino acid composition of potato protein was very slow and practically of the same order as that observed during boiling.[65] With the Danshaku variety, after 15 days of storage following irradiation (7 to 30 krad) certain free amino acids altered significantly, i.e., the concentrations of proline, aspartic acid, and other aliphatic amino acids increased with increasing dose, while those of the basic amino acids and glutamic acid decreased. A study[36] of the free amino acids in the Up-to-date variety showed an increase in aspartic acid, asparagine, threonine, serine, alanine, isoleucine, leucine, lysine, and arginine 24 hr after irradiation and a decrease in glutamic acid, proline, methionine, and phenylalanine. However, the amino acid content of potatoes stored for 105 days after irradiation was similar to that of unirradiated and nonstored potatoes.[66]

These results indicate that changes in the amino acid content of potato tubers by irradiation appeared to be recovered during storage by the biological activity of tubers. No significant difference in the concentration of protein-bound amino acids was found between unirradiated and irradiated (15 krad) tubers.[66] Changes in biological value and net protein utilization of potatoes were not induced by irradiation.[67]

F. Vitamins

Irradiation of potato tubers caused some slight loss of ascorbic acid not exceeding 15% at sprout-inhibiting doses, and since its loss during subsequent storage was comparatively large, no significant difference in ascorbic acid content between unirradiated and irradiated potatoes was found after prolonged storage.[21,45,64,68-71] Pätzold and coworkers described that losses of ascorbic acid as well as dry matter and starch after prolonged storage were less in irradiated tubers than in the unirradiated control owing to germination inhibition.[72,73]

Rubin and Metlitsky[22] observed irreversible oxidation and decomposition of a considerable part of the ascorbic acid in tubers during the first 2 or 3 hr following irradiation. After the initial loss of ascorbic acid and starch, differences between irradiated (10 krad) and unirradiated tubers in the content of ascorbic acid and starch diminished, and the contents of the control fell more rapidly and to a greater extent than those of irradiated tubers as the control tubers emerged from the dormancy state. Thus, eventually, there resulted almost no difference after 8 months of storage between unirradiated and irradiated tubers in the contents of ascorbic acid and starch contents, e.g., 7.0 mg% for unirradiated tubers and 6.8 mg% for irradiated tubers in the ascorbic acid content. The initial loss of ascorbic acid was followed by a buildup of this substance with the onset of warmer temperatures during storage of tubers.[55] A French research group[65] observed that the ascorbic acid content in tubers was significantly reduced by irradiation, but after boiling it was rather higher in irradiated tubers than in the unirradiated control. Since this can be interpreted as the result of liberation of the enol group by boiling, they suggested that one portion of ascorbic acid in tubers may be protected from the action of radiation.

Thiamine, riboflavin, and niacin contents in tubers were unchanged by irradiation,[64,74,75] although the increase in riboflavin content with increasing dose was reported by some authors.[75] The loss of vitamin E by irradiation is higher, but does not exceed the loss due to the common thermal treatment.[76]

With onion bulbs, the ascorbic acid content appeared not to be significantly affected by irradiation.[11,62] Chachin et al.[77] noted that the reduction of effectiveness for sprout inhibition by delayed irradiation of bulbs may result from the increase in ascorbic acid and other reducing compounds which are in the inner buds at the time of emergence from the dormancy state. Ascorbic acid content together with reducing and nonreducing sugar content and amino acid composition of the onion flesh and inner buds was investigated by Mahmoud et al.[78] in the Hungarian program at the final stage of storage of 8 to 10 months after harvest. The results indicated that compositional differences found between irradiated and unirradiated onions appear to be of no nutritional significance.

G. Greening and Solanine Formation

Potato tubers stored at cold temperature (0 to 4°C) under dim light were found to change their skin color into green.[61,79,80] This greening of tubers is due to chlorophyll formation;[81] the bitter principle in green potatoes is the toxic alkaloid, solanine.[82] It has been recently shown that chlorophyll synthesis and photosynthetic activity of chloroplasts are an essential prerequisite for solanine synthesis; some products of photosynthesis may be required for solanine synthesis and some products of photosynthesis may be required for the formation of the solanidine ring of solanine.[83] Under storage at low temperature in continuous dim light, total chlorophyll content increased in the course of storage.

The chlorophyll a content exceeded the chlorophyll b content in the initial stages, but decrease in chlorophyll a and increases in chlorophyll b were found after 4 weeks of storage.[80] Evidence for the direct conversion of chlorophyll a to chlorophyll b was obtained by isolating a cell-free system capable of this reaction in the dark.[80]

Irradiation treatment with sprout-inhibiting doses inhibits the formation of chlorphyll in illuminated tubers progressively with increasing dose.[10,61] The inhibition of chlorphyll formation results in the suppression of solanine synthesis.[84] Difference in the chlorophyll formation between irradiated and unirradiated tubers was greater when illumination was made soon after gamma irradiation.[79] The greening of tubers under light is of concern from the viewpoint of practical marketing.[85] Retail inspectors from New Brunswick, Canada, observed that irradiated potatoes appeared to be more resistant to greening under light than unirradiated potatoes when on display at retail stores. This observation was ascertained by Canadian experiments. For example, both irradiated and unirradiated potatoes were placed in consumer polyethylene packages and exposed to both daylight and direct sunlight in a window. After about 8 days, potatoes that received the irradiation treatment revealed no evidence of greening due to light exposure, whereas unirradiated tubers showed 60% of the stock with an average of 50% light greening.[85] It is noteworthy that tubers irradiated with the lower doses retained their residual ability for cholorphyll formation upon illumination even after a period of dark storage, during which the unirradiated tubers lost significantly their ability to develop chlorophyll.[79]

H. Growth Substances

Brownell et al.[39] indicated that the content of potato hormones was too low to detect the effect of irradiation, but irradiation appeared to increase the hormone content and decrease inhibitor concentration. In a study using the wheat coleoptile test, Trulsen[37] observed no effect of a sprout-inhibition dose on the concentrations of inhibitor in potato tubers and mentioned that radiation-induced sprout inhibition may be considered as an extension of the natural rest period.

The amount of free IAA present in unirradiated resting potatoes was of the order of 0.7 to 1 μg/kg peelings, whereas no detectable amount of IAA was found in peelings from irradiated potatoes. This indicates that the complete destruction of free IAA takes place in potato tubers during storage after irradiation.[29,86] It was observed that the radiation-induced dormancy of potato tubers is reversed by the treatments with IAA and GA after irradiation.[29,86] Only the treatment with 20 ppm IAA 6 hr after irradiation was effective for restoring the enzyme activity of the IAA-synthesizing system and the later treatments were ineffective.[29,87] These results suggest that IAA exogenously supplied may somehow trigger the mechanism for synthesis of the IAA-synthesizing system which has been inhibited by radiation. The ineffectiveness of delayed treatment may mean that the genetic mechanisms for the synthesis of the IAA-synthesizing system is permanently repressed within less than 6 hr after irradiation.[29]

With onion bulbs, inner buds of irradiated onions contained both growth promotors and inhibitors to the same degree as the unirradiated ones 5 days after irradiation[21] and the disks contained growth promotors at the lower content in irradiated onions than in unirradiated ones during the post-irradiation storage.[60] Ogata et al.[21] observed that irradiation does not cause any important alteration in the growth promoting and inhibiting substances in the inner buds of onion bulbs.

I. Volatile Compounds

Both immediately after irradiation with 11 krad and after postirradiation storage for 50 days, no significant difference in total and individual amount of volatile compounds on cooking were found between unirradiated and previously irradiated potatoes by means of gas chromatography and thin layer chromatography. Ethanol, propanal, acrolein, acetone, isobutanol, butanal, isopentanal, hydrogen sulfide, and methyl mercaptan were identified among the volatile compounds from cooked potatoes. Irradiation with 100 krad resulted in the increase in volatile compounds, especially carbonyl compounds.[88]

No significant difference between the control and irradiated onions was found by an electrochemical technique and gas chromatography;[63,89] no effect of irradiation on volatiles was observed with samples tested for dehydration quality.[63] The volatile constituents of onion, displaying the characteristic odor, consist mainly of various sulfur-containing compounds developed by an enzymatic degradation process with alkyl-L-cysteine sulfoxides as precursors. Of the sulfur compounds the major pungent flavor substances are known to be di-*n*-propyl disulfide and propenyl propyl disulfide, while the lachrymatory factor, 1-propenyl sulfenic acid, which occurs spontaneously on cutting of onion bulbs, decomposes rapidly to propionaldehyde.

As for the changes in the volatile flavor components of onion bulbs induced by irradiation, the lyase activity and formation of di-*n*-propyl disulfide were apparently decreased by irradiation with relatively lower doses.[90] The lachrymatory factor and pungency of onions were not significantly influenced by lower doses but decreased markedly at dose levels of 70 to 700 krad.[91] According to Bandyopadhyay et al.,[92] there is no effect of irradiation up to 50 krad on the volatile components of onion and the pungent flavor substances such as di-*n*-propyl disulfide and propenyl propyl disulfide, as well as other volatile compounds, are the same in irradiated red globe onions and the unirradiated control.[92]

IV. FACTORS AFFECTING SPROUT CONTROL BY IRRADIATION

A. Date of Irradiation

So far, many studies have indicated that irradiation during the dormancy period of tubers and bulbs is most effective for sprout control. The length of the dormancy period of potatoes is little affected by storage temperature, but it is significantly dependent upon their variety. For example, the Danshaku and Norin-Ichigo varieties have a dormancy period of about 100 days following harvest, while the dormancy period of Shimabara potatoes is around 45 days.[93,94] If potato tubers of Shimabara were irradiated with 7 and 15 krad 45 days after harvest, when it would be the beginning of emergence from the dormant state, buds on potato tubers (1 to 2 mm in length) would be suppressed with regard to their rate of sprouting and concomitant growth. In this case, the tissue of buds damaged by irradiation caused softening and became susceptible to infection and spoilage, but such spoilage was mostly localized within bud tissues.[93] The effectiveness of irradiation for sprout control is not affected by delay between harvesting and irradiation, if the treatment is practiced during the dormancy period.[95]

The dormancy period of onion bulbs is mostly shorter and not so clear as in potatoes. The length of the dormant state of Sapporoki onions appeared to be about 1 month following harvest.[96,97] Even within the dormancy period, some inner growth of buds in onion bulbs takes place, resulting in the gradual increase in the necessary dose for sprout control. The sensitivity to sprout control by irradiation in bulbs decreases with increasing intervals between harvest and irradiation.[52,62,97-101] Thus, the earlier irradiation after harvest is preferable for the most effective sprout control in onion bulbs.[15,78,89] The results obtained by Mullins and Burr[102] indicated that irradiation 12 days after harvest resulted in almost complete sprout inhibition with a dose as low as 2 krad during 6 month storage, while irradiation 76 to 118 days after harvest was remarkably less effective for sprout inhibition during the concomitant storage of 2 or 3 months. According to Grünewald,[103] the complete inhibition of sprouting in onions was accomplished by the irradiation treatment not later than 4 weeks after the harvest, but irradiation after 8 to 10 weeks of storage was not effective for sprout control.

However, the dormancy period could be extended by low temperature storage of onions after harvest.[110] Growth of the inner bud was suppressed at 3 to 5°C even after 3 months of storage and the length of the inner bud was similar to that of 1-month-old onion. Thus, sprouting was inhibited by irradiation and the operational period of the irradiation plant could be extended to 3 to 4 months from the harvest date when onions were stored at low temperature after harvesting.

Sometimes a transitory stimulation of sprouting in onions is observed with delayed irradiation,[13] although not in shallots. Brownell et al.[7] found that irradiation sometimes appeared to accelerate the first appearance of sprouts, but these early sprouts did not continue to grow. After the initiation of growth of inner buds, irradiation cannot prevent their growth with low doses. Since the reducing compound contents in inner buds rapidly increased at the emergence from the dormancy state, the ineffectiveness of irradiation at this stage was correlated with their increase.[42] The appearance of sprouts and roots in irradiated onion bulbs is not the result of the normal increase in tissue due to cell division but of the increase caused by cell enlargement. The sensitivity of cells which have enlarged to radiation is generally lower than that of cells which have divided.[16]

Another reason to recommend the irradiation treatment of onion bulbs at an early stage of postharvest storage is to minimize the amount of discoloration at inner buds which causes darkening during postirradiation storage. In conclusion, to inhibit sprouting, onion bulbs should be subjected to irradiation at an early stage, at least within 1 or 2 months after harvest if sprout control is required, whereas potato tubers can be irradiated postharvest as late as through the winter for the same purpose.

B. Radiation Dose

In general, it is known that the application of appropriate low doses to tubers and bulbs can be expected to result in excellent sprout control, although too great a dosage induces unfavorable side effects. Very small doses, less than 1 krad (e.g., about 100 R) may stimulate sprouting in tubers.[15,104] The minimum dosage for effective sprout control has been found to be about 3 krad for potato tubers and 2 to 3 krad for onion bulbs in many studies, when irradiation is applied at an appropriate date after harvest, as mentioned above.[9,62,64,97,102,105-111] For example, Kume et al.[112] found that doses of 2 to 5 krad were very effective for Sapporoki onions in the dormant state (percent of sprouting after 7 months storage — 5.8% with 2 krad, 20.8% with 1 krd, and 63.8% without irradiation) and concluded that the effective minimum dose for sprout control in onions is 2 krad.

It should also be noted that too high a dose may induce various types of unfavorable side effects disadvantageous for commercial irradiation treatments. The maximum dose for sprout control of tubers and onions is 15 krad, preferably 10 to 12 krad. With potato tubers, over 15 to 20 krad increases darkening or browning, decreases wound healing ability, increases storage rot, spoilage, sweetening, decreases vitamin content,[5,15,39,59,113-115] and brings about changes in chemical composition which did not disappear during subsequent storage.[15]

With onion bulbs, the overdose brings about an increase in rotting, an increase in losses due to softening, and sometimes the acceleration of sprouting or transitory stimulation of sprouting.[102,116] In Sapporoki onions, sprouting could not be sufficiently prevented at doses exceeding 30 krad regardless of the interval between harvesting and irradiating.[97] With Indian white onions, a dose of 25 krad resulted in increased sprouting when compared to the control and lower doses.[64] A similar increase in sprouting at the same dose was also noted in Denmark.[117]

From developmental studies conducted in many countries, doses between 5 and 15 krad, preferably a dose range from 6 to 12 krad, are recommended for sprout control if tubers and onions are in the dormant state shortly after harvest. There seems to be some difference in sprout inhibiting doses between tubers and bulbs. For the 12 potato varieties in Canada, a dose of 8 krad provided complete control of sprouting for up to 1 year and 7.5 to 8 krad would be optimum for long-term storage at high temperatures. Doses as low as 5 krad may be adequate for short-term or low-temperature storage.[10] In the Canadian potato pilot-plant program, a dose of 8 krad completely inhibited internal sprouting, which can be a very serious problem in potatoes stored at higher temperature for processing.[100] (Figure 6). In the initial petition for the irradiation of white potatoes to the U.S. Food and Drug Administration (FDA), an average dose of 7.5 krad of gamma radiation, with a maximum of 10 krad and a minimum of 5 krad, was proposed for sprout control.[12] Subsequently, the petition was amended to set the upper dose limit at 15 krad.

For Canadian varieties of onion bulbs, doses ranging from 2 to 12 krad produced little difference in the effectiveness for sprout inhibition and these doses also prevented rotting.[11,100]

C. Dose Rate

To some extent, dose-rate effects were similarly observed with tubers and bulbs. Rubin and Metlitsky[22] mentioned that an increase in the dose rate reduced the total dose required for the same biological effect, e.g., after planting out, the tubers indicated at a dose rate of 40 R/min and a total dose of 5kR began to sprout after 28 days, while those irradiated at 600 R/min and with the same dose began sprouting only after 53 days. In addition, the higher dose rate had no harmful effects on the immunity of the tubers from microorganisms. They concluded that a total irradiation dose could be lowered to achieve the same antisprouting effect by increasing the dose rate or using divided doses.

Mathur[118] reported that a dose rate of 3 krad/min is more effective in the prevention

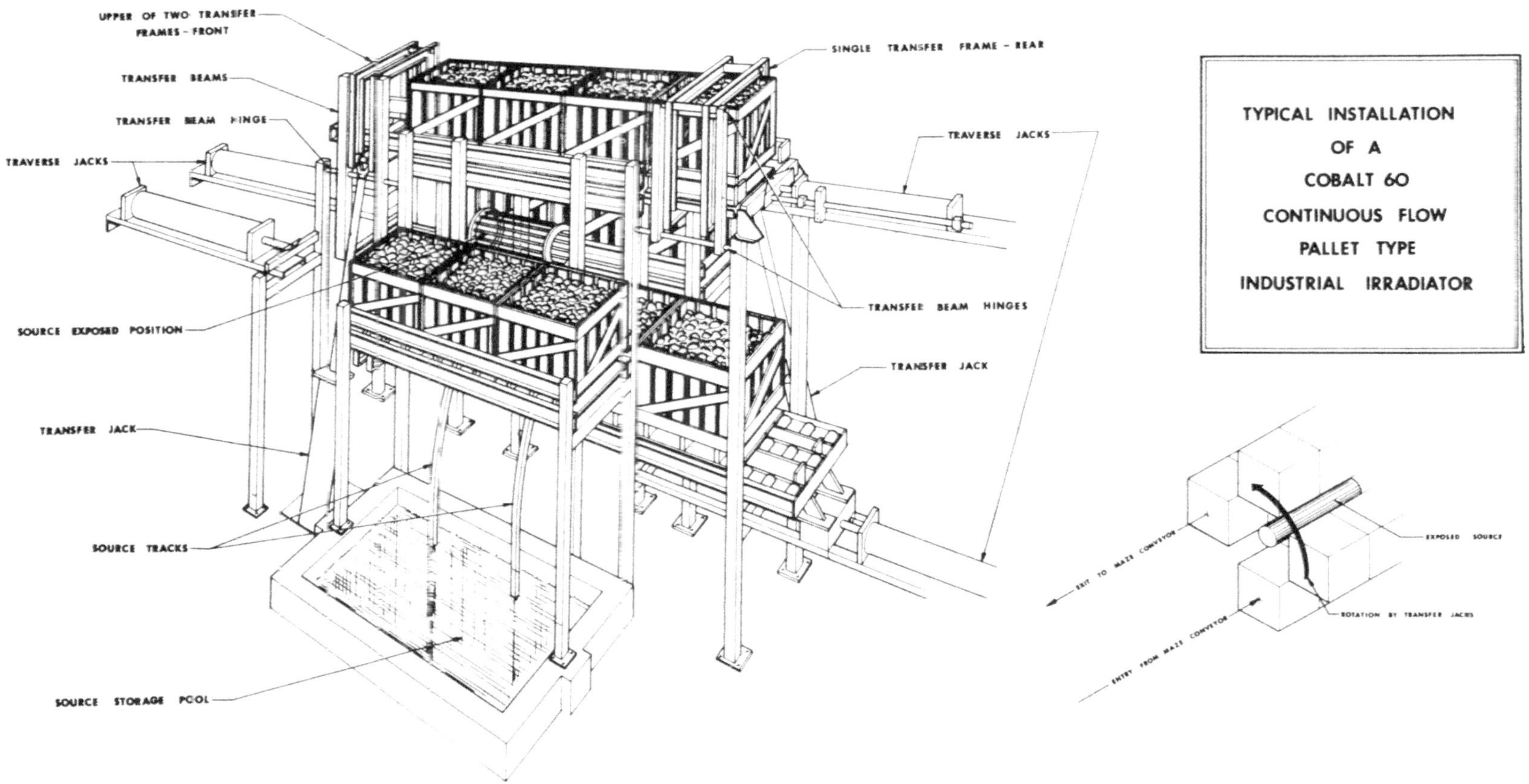

FIGURE 6. Commercial potato irradiator – source pass mechanism. (Courtesy of Atomic Energy of Canada, Ltd.)

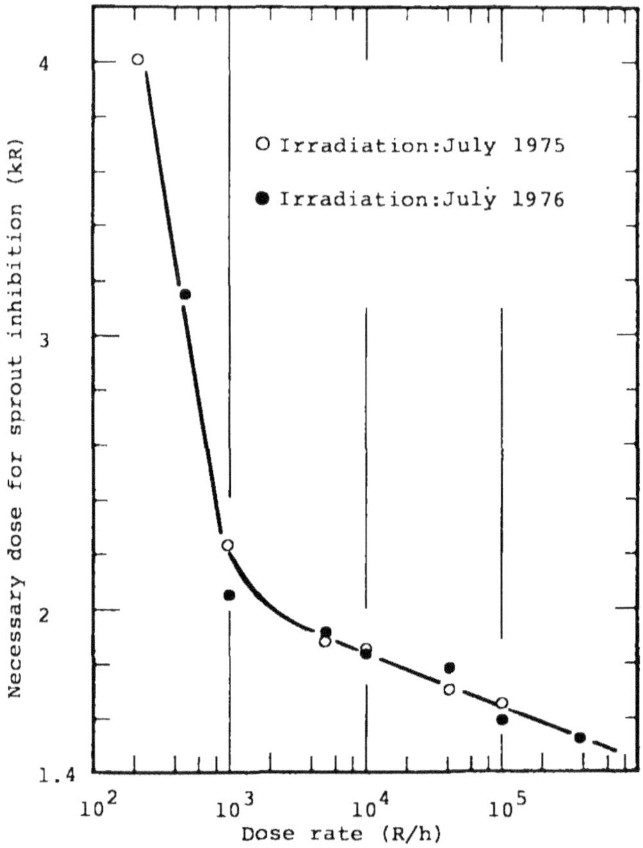

FIGURE 7. Dose rate (R/h) effect on the sprout inhibiting dose for onions. (From Furuta, J., Hiraoka, E., Okamoto, S., Fujishiro, M., Kanazawa, T., Ohnishi, T., Tsujii, Y., Tabata, T., Hori, S., and Ojima, T., *Food Irradiat.*, 12, 1, 1977.)

of sprouting in potatoes than a dose rate of 250 rad/min when the same dosage (6 krad for the Gola variety and 9 krad for the Up-to-date variety) is given. Jaarma[24] compared the sprout inhibiting effects of irradiation at high- (600 krad/hr) and low- (0.1 krad/hr) dose rates with the same dose of 15 krad on the same date and confirmed with 6 varieties that the effectiveness of irradiation for sprout control increases with increasing dose rate.

With onion bulbs, dose-rate effects on sprout control have been similarly observed. Furuta et al.[111] observed that doses required for complete sprout inhibition of onion bulbs, which is about 2 kR at dose rate 1×10^3 R/hr, decreased slowly with increasing dose rates, and increased rapidly with decreasing dose rates (Figure 7).

D. Variety

Differences in the effectiveness of radiation for sprout inhibition in various varieties of tubers and bulbs have been observed during postirradiation storage, the importance of which has been pointed out in both potato and onion irradiation treatments.[54] Rubin and Metlitsky[22] reported the significant difference in Russian varieties of potatoes such as Moscow, New Rose, Peredvik, Epron, Lorkh, and Berlikhingen irradiated with 8 kR after storage for 12 months in an ordinary storage room.

Canadian studies using 12 varieties of potatoes[10,29,100] indicated that varietal difference exhibits a great influence on the storage, chemical, and organoleptic properties of the irradiated tubers. Early U.S. studies[119] demonstrated that sprout inhibition in

Table 1
SPROUTING AND SHRINKAGE OF FOUR POTATO VARIETIES TREATED WITH GAMMA IRRADIATION, 1957 RESULTS

	Katahdin		Green Mountain		Kennebec		Russet Burbank	
Dosage(R)	Sprout.[a]	Shrink.[b]	Sprout.	Shrink.	Sprout.	Shrink.	Sprout.	Shrink.
Check	31.4	10.0	120.4	8.8	38.3	12.2	62.0	12.8
5,000	10.6	7.5	32.5	13.4	30.5	12.2	37.2	12.8
7,500	2.5	9.9	13.6	10.5	26.2	12.2	35.1	10.0
10,000	0.8	8.2	3.2	10.8	10.5	11.8	22.8	8.1
12,500	0.5	4.4	0.0	9.1	3.5	12.0	6.0	6.1
15,000	0.1	6.8	0.0	8.3	0.6	9.9	0.1	5.4

[a] Sprouting given in grams of sprouts per kilogram of tuber.
[b] Shrinkage given in percent.

From Sawyer, R. L. and Dallyn, S. L., *Am. Potato J.*, 38, 227, 1961. With permission.

potatoes was achieved with a dose of 5 krep for White Sebago, Katahdin, and Russet Burbank and with 7.5 krep for Red Pontiac at a storage temperature of 41°F(5°C). At 47°F (8.3°C), for all of these four varieties sprouting was prevented by a dose of 7.5 krep; at 55°F (12.8°C), White Sebago, Katahdin, and Russet Burbank required 7.5 krep for sprout inhibition, while Red Pontiac 10 to 12.5 krep.

Sawyer and Dallyn[120] observed significant variation of sprouting and shrinkage among four potato varieties, Katahdin, Green Mountain, Kennebec, and Russet Burbank, after irradiation with 5 to 15 krep and subsequent storage for 8 months at 10°C (Table 1). Comparing these data with those obtained by Workman et al.,[121] varietal differences appear to be affected by the growing area and postirradiation storage temperature.[122] Six varieties of Swedish potato harvests such as Bintje, Early, Puritan, Eva, President, Primula, and Ulster Chieftain exhibited different sensitivities to the sprout-inhibiting effect of radiation and differences in sensitivities among potato varieties were more remarkable with irradiation at the earlier stage of storage.[31] Grünewald[158] also found that the doses required for sprout inhibition in potatoes are greatly dependent on the variety. In Germany a dosage range from 4 to 16 krad is used owing to the large number of varieties (about 100).

With onion bulbs, some varietal differences in sprout-inhibiting doses were found.[11] Nevertheless, doses required for sprout control in different varieties fall into a dose range of 3 to 15 krad for potatoes or 2 to 15 krad for onions.[64]

E. Storage Conditions

Data on the effects of postirradiation storage temperatures on potato tubers can be summarized as follows: although the sprouting in irradiated potatoes is more effectively prevented during storage at low temperature (4 to 7°C) than at room temperature, the temperature during storage of irradiated potatoes is not so serious for practical purposes.[93,94,123-125] As indicated in Canadian studies,[100,126] storage temperatures appreciably influence chemical, storage, and organoleptic properties and suitability for processing.[10,45,58,93,127] For example, the higher sugar content of irradiated tubers was observed after storage at low temperature as compared with storage at higher temperature,[24,58,61,127] which is disadvantageous for processing of potato tubers. The pilot-plant potato irradiation program conducted in Canada during the 1961 to 1962 potato season indicated that a dose of 8 krad of cobalt-60 gamma rays was completely effective in controlling sprouting in potato tubers regardless of storage temperatures and other factors such as variety, date of irradiation, geographical location, and length of storage.[100,128]

The importance of the preirradiation temperature of potato storage has been stressed by Metlitsky et al.[15] in connection with wound healing after harvest. Based on special research and industrial practice, they recommended the storage at about 15 to 20°C with a rather high relative humidity for the first 2 weeks prior to irradiation for curing mechanical damage by harvesting and handling. Consequently, the most appropriate storage temperature of irradiated potatoes should be set up for each case considering the purpose of irradiation, the length of storage period, and the cost of temperature control.

On the other hand, sprouting losses during storage of onion bulbs are considerably influenced by pre and postirradiation storage temperature.[13] In the case of onion bulbs, effect of storage temperatures on radiation inhibition of sprouting has been differently reported in papers from different countries. With doses of 3 to 15 krad, sprouting in onion bulbs of the Sapporoki variety was almost completely inhibited for about 8 months of storage at 5, 10°C, and room temperature when irradiation was carried out within 1-month storage at room temperature after harvest. Variance analysis of the data clearly indicated that there was no difference in the percentage of sound bulbs between different post irradiation storage temperatures, i.e., 5°C and room temperature.[97] With the Senshuki variety, in contrast, the percentage of sound bulbs was higher after storage at 5°C than at room temperature, although the sprout-inhibition dose was the same;[129] with Indian white onions stored for 5 months, sprouting in bulbs stored at room temperature (27 to 30°C) was 1 to 3% in irradiated groups (6 to 25 krad) as against 8 to 24%, while onions stored at 20°C showed considerable sprouting in irradiated groups.[13]

The stimulation of sprouting in bulbs stored at about 10°C lower than room temperature was reported by several authors,[13,106,130,131] some of whom attributed this stimulation to more vigorous sprouting activity or higher relative humidity. On the other hand, a study conducted at a storage temperature of 12°C by Dallyn and Sawyer[63] with three varieties of onions (Southport, White Globe, Iowa 44, PW 160) indicated that doses from 4 to 12 krep applied within 42 days after harvest almost completely inhibited external and internal sprouting during 5 months' storage.

The preirradiation storage at room temperature longer than about 2 months does not allow effective sprout control by irradiation because of the considerable inner growth of buds. By cold storage at 3 to 5°C during the interval between harvest and irradiation, the growth of inner buds is inhibited and the effective period of irradiation can be extended up to 3 months after harvest.[110] Furthermore, another advantage of cold storage of onions is that inner buds do not exhibit significant discoloration during storage at 3 to 5°C for 8 months after irradiation and for 1 additional month of the marketing period at room temperature.

The higher humidity may prevent the rapid escape of moisture which results in shrinkage, but it may also cause storage rot. Onion bulbs require rather dry storage for optimum quality.[11] The preferable storage humidity would be 85% relative humidity or more for potato tubers but there is no clear recommendation for onion bulbs. With root crops harvested in Pakistan, where the ambient room temperature is 25 to 40°C, it was found that radiation (10 krad) followed by storage at 14 to 16°C is the optimum condition to store potatoes, onions, and garlic; they will keep for more than 5 months, avoiding the serious storage rot problem that occurs under tropical conditions.[132]

V. RADIATION EFFECTS ON CERTAIN TECHNOLOGICAL PROPERTIES

A. Weight Loss

It has been recommended that irradiated Canadian potatoes, Katahdin variety, be stored at 56°F (12.8°C) because weight loss from storage and sprouting plus sugar

Table 2
MEAN PERCENT WEIGHT LOSS UNDER VARIOUS STORAGE CONDITIONS

Storage conditions	No.1, 5°C,RH 90%			No.2, 7°C, RH 90%			No.3, 7°C, RH 95%		
Storage terms[a] (Month)	2(%)	4(%)	8(%)	2(%)	4(%)	8(%)	2(%)	4(%)	8(%)[b]
Irradiated									
Norin-ichigo	3.4	5.0	8.0		4.8	9.4		3.1	10.2
Yukijiro	4.4	6.4	11.1		7.6	18.3		4.8	19.3
Eniwa	3.2	4.9	9.0		4.9	10.0		3.4	16.1
Benimaru	3.6	5.6	10.2		5.9	12.9		2.9	16.2
Dejima	4.5	7.1	12.9		8.2	15.0		4.0	13.8
Nonirradiated									
Norin-Ichigo	2.8	4.9	12.5	3.8	7.2	20.5	2.7	4.9	27.3
Yukijiro	7.0	9.1	16.5	6.2	9.8	21.8	3.9	5.9	26.0
Eniwa	3.0	4.3	8.2	4.1	5.7	13.1	2.3	4.7	20.0
Benimaru	3.2	6.2	14.2	4.0	8.6	26.3	2.7	4.7	31.0
Dejima	5.8	9.4	17.4	6.4	12.2	25.1	4.4	7.0	28.0

[a] Means a storage period after transferring to each storage condition.
[b] Potatoes of No. 3 were reconditioned twice (6/22 to 7/6, 15°C, RH 80% and 7/6 to 7/30, 20°C, RH 80%) before processing; then weight loss showed remarkably.

Umeda, K., Paper presented to FAO/IAEA Res. Coord. Meet. Technological and Economic Feasibility of Food Irradiation, Accra, Ghana, February, 1978.

build-up are minimized at this temperature.[10] In the control tubers the weight loss due to shrinkage increased from 4 to 34% during 5 months following sprout development; in the tubers irradiated at 8.5 krep weight loss increased from 4 to only 9% during the same period. Sawyer and Dallyn[120] showed a similar reduction of weight losses in four varieties of irradiated potatoes during storage at 10°C due to sprouting and shrinkage after exposure to 5 to 15 krad when compared with nonirradiated control potatoes.

With three varieties of Japanese potatoes, the weight loss during storage at room temperature was reduced by 7 and 15 krad, but not during storage at 5°C.[93,127] In another experiment on a commercial scale,[14] five varieties of Japanese potatoes were stored in three different conditions after irradiation (Table 2). After 8 months' storage, the weight losses of potatoes were different depending on the variety and storage condition — from 8.0 to 19.3% for the irradiated group and from 8.2 to 31.0% for the nonirradiated group. In all storage conditions irradiated potatoes remarkably prevented weight loss as compared with nonirradiated potatoes, especially in the case of 7°C storage.

In the 1957 to 1958 Canadian experiment,[10] total weight loss (sprouting plus shrinkage) of onion bulbs irradiated with 6.8 and 8.5 krep was 5.7% as against 23.2% for the unirradiated bulbs after 5 months storage at 55°F (12.8°C). Weight losses of the Senshuki variety of onion bulbs were larger when stored at room temperature than when stored at 5°C.[129] Nuttall et al.[105] reported that no significant differences in weight losses were found between unsprouted control and bulbs irradiated with doses ranging from 3.2 to 9.5 krad.

B. Specific Gravity and Firmness

In a study conducted by Schreiber and Highlands,[55] doses two to six-fold as great as those required for sprout inhibition were found not to affect the specific gravity at the level of commercial significance after storage at 31 to 66°F (−0.5 to 18.8°C). In a Purdue University experiment[119] and subsequent storage, irradiation with doses of 5 to 15 krep at 41°F (5°C) and 55°F (12.8°C) did not alter the specific gravity of the four potato varieties studied. At the end of onion storage, irradiated sound bulbs were firm with skins relatively intact and bright in color.[105]

C. Storage Rot and Wound Healing Ability

During storage, tubers and bulbs are exposed to a risk of infection by harmful microorganisms leading to the development of rotting and decay, i.e., potato blight, wet rot, dry rot, and concomitant creamy decay. Although no significant incidence of storage rot in irradiated tubers and bulbs is reported in some studies.[10,11,100,101,105] the enhancement of storage rot, or decay by irradiation has been observed in numerous experiments.[13,15,97,117,120,129,133,134] However, differences in rotting rate were not so significant and the decay rates of the different varieties were similar.[10] For long-term storage of tubers and bulbs, rotting may become a serious problem. Duncan et al.[135] indicated that rotting in tubers was gradually increased with increasing doses without any definite dose level for occurring storage rot. Bruising of tubers before or after irradiation increased rot incidence over nonbruised tubers, suggesting that careful handling before and after radiation treatment is important to reduce losses due to storage rot.

Curing of potato tubers before irradiation can reduce rotting during subsequent storage, because curing will allow the tubers a chance to heal their harvest wounds. The healing ability of tubers is reduced by irradiation. Consequently, Metlitsky[15] recommended that irradiation of tubers in industrial practice should be done after curing under conditions favorable to rapid formation of wound periderm in mechanically damaged parts; he recommended the doses should not exceed 10 krad to avoid unnecessary side effects of radiation due to damaged healing ability. Susceptibility of potato tubers to rotting has been discussed in connection with their wound healing ability.[15,121] Brownell et al.[136] observed the inhibition of periderm formation and the delay of suberization after irradiation at 19 to 200 krep. Thus, storage rot could occur as a result of infections in the cuts and bruises in the tubers after irradiation treatment. If tubers with bruises were irradiated before they formed periderm, the bruises would then offer entrance for rot-causing organisms. Rubin and Metlitsky[22] observed the temporary reduction of tuber immunity evidenced by the more intensive infestation of tubers with *Fusarium* fungus soon after irradiation, but in the subsequent storage no difference was found in the contamination between unirradiated and irradiated tubers. Metlitsky[15] suggested two possible mechanisms for the increase of storage rot in irradiated tubers, i.e., (1) tubers which for some reason had lowered resistance to disease are further weakened by irradiation and (2) the complete loss of their capacity to form periderm even at doses as low as 8 to 10 krad results in the more rapid infection in irradiated tubers.

In unirradiated tubers the periderm is formed simultaneously with the suberization process, but not in irradiated tubers. The wound-periderm formation is inhibited by a somewhat higher dose, perhaps above 10 to 12 krep,[9,10,22,39] while suberization is not inhibited by sprout inhibiting doses. Thus, irradiated tubers can be protected by the suberous layer which serves as a barrier to infection. However, the suberization process is slow. Metlitsky et al.[15] noted that potatoes should be held for the first 2 weeks at about 15 to 20°C with a high relative humidity (not less than 85%) and good air ventilation to ensure the healing of tissue damage. Ogawa et al.[47] also reported that the defense action of tubers against infection by the black rot fungus was enhanced by irradiation in a dose range of 5 to 12.4 krad, whereas it was rather reduced at 22.4 krad. Thus, the extent of storage rot is of little significance in cured and properly handled tubers irradiated with sprout-inhibiting doses.[12,137] There is neither a harmful nor beneficial effect in respect to physiological and pathological breakdown due to sprout inhibiting irradiation.[10]

With onion bulbs, Dallyn and Sawyer[138] and Umeda et al.[97,129,134] observed no consistent effect of irradiation at the dose levels of sprout inhibition on the development of storage rot. However, the percentages of sound bulbs after long-term storage were higher with irradiated lots than with the unirradiated control, being highest at 7 krad. Nuttal et al.[105] observed that the varieties and bulb sizes of onions differ in susceptibility to storage rot and that irradiation somewhat increased rotting. Subsequently, in the case of the Autumn Spice where rotting was least, the amount of sound bulbs in the irradiated lot was markedly greater than that in the unirradiated control. In the Netherlands there is an unsolved storage problem, i.e., the incidence of neck rot was partially solved on storage of onion bulbs, but irradiation was not effective for preventing storage rot.[139]

D. Peeling and Trim Losses

The reduction of peeling and trim losses would be advantageous for consumers as well as for the prepeeled potato industry. In a study of Sebago and Russett Rural potatoes,[140] it was found that peeling and trim losses were lowest in the lots of potatoes irradiated with 5 and 10 krep. For example, after 2 months of postirradiation storage at 22°C, the total peeling and trim losses in tubers of the Sebago variety irradiated with 10 krep were 22% compared to 32% in the unirradiated tubers. One of the favorable side effects of the irradiation treatment would be the inhibition of greening and solanine formation in the surface layer of tubers stored at sub-room temperatures[10,61] and this may serve for reducing the peeling losses.[84]

E. Discoloration

In some cases of potato irradiation, the enhancement or suppression of discoloration in tubers induced by radiation has been noted. One case is raw tissue browning or darkening observed with uncooked tubers; the other occurs after heat treatments such as boiling, steaming, and frying. The former appears to involve the enzymatic process, whereas the latter would be nonenzymatic.

Concerning discoloration of raw potatoes, there are several different observations. Some authors described an increase in black spot or darkening of potatoes by irradiation.[38,120,141-143] Trulsen[37] observed that the degree of darkening of the central parts of irradiated potato tubers was so serious that these potatoes were considered unacceptable and that this enhancement of darkening of tubers after irradiation might present a serious obstacle to the practical application of irradiation as a sprout-inhibiting method even though it occurred only in a few lots. Enhanced darkening following irradiation was observed in all of the varieties used, i.e., Alpha, Bintje, Dianella, Kennebec, and Up-to-date after storage from harvest in autumn to June. It occurred in some instances at only 5 krad. Discoloration of this type increases with increasing dose in the early stage of storage in potatoes cultivated with intense nitrogen fertilizer. This darkening process appears to be independent of boiling, storage temperature (in a range of 5 to 10°C), and culture locality.

Usually the radiation enhancement of darkening became moderately obvious only after a few months of storage. In contrast, Metlitsky[15] reported with Russian potatoes that, as compared with the unirradiated control, irradiated tubers were less affected by a physiological disease, i.e., variety dependent darkening of the tubers centers, which might occur in some years during storage. Although the cause of such darkening is not well known, it appears that the darkening processs may involve the accumulation of oxidation products of phenolic substances, especially tyrosine, and a decrease in dehydrogenase activity. Irradiation, which increases the dehydrogenase activity without increasing polyphenoloxidase and peroxidase, favors the suppression of darkening in the tuber center.

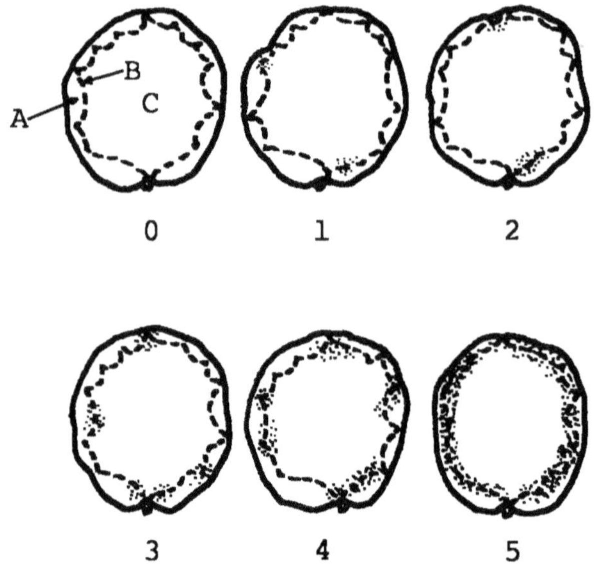

FIGURE 8. Index and scores of the browning of potato tubers. (0) None — Nonindication of browning; (1) very trace — beginning of browning not apparent; (2) trace — beginning of browning; (3) slight — browning in localized areas; (4) medium — browning more widely distributed around the vascular bundle; (5) severe — browning in an advanced stage, with severely colored areas around the vascular bundle. (From Otaga, K., Tatsumi, Y., and Chachin, K., *J. Food Sci. Technol. Jpn.*, 17, 298, 1970. With permission.)

With potato tubers grown in Japan, it has been observed that irradiation-enhanced browning in raw tissues developed around the vascular bundle of the tubers stored at room temperature followed by cell death.[49,89,144-148] The browning became obvious 3 to 7 days after irradiation[144,147] (Figure 8). Occurrence of browning in the raw state appears to be dependent on the cultural locality and condition, variety, maturity, radiation dose, and date of irradiation.[147,149,150] Since browning was also observed in most cases with the unirradiated control, the browning observed with irradiated tubers is radiation stimulated or enhanced discoloration (Figure 9). Tubers from the main potato-growing district in Japan, Hokkaido, showed a lesser extent of browning in the raw state after irradiation.[1,149] Browning in the raw state was observed in tubers when irradiated with 10 to 40 krad 1 month or less after harvest, whereas only a little browning takes place in tubers irradiated with 10 and 20 krad when irradiated 2 months after harvest. However, at a dose of 40 krad they defintely became brown.[147] As a result of the examination of the dose-dependence phenomenon, it was found that browning in raw potatoes irradiated by sprout-inhibiting doses exceeding 5 krad tends to occur.[147] It was concluded that the browning of raw potatoes in Japan occurs in immature, small potatoes of short cultivation period.[150] Development of browning in raw irradiated potatoes during storage was greatly suppressed either by allowing an interval of about 1 to 2 months between harvest and irradiation or by low-temperature storage at 5 to 7°C after irradiation.[1,144,147,149] Lower doses (less than 10 krad) are recommended for avoiding browning in the raw state. Thus, the discoloration problem does not offer any practical difficulty for commercial-scale production of potatoes by irradiation.[151]

Although the biochemical basis of browning in the raw state is not well known, the correlation between the development of browning and both polyphenol content and

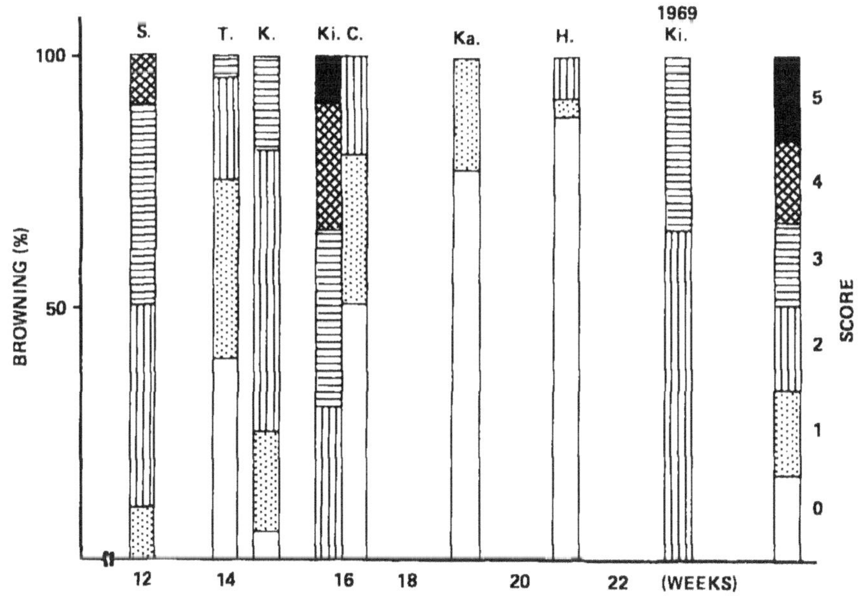

FIGURE 9. Relationship between the browning of potato tubers irradiated and term of cultivation. The tubers harvested at seven localities were irradiated 2 weeks after harvest with doses of 10 krad. Browning of tubers was compared on 7th day after irradiation (harvested 1970). S. — Shikoku; K. — Kyushu; C. — Chubu; H. — Hokkaido; T. — Tohoku; Ki. — Kinki; Ka. — Kanto; 1969 Ki. — harvested in 1969 Kinki. (From Tatsumi, Y., Chachin, K., Matsuzuka, M., and Ogata, K., *J. Food Sci. Technol. Jpn.*, 20, 132, 1973. With permission.)

oxidizing enzyme activities was examined.[49,50] When Danshaku potatoes were irradiated soon after harvest and developed browning, increase in o-diphenol content and the activity of polyphenol oxidase and peroxidase and a decrease in ascorbic acid content were found.[50] In contrast, these changes were not observed with tubers irradiated about 3 months after harvest in which browning did not occur. An increase in o-diphenol and decrease in ascorbic acid were noted in the cortex and vascular bundle until 7 days after irradiation; polyphenol oxidase activity increased in the cortex. These parts of tubers were similar to those developing browning. Tatsumi et al.,[50] who obtained these results, suggested that such an enzymatic browning process may involve the oxidation of polyphenol into quinones by polyphenol oxidase followed by oxidative polymerization of quinones into browning substances through the electron transport system.

After-cooking darkening is one of well-known discolorations which develop in many potato varieties grown in different parts of the world.[152] Discoloration of this type appears to be due to the interaction of ferrous salts in potatoes with polyphenols such as chlorogenic and caffeic acids resulting in the formation of complexes which are oxidized to black pigments.[153,154] After-cooking darkening of potatoes is enhanced by irradiation with sprout-inhibiting doses[155-157] and its intensity increased with increasing dose.[120] The details of the radiation enhancement of after-cooking darkening have not yet been known. No significant difference in the ferrous content of the Kennebec variety was found between unirradiated and irradiated tubers.[156]

The Bintje variety, which is the predominant variety in the Netherlands and Germany owing to its excellent technological qualities and high yields (55 to 60 t/ha), is most sensitive to radiation and causes enhanced discoloration by heat used in cooking and prefrying.[158] See Figures 10A and B for equipment used in some of the German experiments. Grünewald[158] commented that discoloration of irradiated potatoes could

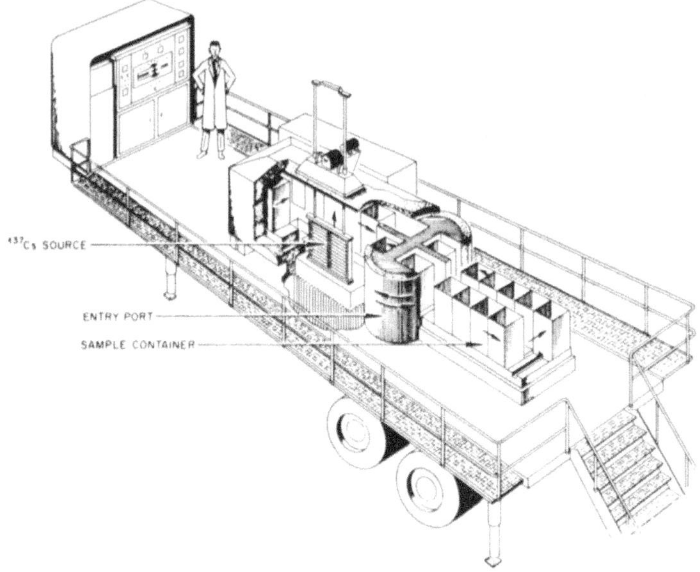

A

FIGURE 10. French mobile irradiator "IRMA" used by the German group for irradiation of potatoes in a warehouse. (Courtesy of Dr. Th. Grünewald.)

be overcome and discoloration of irradiated cooked potatoes or potato puree is not the enzymatic browning occurring in fresh-cut tubers. Even within the same variety there are differences in the intensity of discoloration. After-cooking darkening occurred to some extent with all treatments including the use of chemicals, but irradiated potatoes showed most severe darkening when compared with that of the untreated or chemically treated ones.[159] Sparenberg and Buitelaar[160] reported that a 2- or 4-week reconditioning at 15°C before the cooking or prefrying of the Bintje variety originating from 5 different soil types showed no reduction of discoloration of cooked potatoes. Studies conducted in Canada[161] revealed that irradiated and stored tubers, especially after a dose of 8.5 krep or more, showed considerable after-cooking darkening and there were varietal differences in the extent of darkening (more darkening in Katahdin than in Netted Gem).

Thomas and Joshi[157] reported that potatoes grown in India, irradiated with 10 krad and stored for 8 months at 15°C, developed discoloration after boiling. The discoloration turned from gray to dark gray with increasing time of post-cooking exposure to air. On the other hand, such discoloration was not observed with potatoes of many varieties grown in several countries. There are some reports from different countries on observations of no or less discoloration after cooking and processing with a number of potato varieties.[159] Therefore discoloration on cooking or processing appears to be dependent upon variety and other agronomic and climatic factors such as cultivating district, soil, fertilizer, and the planting period.

It is known that the high concentration of reducing sugars in stored tubers is an important cause of browning. In fact, radiation-induced or enhanced discoloration on frying and chipping potatoes has been found, especially at higher reducing sugar contents.[158,159] In some instances, however, irradiation does not increase the browning of products.[162] Since intensified browning has been observed with chips prepared from potatoes with the lower content of reducing sugars, discoloration of chips is not necessarily dependent upon the reducing sugar content.[94]

FIGURE 10B.

The chemical and biochemical processes for discoloration in irradiated potato tubers are considered to be these: (1) enzymatic browning or darkening as found in the raw tissue, (2) nonenzymatic darkening after cooking due to oxidation into black pigments of complexes resulting from the interaction of ferrous salts in the tubers with polyphenols, and (3) the Maillard reaction, a heat-induced reaction between reducing sugar and amino acids. This reaction would be responsible for discoloration on frying or chipping. In any event, it has been noted that in some cases irradiation increases these discolorations. This radiation effect is a problem that should be solved before the commercialization of potato irradiation.

Several methods have been proposed to prevent or reduce discoloration: (1) delayed irradiation, 1 to 2 months after harvest, to preclude browning development around a vascular bundle of tubers during storage; (2) the reduction of the reducing sugar content by reconditioning at about 15°C or room temperature for 2 weeks[94,162] or at 70°F (21.1°C) for 3 weeks[157,161] prior to processing if tubers are to be stored in the cold (5 to 10°C); (3) prepeeling of tubers before cooking, (4) reconditioning at 34 to 35°C for 4 to 6 days, or (5) soaking and cooking in ethylenediaminetetraacetic acid (EDTA) or citric acid solution, which has been proposed as a deterrent to after-cooking darkening.[157]

Reconditioning after irradiation treatment also suppressed the increase in reducing sugar.[14] In this experiment, Japanese potatoes (var. Yukijiro) were cured after harvest at 15°C, relative humidity 80% for 1 month and then transferred to the preirradiation

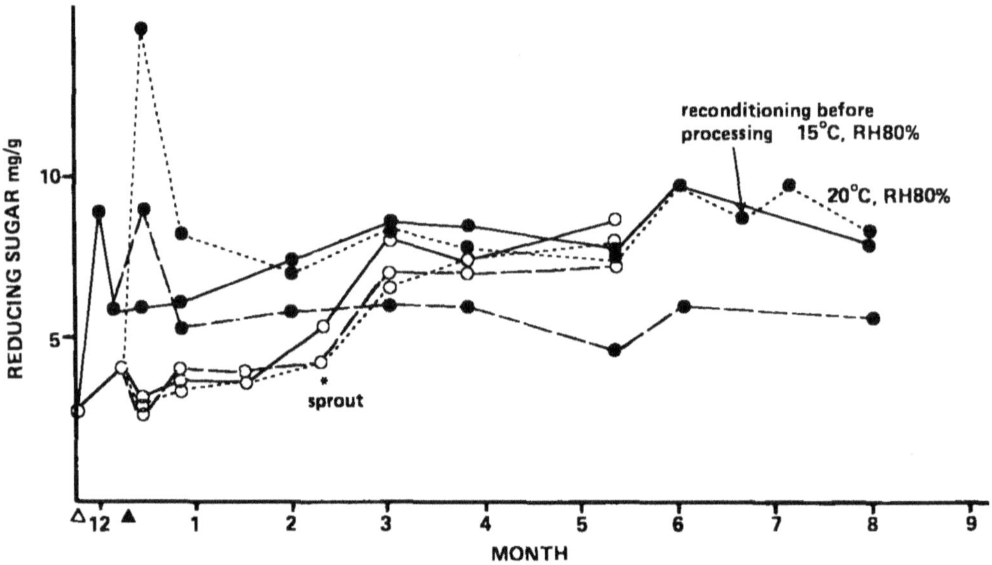

FIGURE 11. Reducing sugar content of potatoes during storage.

(Yukijiro)

○ ; Non-irradiated — No. 1, 5°C, RH90% (solid line)

●; 15 krad — No. 2, 7°C, RH90% (dashed line)

△ ; Reconditioning after irradiation (15°C, RH80%, 2 weeks) — No. 3, 7°C, RH95% (dotted line)

▲ ; Non-reconditioning after irradiation

storage condition (No. 1: 5°C, relative humidity 90%; No. 2: 7°C, relative humidity 90%; No. 3: 7°C, relative humidity 95%). Figure 11 shows the results during 9 months after irradiation. Irradiated and nonirradiated samples of No. 1 and No. 2 were reconditioned for 2 weeks at 15°C, relative humidity 80% after irradiation date and transferred to the long-term storage condition which was the same condition as the preirradiation. On the other hand, No. 3 samples were transferred to the long-term storage condition without reconditioning. As a result, samples which were irradiated, reconditioned, and transferred to 7°C, relative humidity 90% (No. 2) maintained the lowest reducing sugar content from March to July compared with other irradiated and nonirradiated samples. To decrease the reducing sugar content of nonreconditioned samples after irradiation, irradiated potatoes stored under No. 3 conditions have been reconditioned at 15°C, relative humidity 80% for 2 weeks. In the latter part of the storage period which occurred in June, usual reconditioning contrary to expectation did not decrease the reducing sugar content of potatoes but actually increased it. However, under the much more severe reconditioning at 20°C, relative humidity 80% for 25 days encountered in the beginning of July, the reducing sugar content of potatoes decreased rapidly. However, this severe reconditioning caused shrinkage and quality deterioration of potatoes.

Discoloration observed with irradiated onions is the darkening or browning of inner buds.[13,63,105,109,110,124,129,138,163-165] However, it is not a serious problem with regard to marketing value. Sprouts in irradiated bulbs withered and died after some growth.[62,63,105,134,138] Dallyn and Sawyer[63] mentioned that the dark centers in irradiated onion bulbs may not be objectionable to users if the alternative is a sprouted onion. Color of inner buds of onions turned into a brownish or dark color; this discoloration is apparently due to death or injury of the growing points.[11] Discoloration of the interior of bulbs due to injury or cell death in growing points was observed even in longitudinal sections of apparently sound marketable bulbs. In a Canadian study, the percentage of internal discoloration increased with radiation dose[105] and the length of a brownish inner bud decreased with increasing dose and dose rate.[111] Discoloration of the inner buds was in most cases accompanied by internal decay.[13,105] The Canadian study[105] suggested that irradiation may merely intensify an already existing condition in untreated onion bulbs.

The discoloration of the growth center in irradiated onions may be due to cell death in meristem tissues which increase their susceptibility to microbial attack.[13] On the other hand, Brunelet and Vidal[163] suggested that discoloration might be caused by enzymatic activity.

However, the browning of buds observed in irradiated onions during the storage period was prevented by low-temperature storage after irradiation. Onions so treated have retained good market quality for 1 month when transferred from low temperature to ambient temperature from spring to summer. (Figure 11).[110]

F. Organoleptic Qualities

The results of analyses of cooking qualities such as color, texture, flavor, off-flavor, and preferences have indicated neither significant change in culinary properties nor difference in a conclusive evaluation between unirradiated and irradiated potato tubers and onion bulbs during storage over several months (4 months or longer).[5-7,9,12,21,22,62,63,94,105,117,140,156,166-168] A better quality of irradiated potatoes was noted during the off-season.[169] Deterioration changes observed in the course of post-irradiation storage of potatoes are sweetening and enhanced darkening on cooking.[156,161] A sweet taste in potatoes was noted after higher radiation doses.[5,123,170]

In the case of onions, the characteristic flavor was not seriously altered by irradiation,[62] and no off-flavor developed after irradiation.[105] However, it has been reported that irradiated bulbs lost (somewhat) their sharper astringent qualities and lachrymatory character, resulting in milder flavor, being slightly sweeter, and requiring a shorter cooking time for the same degree of the tenderness as compared with the unirradiated bulbs.[1,60,100,105,130]

Irradiation treatment did not have any significant effect on the qualities of potatoes for processing into chips, instant mashed potato flakes, frozen french fries, and fresh prepeeled boilers.[85,94,100,167] The internal firmness displayed by irradiated potatoes after prolonged storage at 60 to 70°F (15.6 to 21.1°C) was considered an advantage for processing.[85,100]

Close correlation between the reducing sugar content and whiteness of processed chips was found, although the conclusive organoleptic evaluation was not solely dependent upon the reducing sugar content.[156,162] Selection of good chipping varieties and careful control of temperatures appeared to be needed for chipping purposes on irradiation treatment of potatoes with sprout-inhibiting doses.[88] As mentioned above, reconditioning to decrease the reducing sugar content prior to frying resulted in improved chips and crisps with respect to color, texture, flavor, and crispness; they were better in quality compared to products prepared from unirradiated potatoes.[161,162] After 9 months of storage at 68°F (20°C) the chips made from irradiated potatoes became generally unacceptable, being soggy, often dark, and of bitter flavor; however, the unirradiated potatoes could not be used for chipping.[161]

As for onion processing, discoloration of inner buds was considered not to be acceptable for the dehydrated product.[11] Cold storage (at 3 to 5°C) after irradiation may be effective to abolish the disadvantage due to the darkening of inner buds.[10]

VI. IRRADIATION TECHNOLOGY FOR SPROUT CONTROL

A. Materials to be Irradiated

To take advantage of radiation effects on tubers and bulbs attention should be paid to variety, maturity, size, and soundness of materials to be irradiated.

The importance of varietal difference in the effectiveness of irradiation for sprout control has been stressed with tubers and bulbs.[54] Varieties of poor storage quality are not suitable in principle, e.g., the Yellow Sweet Spanish onion variety appears to be little improved by irradiation treatment because of rapid rot development in both unirradiated and irradiated bulbs.[100] In the Canadian potato study, the Netted Gem was found to be the best stable-stock variety and the Kennebec best for chipping.[100] The smaller size of onion bulbs gave lower incidence of rotting and a higher percentage of sound bulbs.[105] Smaller bulbs sprout faster than larger ones of the same variety.[8,105,117,171]

With long-term storage, potato varieties of long-keeping quality may in general be suitable for irradiation treatment. However, although Yellow Sweet Spanish is a soft and relatively poor-keeping onion, it would appear that irradiation may be of some use in extending its storage life if the characteristic quality of this variety is desired.[105] On the other hand, potato tubers of immaturity and small size are prone to display discoloration in their raw tissues during storage.[49,172]

B. Pretreatment

Careful handling and grading according to appearance, wound damage, and quality are recommended. Mechanical injury during harvest as well as preirradiation handling and storage increases the incidence of rotting. To complete the wound-healing process, curing at about 15 to 20°C or room temperature with high humidity under good air circulation (with 95% or higher rather than 85% relative humidity) for 1 to 2 weeks is necessary. The most desirable curing is one that occurs before irradiation.[121]

If there were a discoloration problem in the raw state, the storage interval between harvest and irradiation would be, preferably, 1 to 2 months at room temperature. Onion bulbs should be dried for about 2 weeks following harvest to protect them against rotting.

C. Irradiation

Irradiation should be practiced during the dormancy period of tubers and bulbs to minimize the necessary dose. Low-temperature storage (at up to 5°C) usually prolonged in some measure the effective dormancy period of tubers and bulbs, especially in the case of bulbs.[110] On the other hand, the results obtained with Indian onions[13] indicated that room-temperature storage at 27 to 30°C yielded the smallest sprouting rate and suggested that storage at 3 to 5°C may shorten the state of rest while storage at 10°C accelerates sprouting.[13,173,174] The effects of irradiation on sprout control and internal discoloration in Japanese onion bulbs were found to be greatly affected by the extent of internal sprouting.[97]

The suppression of internal sprouting by low temperature (3 to 5°C) reduced the dose required for sprout inhibition. As mentioned earlier, data on the suitable dose for sprout inhibition obtined so far in different countries fall into a range of 3 to 15 krad for potatoes[10,12,15,64,72,89,107,148,175-182] and 2 to 15 krad for onions.[11,64,89,101,105,107,176-178,183-186] Since the necessary dose is influenced by several factors such as agronomic and climatic factors, variety, date of irradiation, storage temperature, and the intended period of extension, the most appropriate dose should be determined for each case under the specific area conditions.

In the Canadian study on potato irradiation,[10] it was concluded that doses of 7.5 to 8 krad would be optimum for long-term storage at high temperature and doses as low as 5 krad may be sufficient for short-term or lower temperature storage. A dose range between 2 and 12 krad was found sufficient for sprout control in onion bulbs.[11] From the economics point of view the minimum dose of 2 krad could be used for onion irradiation commercially but a higher level of 4 to 8 krad[54] or 8 to 12 krad[11] was recommended to minimize discoloration in growing points. For the same purpose, irradiation of onions at the early postharvest stage is effective.

In hot countries, a little higher dose seems to be required for sprout control, e.g., 5 to 15 krad for Indian onions,[13,187] 10 krad for Indian potatoes,[13] and 8 to 10 krad and 12 to 15 krad for potato tubers produced in Bangladesh[31] and Iran,[181] respectively. Khan and Wahid[132] reported that onions, shortly after harvest in Pakistan, receiving doses of 5 to 15 krad, showed excellent sprout inhibition.

Problems of potato storage in hot countries relate the control of tuber moth and rotting at higher temperature. Adult emergence was completely inhibited by 10 krad when tubers were infested with eggs and early larval instars, but for those infested with late larvae a dose of 20 krad was required for the same effectiveness.[13]

D. Postirradiation Storage

In the case of irradiated potatoes, it is not necessarily required that tubers be kept at low temperature for culinary use or chipping. However, if a long period is required for storage, low temperature (5 to 10°C) is employed. Humidity should be kept high with good ventilation during storage, e.g., at a relative humidity 95% or higher rather than at relative humidity 85% at 7°C.[188] However, in the case of injured tubers somewhat lower relative humidity, for example 70 to 80%, relative humidity is recommended to reduce storage rot.[188] When tubers are processed into chips after storage at low temperature, reconditioning should be performed to decrease the reducing sugar content. Reconditioning is effective for lowering the content of reducing sugars which increased during cold storage and for yielding chips of improved quality as described before.[162] Rumpf[189] mentioned that only the irradiated potatoes containing fructose and glucose at not more than 0.07% each of the fresh weight yielded marketable chips of good quality, and that such lower content of reducing sugars resulted in a very low rate of rotting even in the irradiated batches.

Good ventilation and the use of pallet boxes as containers for tubers offer advantages in storage and handling.[190] Postirradiation storage of onion bulbs at low temperature is recommended when discoloration of growth centers must be suppressed. High humidity in the storage room appears to be largely responsible for rotting.[105]

VII. SPROUT CONTROL OF OTHER ROOT CROPS

Inhibition of sprouting by low-dose irradiation has been reported for various types of root crops, such as carrots, sweet potatoes, ginger roots, garlic, shallots, red beets, turnips, Jerusalem artichoke tubers, and chestnuts.[15,54,73,191-196]

Irradiation of carrots to inhibit sprouting has received attention. Mikaelsen et al.[197] exposed carrots to 5, 10, and 15 kR of gamma rays. After 8 months of storage, doses of 10 and 15 kR were found to prevent sprouting completely. Although at 5 kR some sprouting was noticed, there was less sprouting and growth in comparision with the unirradiated control. At the same storage time, new-root hair appeared in the unirradiated carrots but not in irradiated ones. Rotting was not affected by 5 kR, but at the higher doses increase in rotting was observed after 8 months of storage. In a carrot irradiation experiment, Madsen et al.[198] observed fairly effective inhibition of sprouting at 9.3 krad and almost complete inhibition at 18.6 krad. No dose-rate effect was found with sprout control by irradiation. The length and weight of carrot sprouts increased as the temperature of preirradiation thermal treatment increased from 40°F (4.4°C) to 120°F (49°C).

Taste quality of carrots did not significantly change immediately after these irradiation treatments, but the carrots irradiated with 18.6 krad and stored at 50°F (10°C) for 3 to 4 months showed significantly better flavor than the unirradiated carrots. The Canadian carrot irradiation program[100] indicated that a dose of 4 krad was sufficient to inhibit root-hair growth and the effective inhibition of sprouting at the crown required 12 krad. Irradiation did not affect the flavor of carrots on cooking tests. Irradiation treatments showed no consistent effects on the ascorbic acid, carotene, and soluble-solids values of carrots. In Canada, commercial irradiation of carrots for sprout control could be economically performed using a potato irradiator, since the highest concentration of carrot production in Canada is concentrated in a potato and onion-growing area.[100]

Sprouting in sweet potato roots was inhibited appreciably by doses of 5 to 8 kR and completely by doses of 8 to 20 kR without any significant changes in appearance, taste, and vitamin content.[199] Irradiation for sprout control in sweet potato roots was more affected just before sprouting in the control (6 months after harvest) than on those immediately after harvest.

Sweet potatoes showed a lower sensitivity to gamma radiation than potatoes with respect to respiratory amount, polyphenol content, and peroxidase activity.[40] The effects of gamma irradiation on metabolic changes in the sweet potato root tissue in response to cutting were studied.[200,201] In both unirradiated and irradiated samples after cutting, the polyphenol content and activities of o-diphenol oxidase, peroxidase, and phenylalanine ammonialyase increased remarkably. The polyphenol content and peroxidase activity of samples receiving 60 krad or more in response to cutting was significantly stimulated over the unirradiated samples, but formation of a lignin-like substance was inhibited. No significant effect on o-diphenol oxidase was found when doses of 5 to 12 krad were applied.[200]

Irradiation significantly stimulated the development of peroxidase as well as the production of ethylene in sweet potato root disks. Exogenous ethylene stimulated the peroxidase development in both unirradiated and irradiated disk tissues in the additive mode of action.[2]

The results of kinetic experiments indicated that the stimulation of peroxidase development by irradiation is mainly mediated by an unknown mechanism independent of the ethylene effect, but it may be partly mediated through the enhanced amounts of ethylene produced in disks by irradiation.[201,202]

Irradiation was found to be effective for the inhibition of sprouting in yams for a storage period of 6 months at doses between 7.5 and 15 krad without any adverse changes in acceptability or physiological properties.[172] Doses higher than 15 krad tended to cause damage to internal tissues of the yam tubers. On the other hand, the chemical treatments with maleic hydrazide and three other inhibitors were not effective for sprout control of stored yams.

VIII. COMPARISON OF IRRADIATION WITH THE ALTERNATIVE METHODS

For chemical control of sprouting in potatoes, maleic hydrazide (MH-30), methyl ester of α-naphthalene acetic acid (M-1), chloroisopropyl N-(3-phenyl) carbamate (CIPC), tetrachloronitrobenzene (Fusarex®), and nonyl alcohol have been used in several countries. These sprout-inhibiting chemicals are applied in different procedures. For instance, MH-30 is sprayed on the potato plants in the field 4 to 6 weeks before harvest; Fusarex® is applied 4 weeks postharvest by a dust treatment as fogging dosage. With onions, most of the chemical sprout inhibitors used for potatoes are not effective[203] because the chemicals cannot penetrate the tough outer skin of onion bulbs to reach the growing points inside onion bulbs.

The only recommended chemical for sprout inhibition in onions is MH-30, which must be applied at least 24 hr before rain at the right stage in maturity.[11] As compared with irradiation treatment, chemical methods have some benefits, i.e., cheaper, no need for special irradiation facilities and technology, no need to transport the crops to an irradiator for treatment, and a longer experience relevant to usage. However, the use of chemical inhibitors suffers from numerous limitations, e.g., a degree of ineffectiveness, need for legal permission in Japan for practical use, and risks of toxicity. The toxicology problem of MH-30 and other chemical inhibitors should be carefully studied including the toxicology of hydrazine as an impurity or metabolic product. Hydrazine is a known carcinogen. MH-30 has shown mutagenic and carcinogenic activity in plants and animals. Nishi et al.[204] reported positive cytogenic effects of MH-30 and its salts on cultured V79 cells in vitro. No evidence is available on human carcinogenicity at present.[205] Considering such wholesomeness aspects, it would not be expedient to use chemicals in combination with radiation to reduce the dose required for sprout control.

An alternative method for sprout control is refrigeration. Because this method is very expensive, it has found very limited use in practical application. After shipping from refrigeration storage, tubers and bulbs tend to sprout faster. Since the sugar content of potato tubers increases during storage at refrigeration temperature, reconditioning is required to lower the sugar content.

In evaluating these two methods, the irradiation method will be seen to have several advantages. The technological procedures of irradiation treatment have been well established and provide a product of better quality except for after-cooking darkening. The most suitable conditions for radiation inhibition of sprouting in tubers and bulbs should be examined prior to application and storage. The variety of the potato to be irradiated and agronomic and other factors in different countries and potato growing districts should be defined in depth. In a large-scale operation irradiation can be applied with uniform effect. Taking all the factors into account, it is concluded that the sprout-inhibition effect is more effectively achieved by irradiation at low doses without unfavorable side effects, except for after-cooking darkening, as compared with alternative methods. The only disadvantage of the irradiation method is its cost, which is somewhat higher than that of the chemical method. However, it should be noted that irradiated potatoes and onions produced under the specified condition have been shown to be safe for human consumption, whereas the chemical method provokes questions with respect to its wholesomeness aspects.

IX. IDENTIFICATION OF IRRADIATED TUBERS AND BULBS

Shimomura et al.[19,206] state that irradiated potatoes can be detected by the observation of the bud shape. They mentioned the remarkable difference in the growth of buds after the treatment with gibberellin or kinetin, since the dormant buds in irradiated potato tubers show no visual change during 14 days following the treatment with gibberellin in the dark or kinetin in the light, whereas those in the unirradiated control similarly treated developed at this time into a stem or young leaves in the particular shape and color (Figure 12). As a biochemical parameter for identifying the irradiated potatoes, Thomas et al.[207] have proposed measurement of the radiation induction of PAL activity. The capacity of irradiated tubers inducing PAL activity by second irradiation decreases, while the unirradiated tubers exhibit the full response to irradiation in this capacity even after 3 months of storage. Penner and Fromm[208] noted that since the content of chlorogenic acid in tubers immediately rises after irradiation and decreases down to the normal level within several weeks, the identification of irradiated potatoes was possible by determination of chlorogenic acid content only within a few days after irradiation and when compared with the unirradiated potatoes of the same lot.

With irradiated onions, Hori et al.[209] proposed a method for the early detection of the irradiation treatment, namely, by observing whether or not root elongation from

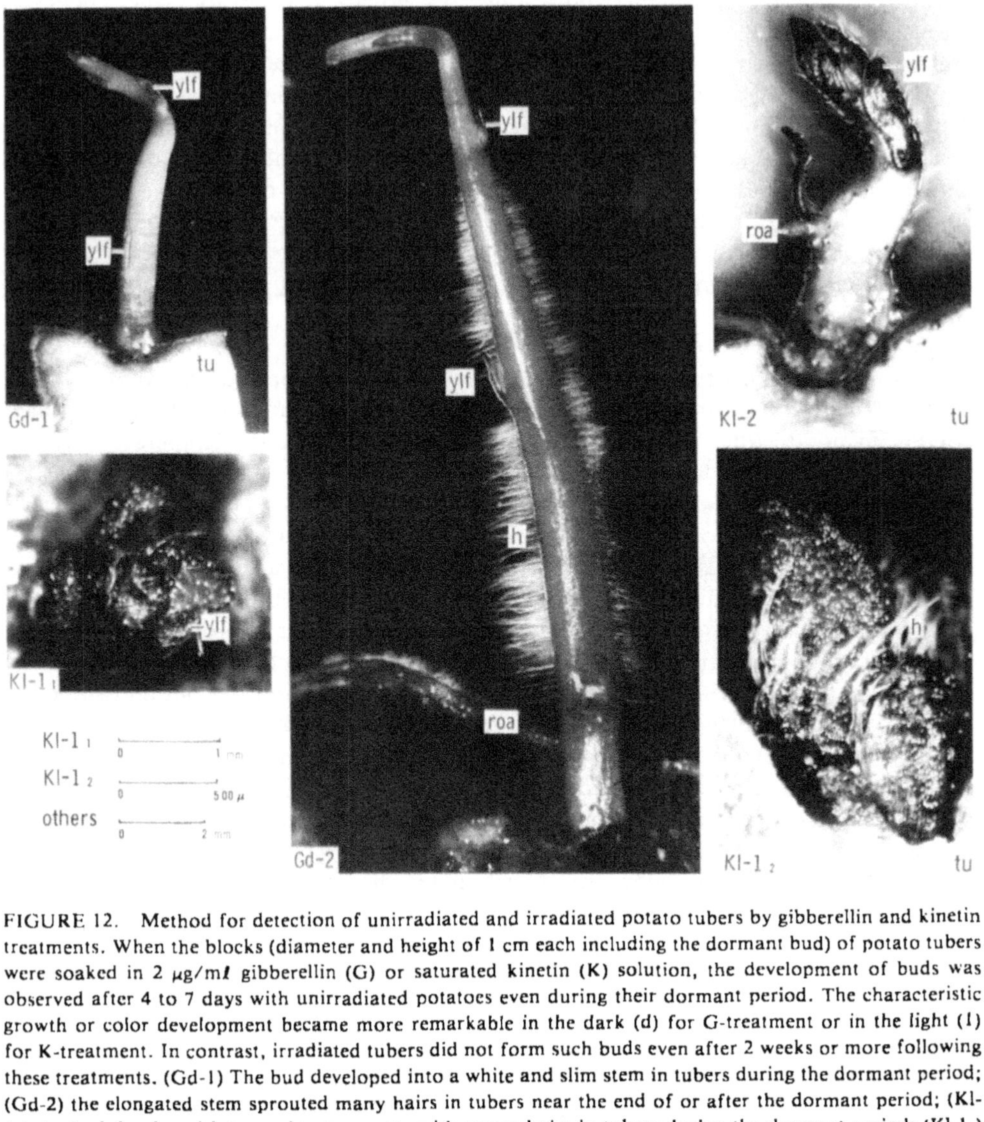

FIGURE 12. Method for detection of unirradiated and irradiated potato tubers by gibberellin and kinetin treatments. When the blocks (diameter and height of 1 cm each including the dormant bud) of potato tubers were soaked in 2 μg/m*l* gibberellin (G) or saturated kinetin (K) solution, the development of buds was observed after 4 to 7 days with unirradiated potatoes even during their dormant period. The characteristic growth or color development became more remarkable in the dark (d) for G-treatment or in the light (l) for K-treatment. In contrast, irradiated tubers did not form such buds even after 2 weeks or more following these treatments. (Gd-1) The bud developed into a white and slim stem in tubers during the dormant period; (Gd-2) the elongated stem sprouted many hairs in tubers near the end of or after the dormant period; (Kl-1_1) the bud developed into a crimson rosette with sparse hairs in tubers during the dormant period; (Kl-1_2) outer young leaves with many hairs in tubers during the dormant period; (Kl-2) white burly stem with a crimson bud in tubers near the end of or after the dormant period. Abbreviations: h, hair; roa, adventitious root; tu, tuber; ylf, young leaf. (The authors thank Dr. H. Shimomura for photographs and legends.)

irradiated bulbs immersed in tap water at 20 to 25°C has occurred. Since the unirradiated control allowed rooting within 1 to 4 weeks, no rooting indicates irradiation action. Münzner[210] reported a rapid method to identify irradiated bulbs by failure of root formation. After cutting off half of the onion disk and holding the bottom part of the onion bulb with the other half of its disk in tap water at 22 to 24°C, bulbs not treated with radiation as well as those treated with chemicals developed roots 2 to 6 mm in length 16 to 20 hr after the treatment; irradiated onions did not form roots. The discoloration of inner buds of onion bulbs would be also helpful to detect the irradiation treatment. Considering the radiobiological approach, the examination of chromosomal aberration in cells of the growing points might provide a useful identification method for irradiated tubers and bulbs, but no work on this has been reported so far.

X. THE COMMERCIAL IRRADIATION PLANT IN JAPAN[1]

A. Fundamental Concept

Any nation contemplating building an irradiation facility should consider its adaptability and contributions to that nation's industrial system, its economy, and its public health services. In other words, the mere comparison of the cost of irradiation with other food processes is insufficient grounds for a decision. It would be illogical, for example, to consider food problems from the same economic perspective as would be used to evaluate a textile manufacturing technique.

Every country has its own national industrial system and if that country belongs to the Economic Co-operation Organization it will be useless for that nation alone to introduce irradiation preservation into its food economy. On the other hand, we can understand the negative attitude of the U.S. industry toward the radiation preservation of potatoes. The existing manufacturing and distribution system seems adequate, at least for the present. For instance, the cost of the irradiation treatment of potatoes to prevent sprouting compared with the cost of the chemical treatment to inhibit sprouting shows no great advantage of one over the other in the opinion of many people in the U.S. potato industry.

Japan, on the other hand, has two situations in regard to its domestic potato distribution system that differ from that of the U.S. and the European countries. The big difference is the prohibition of chemicals under the food additive laws of Japan for sprout inhibition of tubers and bulbs. The second difference is the absence of any large potato processing industries which have their own potato distribution system. These two differences plus government subsidies for stabilizing the supply and market price of potatoes have accelerated the application of the irradiation process to achieve sprout inhibition. Japanese experience in establishing the first commercial food irradiation plant will be useful to countries that have a domestic potato distribution system similar to that of Japan.

B. The Shihoro Experience

An irradiation plant with a monthly capacity of irradiating 10,000 tons of potatoes was established in 1973 at Shihoro, located in the central part of Hokkaido. The potato season starts at the end of April in Kyushu in the southern part of Japan; it begins in June in Honshu. As a rule, the potatoes are sent directly to market without storage. In Hokkaido in northern Japan, potato harvesting ends in mid-October. The yield amounts to about 2 million tons/year, which represents 65% of the total potato production in Japan (Figure 13). Potatoes from Hokkaido are supplied to the market from October to April. Until February these potatoes can be successfully stored at low-temperature storage, but after February and until April, sprouting, rotting, and weight losses occur and transportation conditions are often bad due to the heavy snow fall. As a result prices are adversely affected.

The national project for food irradiation in Japan was initiated in 1967 with the main objective of stabilizing the supply and market price of potatoes. The potato irradiator in Hokkaido was started with the object of keeping potatoes for nearly 7 months after harvest. However, if Hokkaido potatoes are stored for a longer period by means of irradiation, this means that the production of potatoes in other areas will be reduced.

Production of potatoes in Shihoro is 250,000 tons/year. Most of it is used for starch; household use is small. The domestic use of potatoes is mainly confined to the southern part of Hokkaido, because of the preferable sizes and color.

Irradiated potatoes were meant to stabilize prices and prevent the wide fluctuations during the off season (February to April). The price fluctuations of agricultural prod-

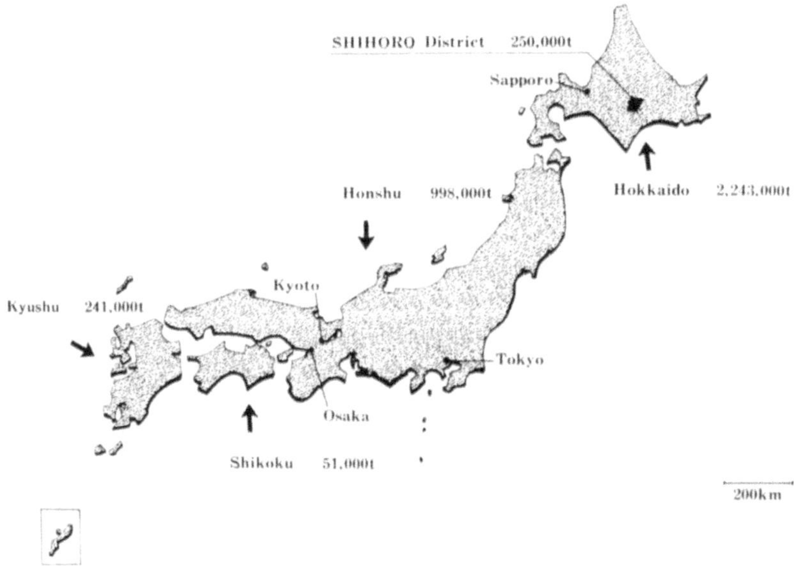

FIGURE 13. Potato yield in Japan (total 3,200,000t).

ucts in markets are basically due to shortage or oversupply of products. However, in the case of potatoes and onions, the fluctuation of prices cannot be completely explained by these reasons. The speculative purchases of potatoes could be avoided by having available several thousands of tons of irradiated potatoes. Thus, before the demand and supply of potatoes could become uncontrollable, the irradiated potatoes would help stabilize prices.

Although irradiated potatoes help stabilize prices when the potatoes from the new harvest come to market from Kyushu and the southern part of Honshu, irradiated potatoes do not sell well in the market. However, when an irradiation plant is established in areas where potato processing plants are available, it is possible during off-seasons to control the marketing of domestically consumed potatoes and the use of the remainder of irradiated potatoes.

Before the potato irradiator was established in Shihoro, there already existed both public and private potato processing plants, some with the capacity for manufacturing starch at the rate of 180,000 tons/year, mashed potato plants with capacities of 17,000 tons/year, and in addition there were plans for establishing processing centers. Thus, by using these plants, the remaining potatoes could not only be efficiently utilized but the cost of irradiation could be absorbed by the several plants and the operation period of the potato processing center could be extended. There was, therefore, no necessity to increase prices of both domestically used potatoes and potatoes for processing. These then are the reasons for the construction of the first potato irradiator in Shihoro, Hokkaido.

C. Description of the Irradiation Plant

The radiation plant established at Shihoro in Hokkaido has a ^{60}Co source of 300 kCi and can irradiate about 350 tons of potatoes per day for a period of 3 months in a year, thereby irradiating over 30,000 tons of potatoes per year. As shown in diagrams of the plant (Figure 14), the radiation shielding consists of a 2-m thick concrete wall and ceiling. This provides full shielding even if the strength of the ^{60}Co source is increased to 1 million Ci. The radiation plant consists of three areas: an irradiation chamber with source pool and circular conveyor; a control area with operational, ex-

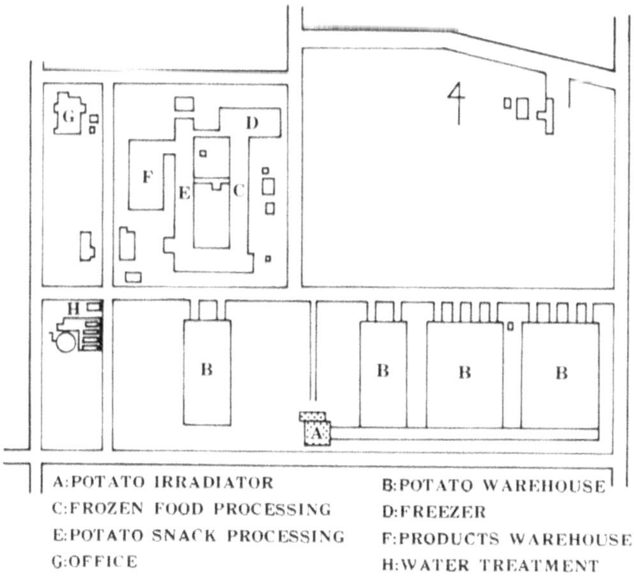

FIGURE 14. Layout of Shihoro potato processing center.

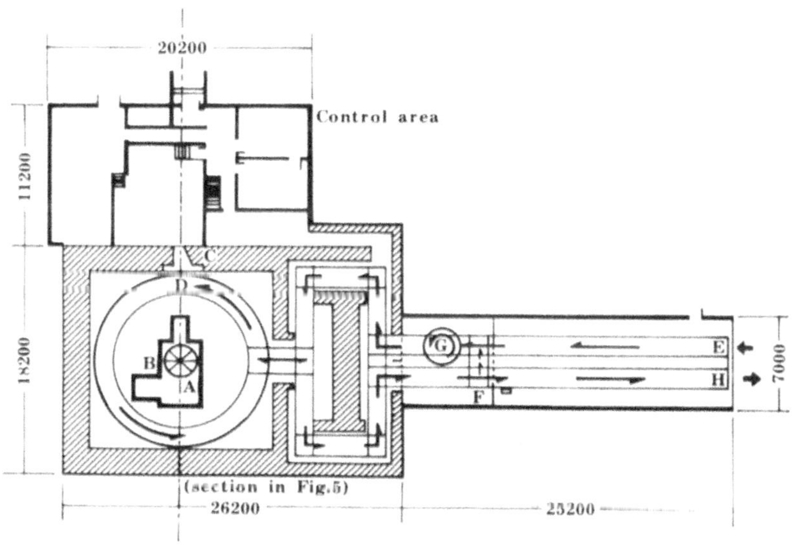

FIGURE 15. Shihoro Agricultural Cooperative Association potato irradiator (plane view): (A) source; (B) water pool; (C) window; (D) irradiation conveyor; (E) entrance line; (F) line transfer; (G) turntable; (H) exit line.

perimental, and analytical rooms; and a conveyor building with a carrying conveyor and turntable. Figures 15, 16, and 17 depict the radiation facility at Shihoro.

Figure 15 shows (A) the cyclindrical-shaped source with a diameter of 1 meter; (B) the water pool with a depth of 6.8 m; (C) tbe viewing window, and (D) the conveyor. The conveyor can carry 19 containers. Potatoes to be irradiated are introduced in the conveyor at E and pass through D. They are circulated for about an hour. They come out at F. At G the containers do a turn around, passing back to conveyor D, which circulates around them for another hour. Finally they come out at H and from there pass to the warehouse. This system irradiates about 10 containers in 1 hr.

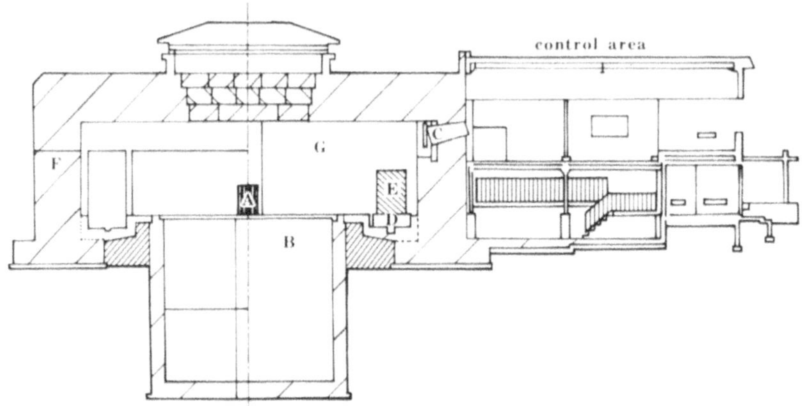

FIGURE 16. Shihoro Agricultural Cooperative Association potato irradiator (section): (A) source; (B) water pool; (C) window; (D) irradiation conveyor; (E) potato container; (F) shield; (G) irradiation chamber.

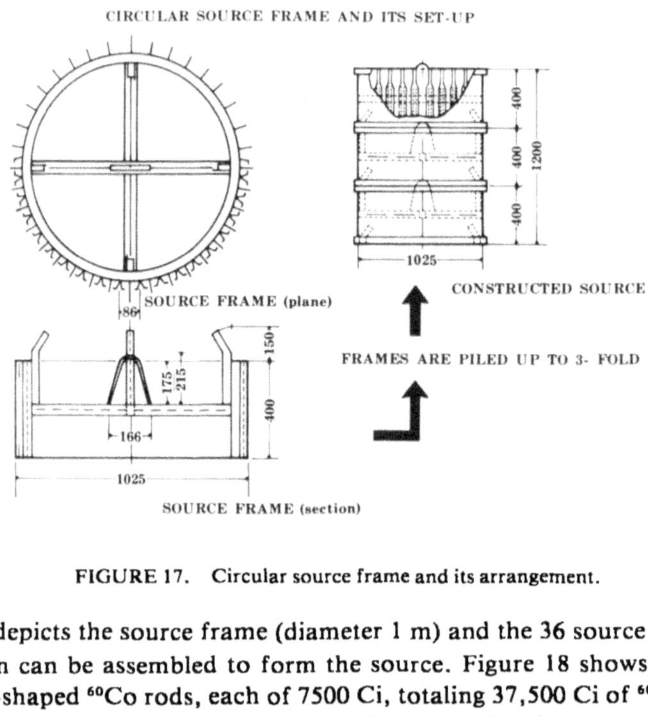

FIGURE 17. Circular source frame and its arrangement.

Figure 17 depicts the source frame (diameter 1 m) and the 36 source units. The three frames shown can be assembled to form the source. Figure 18 shows the source unit with 5 pencil-shaped ^{60}Co rods, each of 7500 Ci, totaling 37,500 Ci of ^{60}Co. The assembled source frame without the source unit is illustrated in Figure 19. Figure 20 shows the conveyor traveling to and from the warehouse. A complete view of the irradiation chamber, taken by a wide angle lens, can be seen in Figure 21. In the center is the source and at the far left the window leading to the control room. In Figure 22 the potato container is shown on the round turntable, which is equipped with a device to hold alternate containers as partly irradiated and nonirradiated potatoes enter the chamber.

An overall view of the Shihoro facility is shown in Figure 23. This plant costs \$1.3 million (approximately in 1973; 1 dollar = 300 Japanese yen) of which \$866,000 was provided by the Japanese Ministry of Agriculture and the Hokkaido Prefectural Government. The rest was provided by the Shihoro Agricultural Co-operative Association

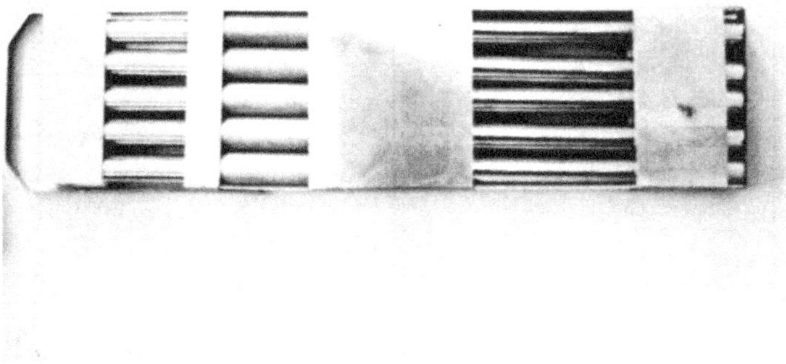

FIGURE 18. Source unit with five pencil-shaped ^{60}Co rods. Commercial Irradiation Plant at Shihoro.

FIGURE 19. Commercial Irradiation Plant at Shihoro. Cylindrical source frame.

(SACA). Annual operation costs, calculated from the period 1974/75 and 1975/76, were approximately $200,000. When the plant is operating at full capacity (3 months), the operational cost of the irradiated potatoes is estimated at $7 to $10/ton. If onions are to be irradiated in this plant, because of their lower dose requirement (half that of potatoes), the only adjustment needed is to double the conveyor speed. Because of the lesser bulk density of onions, there is more uniformity with regard to the absorbed dose.

D. Merits of the Irradiation Process

The SACA facility is also designed to process potatoes. Of the 250,000-ton annual potato production, 180,000 tons or less were used for starch manufacture. The rest was marketed for general distribution. In the same year as the founding of the irradiation plant, a $6.7 million potato-processing center was established in Shihoro by the SACA.

FIGURE 20. Commercial Irradiation Plant at Shihoro. Conveyor to and from warehouse.

FIGURE 21. Commercial Irradiation Plant at Shihoro. Irradiation chamber.

The SACA potato snack plant (chips and shoe strings) has an output of 60 tons/day with an annual operation of 200 days. The frozen food plant (pommes frites and diced potatoes) has a capacity of 20 tons/day with an annual operation of 250 days. The capital investment for both of these plants was $2.5 million. In addition, a warehouse for raw material storage with a capacity of 22,000 tons was built at a cost of $2 million. The plant for waste-water treatment cost $0.33 million. Out of the total $6.7 million, over half was financed by the Ministry of Agriculture and Forestry and the Hokkaido Prefectural Government.

In this processing center 17,000 tons of potatoes can be handled each year. For nonirradiated potatoes, if suitable for processing purposes and if economically feasible, the operation period could be extended to 10 months, the last 5 months of which the center would be supplied with irradiated potatoes.

Up to now, agricultural product processing and food processing have been difficult to carry out in parallel because of their short shelf life and the short operational period. This problem is now solved for potatoes. These plants would profit the grower by producing processed foods and benefit him also by the regulation of potato shipments

FIGURE 22. Commercial Irradiation Plant at Shihoro. Potato container on turntable.

FIGURE 23. Commercial Irradiation Plant at Shihoro. Front view.

to the market when the market price fluctuates. As mentioned earlier, the radiation treatment costs about $7 to $10/ton, but the SACA intends to sell the irradiated potatoes at the same price as nonirradiated potatoes since they can recoup the irradiation costs from the profits on processed foods. Thus the radiation plant benefits both the producer and consumer.

The most important objective of potato irradiation is the control of market price fluctuation and it can be reported that this objective has been attained. Figure 24 shows the range of potato price fluctuation for the 3 years prior to the Shihoro radiation facility becoming operational. The irradiation plant started its operation on January 12, 1974; the output of irradiated potatoes during 1973/74 was small. During 1974/75 the Government controlled Vegetable Price Stabilization Fund bought 5500 tons of irradiated potatoes in expectation of price fluctuations from the end of February to April. During this period they were able to control the price fluctuation of potatoes.

To prevent price fluctuation, the Vegetable Price Stabilization Fund and Tokyo Metropolitan Government have also been buying several thousand tons of potatoes. They were successful in overcoming the price fluctuation until February. During March and April, however, these potatoes started sprouting and they were unsuccessful in accomplishing their objective.

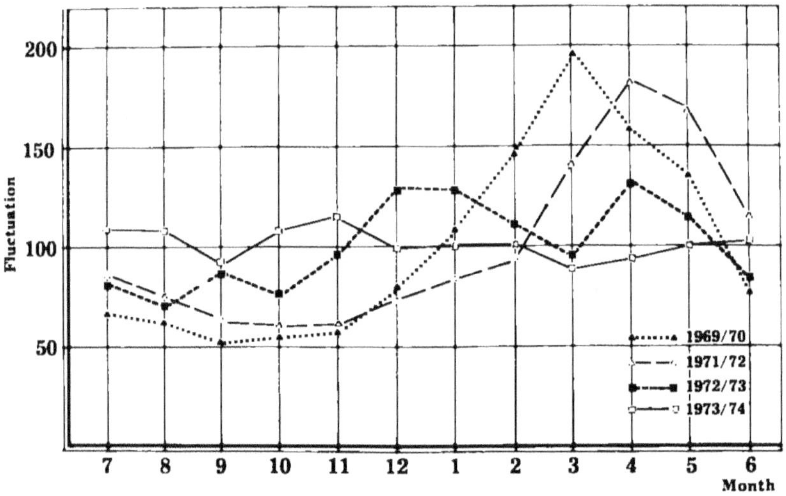

1967/70: Most fluctuated season during last 10 years.
1974/75: In this season Vegetable price stabilization Fund started to buy the irradiated potatoes for controling the wholesale price.
: Average price of each season is indicated as 100.

FIGURE 24. Price fluctuation of potatoes in Tokyo wholesale market.

During 1975/76 the Vegetable Price Stabilization Fund bought 3700 tons of irradiated potatoes and was successful in controlling the price fluctuation. Although every month 12,000 tons of potatoes enter the Tokyo market, the fund was able to control the price fluctuation with 3700 tons. This is incredible, especially since it stopped speculative buying. Out of the 3700 tons, the fund sold 2000 tons in March, 400 tons in April, and 550 tons in May. The remainder was used for starch manufacture. In April and May the irradiated potatoes were sold very cheaply because by this time the newly harvested potatoes started coming onto the market from Kyushu and the southern part of the main island.

The remainder of irradiated potatoes (those left after serving to control price fluctuation) has been used in the manufacture of starch. It is necessary, of course, to use them in more profitable ways, such as to make potato chips and pommes frites. Up to now potato processors have used potatoes from Hokkaido from autumn to spring; in the summer season they use potatoes from Kyushu and Honshu. For processing purposes Hokkaido potatoes are the best in quality and so, if Hokkaido potatoes could be made available to processors all year round, it would be ideal. Furthermore, if this were so, the processing plants in growing areas could extend their period of manufacture from seasonal to all year round.

As mentioned above, the objective for irradiated potatoes is twofold: (1) to control price fluctuation until April and (2) to supply Hokkaido potatoes to the potato processors during March to July. But the problem of using irradiated potatoes during the period from March to July has not yet been solved.

E. Present Technological Problems

The Shihoro district potato harvesting period is rather short, only 1.5 months, from September 15 to the end of October. During this short period 60,000 to 70,000 tons of potatoes must be harvested. Harvesting is done by potato harvesters and the harvest is put into plastic minicontainers of about 25 kg, and air-dried in the field for 1 to 2 days. The containers are then loaded into trucks carrying about 6 to 7 tons and sent to the receiving station and weighed. Thereupon, 50-kg samples are taken out and submitted to preliminary quality and grading tests.

The potatoes are then manually graded and the rotted, damaged, irregularly shaped potatoes are rejected. The potatoes are then automatically classified in M, L, and LL sizes. The potatoes are then bulk stored or stored in containers. Because large quantities are handled in a short period of time, it is almost impossible to grade them perfectly. As a result, many damaged potatoes are included among the irradiated potatoes resulting in rotted potatoes after long storage.

At present there are facilities for storing 24,000 tons in bulk storage and 18,000 tons in container storage facilities. The temperature of these warehouses during the curing period of September to October for bulk storage is maintained at 10 to 15°C with 70 to 80% relative humidity; for container storage the temperature is kept at 10 to 15°C and relative humidity at 80 to 90%. After curing, to prevent the increase of reducing sugar content, potatoes for processing are stored at 7°C, relative humidity 90% whereas those for household use are maintained at 5°C.

There are both advantages and disadvantages in bulk and container storage, but at present both types are used. Container storage, because of easy handling methods made possible by the fork lift, results in much less damage and is the ideal method. During storage in containers, however, the air cannot penetrate sufficiently into the container and some rotting results.

By contrast, potatoes in bulk storage are kept in a space 1600 m² by 4.5 m. Air enters from below, providing good aeration and, therefore, prevents rotting. But unfortunately these potatoes have to be put into containers for irradiation and during this process damage to the extent of about 5% cannot be avoided. For example, in the case of potatoes for household use, (1) there is a rough inspection of the potatoes, (2) they are put into containers, (3) cured (4) irradiated (5) taken out of containers for bulk storage, (6) put into containers again for transportation, (7) taken out of the containers for inspection (which is rather severe), and (8) again put into carton boxes (20 kg each) for final marketing. Thus, it can be seen that there are four times when loading and unloading into containers is carried out. This excessive handling is the source of much damage.

It is important to develop methods to reduce the number of times that loading and unloading into containers is done. As countermeasures, the transportation of potatoes in April for household use and potatoes for processing from April to July are treated after reaching receiving stations and cured in bulk storage. Potatoes for domestic use are bulk stored at 3 to 5°C and those for further processing stored at 7°C. By the end of February they are loaded into containers, irradiated, and stored until transportation. For proper effect, irradiation of potatoes must be done as early as possible after harvest but in this method, since irradiation is done in February, the risk involved is greater, which contradicts an earlier statement. At the end of February, when the dormancy period is over and the irradiation effect is less, at even the lowest dose sprouting will take place. During storage and distribution rotting sometimes takes place.

From the results of the last 4 years of experience, the storage method of choice for irradiated potatoes can be stated as follows. Potatoes for processing should be stored at 7°C, relative humidity 90%, both before and after irradiation, in order to keep the reducing sugar low and to prevent shrinkage. Potatoes with high reducing sugar content on arrival at receiving stations should be reconditioned immediately after irradiation. On the other hand potatoes for household use should be stored at 3 to 5°C, and before transportation the temperature should be brought gradually to the ambient stage.

At present, the most important improvements technically are improvements in gathering and inspection of potatoes at harvest time and the establishment of a container storage system. Potatoes to be irradiated should be mature and sound. Farmers should be paid a premium for the best potatoes. It is recommended that another line be added to the existing inspection line. Another problem is container storage. If proper air circulation is carried out, long-term storage becomes possible without rotting. Irradiation could also be applied during the dormancy period. An urgently needed improvement is the design of each container so that proper air circulation is assured.

XI. CONCLUSION

During the previous decade, many countries have conducted investigations on consumer acceptance or test marketing of irradiated potatoes and onions[180,211-215] to assess the feasibility of the irradiation process. On the basis of the encouraging scientific data, three commercial or semicommercial irradiators mainly for tubers and bulbs are scheduled to be established in 1 or 2 years: (1) a ^{60}Co, 150-kCi semicommercial irradiator of the Institute of Nuclear Energy Research, Atomic Energy Council in the Republic of China,[216] (2) a ^{60}Co, 30-kCi prototype onion irradiator in Hungary,[214] and (3) a ^{60}Co, 180-kCi Italian commercial irradiator.[215] The Italian irradiator, financed entirely by the government (U.S. $3.5 million), is planned to be installed in the Fucino area (100 km east of Rome). It will be devoted mainly to inhibit sprouting of potatoes, onions, and garlic.

As mentioned above, the irradiation process for sprout inhibition of potatoes and onions is now coming to the stage of commercialization and many countries are conducting their own projects to bring this application to realization. However, it should be recognized that there are several fundamental problems which must be solved in each country and region to make sure of the success of the project as a whole.

A. Essential Requirements in Planning and Research

Considering the market structures of each country in regard to its distribution scheme and consumer demands, appropriate amounts of irradiated potatoes or onions to be provided for the market should be determined according to such criteria as objectives of irradiation control of market prices, steady supply to the market, or use of the product (household consumption or processing industries).

The cost structure of irradiation should be analyzed in assessing the functions of the irradiation plant in the total system of potatoes and onions in the food chain including distribution and consumption. Essentially, it is not just building an irradiation plant per se, but building an effective plant within an integrated system that makes it possible to provide a sufficient amount of potatoes and onions to control market prices.

The operational conditions for irradiating, curing, reconditioning, and storing potatoes and onions are greatly dependent not only on varieties and objectives, but also on different parameters of each region in agronomic, climatic, weather during the harvesting period, and soil conditions. For example, potato varieties of the deep-end type, when grown in black soil, tend to carry soil on the tuber in storage after being harvested, cured, and irradiated. In this case, low storage moisture is required to avoid partial overhumidification and spoilage.

B. Essential Requirements in Plant Construction and Operations

Irradiation facilities should be simple in mechanical design and easy to handle during operation. For instance, an S-shaped conveyor system with adequate incline from inlet toward outlet of irradiation room facilitates, by its own weight, movement toward the outlet. Whatever conveyor-driving system is selected, it should be simple.

An irradiation system needs to be linked up with curing, reconditioning, and storage houses. The success of an irradiation operation requires proper treatment of the product before and after irradiation, and complete fulfillment of storage requirements after irradiation.

A sorting system should be established in order to select sound potatoes and onions for irradiation. In prolonged storage, i.e., for more than 6 months, sound tubers and bulbs with no injury are essential.

C. Marketing and Aftercare

An appropriate public information system should be provided. Through acceptance tests or other means, public understanding of food irradiation can be advanced. Irradiated potatoes and onions were the first to come to the market as irradiated food; these products are mainly responsible for the research and development of other irradiated foods that have followed or will follow after them. They have played an important role in furthering the peaceful use of atomic energy.

In commercial operation of the irradiation process many unexpected problems, never before seen on the research level, may arise. Administrative action of some kind to cope with such situations may be quite helpful to overcome the obstacles.

The success of irradiated potatoes and onions will accelerate the acceptance of other irradiated foods and it is a milestone in establishing the radiation process as one of the food preservation technologies, equivalent to heat and freezing methods.

REFERENCES

1. **Umeda, K.,** Background to the establishment of the first food irradiation plant in Japan, in *Requirements for the Irradiation of Food on a Commercial Scale,* Proc. of a Panel, International Atomic Energy Agency (IAEA), Vienna, 1975, 113.
2. **Siu, R. G. H. and Mandels, G. R.,** Beginning of radiobiology, in Radiation Preservation of Food, U.S. Army Quartermaster Corps, Government Printing Office, Washington, D.C., 1957, 31.
3. **Metlitsky, L. V.,** X-irradiation of vegetables in storage, *Konservn. Promst.,* 4, 1936.
4. **Sparrow, A. H. and Christensen, E.,** Effects of X-ray, neutron, and chronic gamma irradiation on growth and yield of potatoes, *Am. J. Bot.,* 37, 667, 1950.
5. **Sparrow, A. H. and Christensen, E.,** Improved storage quality of potato tubers after exposure to Co^{60} gammas, *Nucleonics,* 12(8), 16, 1954.
6. **Sparrow, A. H. and Schairer, L.,** Effect of X rays, gamma rays, fast electrons and fast neutrons on inhibition of growth and sprouting in potatoes, in Conf. on Biological, Physical and Industrial Aspects of Potato Irradiation, Brookhaven National Laboratory, Upton, N.Y., 1955, 7.
7. **Brownell, L. E., Anderson, L. C., Gomberg, H. J., Kempe, L. L., Martin, J. J., Nehemias, J. V., and Solfc, R. A.,** Irradiation-Utilization of gross fission products, University of Michigan Eng. Res. Inst. Prog. Rep. No. 7, Ann Arbor, 1954, 64.
8. **Dallyn, S. L., Sawyer, R. L., and Sparrow, A. H.,** Extending onion storage life by gamma irradiation, *Nucleonics,* 13, 48, 1955.
9. **Burton, W. G. and Hannan, R. S.,** Use of γ-radiation for preventing the sprouting of potatoes, *J. Sci. Food Agric.,* 8, 707, 1957.
10. **Atomic Energy of Canada Ltd.,** An Application to the Food and Drug Directorate, for the approval of the use of gamma radiation from Cobalt-60 for the prevention of sprouting in potatoes, Department of National Health and Welfare, Ottawa, 1962.
11. **Atomic Energy of Canada Ltd.,** An Application to the Food and Drug Directorate, for the approval of the use of gamma radiation from cobalt-60 for the prevention of sprouting in onions, Department of National Health and Welfare, Ottawa, 1962.
12. **U.S. Army Natick Laboratories,** Irradiation of white potatoes for sprout inhibition with cobalt-60, the Army Petition for Clearance and Approval to the Food and Drug Administration, Natick, Mass., 1963.
13. **Nair, P. M., Thomas, P., Ussuf, K. K., Surendranathan, K. K., Limaye, S. P., Srirangarajan, A. N., and Padwal-Desai, S. R.,** Studies of sprout inhibition of onions and potatoes and delayed ripening of bananas and mangoes by gamma irradiation, in *Radiation Preservation of Food,* Proc. Symp., Bombay, 1972, IAEA-SM-166/11, IAEA, Vienna, 1973, 347.

14. **Umeda, K.**, Effects of temperature and humidity on the reducing sugar, poly-phenol, organic acid, free amino acid and ascorbic acid content of irradiated potatoes, paper presented to FAO/IAEA Res. Coord. Meet. Technological and Economic Feasibility of Food Irradiation, Accra, Ghana, February 1978.
15. **Metlitsky, L. V., Rogachev, V. N., and Krushchev, V. G.**, *Radiation Processing of Food Products*, (Engl. transl.) Moscow, 1967. Gerrard, M., McKinney, F. E., Baker, P. S., and Wierbicki, E., Eds., ORNL-IIC-14, Oak Ridge National Laboratories, Oak Ridge, Tenn., July 1968.
16. **Shimomura, H. and Kuriyama, E.**, The morphological and histological effects of ^{60}Co γ-rays. I. Dormant buds of potato tuber, *Jpn. J. Bot.*, 45, 343, 1970.
17. **Shimomura, H. and Kuriyama, E.**, The morphological and histological effects of ^{60}Co γ-rays. III. The developmental structure of buds of potato tuber, *Jpn. J. Bot.*, 51, 303, 1976.
18. **Kuriyama, E. and Shimomura, H.**, The morphological and histological effects of ^{60}Co γ-rays. IV. The deformed buds on the irradiated potato tuber, *Jpn. J. Bot.*, 52, 86, 1977.
19. **Shimomura, H., Kuriyama, E., and Takai, Y.**, A method for early detection of irradiated potato tubers. I. Morphological changes by ^{60}Co γ-rays on dormant buds, *J. Food Hyg. Soc. Jpn.*, 13, 531, 1972.
20. **Shimomura, H. and Kuriyama, E.**, The morphological and histological effects of ^{60}Co γ-rays. II. Dormant buds of onion bulb, *Jpn. J. Bot.*, 48, 295, 1973.
21. **Ogata, K., Iwata, T., and Chachin, K.**, The effect of gamma radiation on sprout prevention and its physiological mechanism in the potato tuber and the onion bulb, *Bull. Inst. Chem. Res.*, 37, 425, 1959.
22. **Rubin, B. A. and Metlitsky, L. V.**, A study of the action of ionizing radiation on the metabolism of potato tubers in relation to the problem of their all-year-round storage, *Proc. 2nd Int. Conf. Peaceful Uses At. Energy*, 27, 437, 1958.
23. **Korableva, N. P. and Metlitsky, L. V.**, *Studies on a Role of Nucleic Acid Metabolism and Its Regulation in the Dormant State of Plant Storage Organs* (Jpn. transl.), Japanese Research Association for Food Irradiation, Tokyo, 1967.
24. **Jaarma, M.**, Effects of ionizing radiations on inhibition of sprouting and biochemical and physiological changes in potato tubers, Risö Rep. No. 16, 70, 1960.
25. **Skoog, F.**, The effect of X-irradiation on auxin and plant growth, *J. Cell. Comp. Physiol.*, 7, 227, 1935.
26. **Gordon, S. A.**, Occurrence, formation, and inactivation of auxins, *Annu. Rev. Plant Physiol.*, 5, 341, 1954.
27. **Gordon, S. A.**, Studies on the mechanism of phytohormone damage by ionizing radiation, *Proc. Int. Conf. Peaceful Uses At. Energy*, 11, 283, 1956.
28. **Mathur, P. B.**, Reversal of γ-ray-induced dormancy of potato tubers by gibberellic acid, *Nature (London)*, 190, 547, 1961.
29. **Ananthaswamy, H. N., Ussuf, K. K., Nair, P. M., Vakil, U. K., and Sreenivasan, A.**, Role of auxins in the reversal of radiation-induced growth inhibition in plants, *Radiat. Res. Rev.*, 3, 429, 1972.
30. **Mathur, P. B.**, Reversal of gamma-ray-induced susceptibility to decay of potato tubers and tomato fruit by methyl ester of indolyl-3-acetic acid, *Nature (London)*, 199, 1007, 1963.
31. **Kuzin, A. M., Plyshevskaya, E. G., Kopylov, V. A., Ivanitskaya, E. A., Lebedeva, N. E., Kolomijtseve, I. K., Tokarskaya, V. I., and Melnikova, S. K.**, On the role of the orthophenol-orthoquinone system in the initial effect of ionizing radiation on the organism, *Int. J. Radiat. Biol.*, 10, 1, 1966.
32. **Kopylov, V. A., Osipova, I. N., and Kuzin, A. M.**, The mutagenic effect of extracts from gamma-irradiated potato tubers on the sex cells of male mice, *Radiobiologija*, 12, 524, 1972.
33. **Levinsky, H. V., Wilson, M., and MacFarland, H. N.**, A study of the mutagenic effects of an alcoholic extract of irradiated potatoes in the mouse, Int. Proj. in the Field of Food Irradiat., IFIP-R9, Karlsruhe, 1973.
34. **Hossain, M. M., Huismans, J. W., and Diehl, J. F.**, Mutagenicity studies on irradiated potatoes and chlorogenic acid; micronucleus test in rats, *Toxicology*, 6, 243, 1976.
35. **Gustafson, F. G., Brownell, L. E., and Martens, R. A.**, Influence of gamma irradiation of potato tubers on the rate of respiration, *Am. Potato J.*, 34, 177, 1957.
36. **Ussuf, K. K. and Nair, P. M.**, Metabolic changes induced by sprout inhibiting dose of γ-irradiation in potatoes, *J. Agric. Food Chem.*, 20, 282, 1972.
37. **Truelson, T. A.**, Irradiation of potatoes. Effects on sugar content, blackening, and growth substances, *Radiation Preservation of Foodstuffs*, 2nd Scand. Meet. Food Preserv., 138, 55, 1963.
38. **Sussman, A. S.**, The effects of ionizing radiation upon respiration and oxidase of the potato tuber, *J. Cell. Comp. Physiol.*, 42, 273, 1953.
39. **Brownell, L. E., Gustafson, F. G., Nehemias, J. V., Isleib, F. R., and Hooker, W. J. M.**, Storage properties of gamma irradiated potatoes, *Food Technol. (Chicago)*, 11, 306, 1957.
40. **Ogawa, M., Hyodo, H., and Uritani, I.**, Biochemical effects of gamma radiation on potato and sweet potato tissues, *Agric. Biol. Chem.*, 33, 1220, 1969.

41. **Ojima, T., Hori, S., Toratani, H., Fujimoto, H., Katayama, T., Kitoh, S., Mori, S., Shiomi, N., and Kowyama, Y.**, The effect of ionizing radiation upon the sprout inhibition, loss of weight and respiration in potato tubers, *Annu. Rep. Radiat. Cent. Osaka Prefect.*, 11, 139, 1970.
42. **Chachin, K., Ogata, K., and Honjo, H.**, Relationships between some reducing compounds levels in the inner buds of onions and the effect of sprout-inhibition by gamma irradiation, *J. Food Sci. Technol.*, 23, 351, 1976.
43. **Chachin, K., Honjo, H., and Ogata, K.**, Oxidative capacity of some organic acids by mitochondria of irradiated potatoes and onions, *Food Irradiat.*, 9, 76, 1974.
44. **Chachin, K. and Ogata, K.**, Alternation of respiratory chain in some fruits and vegetables irradiated by gamma radiation, *Food Irradiat.*, 11, 13, 1976.
45. **Schwimmer, S., Weston, W. J., and Makowe, R. U.**, Biochemical effects of gamma radiation on potato tubers, *Arch. Biochem. Biophys.*, 75, 425, 1958.
46. **Jaarma, M.**, Influence of γ-irradiation on β-fructofuranosidase in potato tubers, *Acta Chem. Scand.*, 20, 592, 1966.
47. **Ogawa, M., Majima, R., Uritani, I., and Namiki, M.**, Effects of gamma-ray irradiation on metabolic changes in potato tubers in response to cutting, *Plant Cell Physiol.*, 9, 511, 1968.
48. **Pendharker, M. B. and Nair, P. M.**, Induction of phenylalanine ammonia lyase (PAL) in gamma irradiated potatoes, *Radiat. Bot.*, 15, 191, 1975.
49. **Ogawa, M. and Uritani, I.**, Tissue browning of potato tubers induced by gamma irradiation, *Agric. Biol. Chem.*, 34, 870, 1970.
50. **Tatsumi, Y., Chachin, K., and Ogata, K.**, Studies on the browning of potato tubers by gamma irradiation. II. The relationship between the browning and the changes of o-diphenol, ascorbic acid and activities of polyphenol oxidase and peroxidase in irradiated potato tubers, *J. Food Sci. Technol.*, 19, 508, 1976.
51. **Berset, C. and Sandret, F.**, Sprout radio-inhibition of potato tubers. Comparison of β or γ radiations after-effects on polyphenol content, polyphenoloxydase activity and after-cooking blackening, *Lebensm.-Wiss. Technol.*, 9, 85, 1976.
52. **Chachin, K., Matsuzuka, M., Honjo, H., and Ogata, K.**, Sprout-inhibition of Sapporo-ki onions and the changes of nucleic acids in their inner buds by gamma radiation, *J. Food Sci. Technol.*, 20, 158, 1973.
53. **Chachin, K., Tatsumi, Y., and Ogata, K.**, Effects of gamma radiation on morphological changes and nucleic acid content in buds of potato tubers, *J. Jpn. Soc. Hortic. Sci.*, 43, 194, 1974.
54. *Training Manual on Food Irradiation Technology and Techniques*, STI/DOC/10/114, IAEA, Vienna, 1970.
55. **Schreiber, J. S. and Highlands, M. E.**, A study of the biochemistry of irradiated potatoes stored under commercial conditions, *Food Res.*, 23, 464, 1958.
56. **Takano, H., Tanaka, Y., Umeda, K., and Sato, T.**, Sprout inhibition of potatoes by ionizing radiation. III. Dose to inhibit sprout and the change of sugar content during the low temperature storage, *Rep. Natl. Food Res. Inst.*, 27, 64, 1972.
57. **Moore, M. D., van Blaricom, L. D., and Senn, T. L.**, The effect of storage temperature of Irish potatoes on the resultant chip color, *Clemson Coll. Res. Ser.*, 43, 1, 1963.
58. **Burton, W. G., Horne, T., and Powell, D. B.**, The effect of γ-irradiation upon the sugar content of potatoes, *Eur. Potato J.*, 2, 105, 1959.
59. **Cloutier, J. A. R., Cox, C. E., Manson, J. M., Clay, M. G., and Johnson, L. E.**, Effect of storage on the carbohydrate content of two varieties of potatoes grown in Canada and treated with gamma radiation, *Food Res.*, 24, 659, 1959.
60. **Ogata, K.**, Improved storage life of fruits and vegetables by ionizing radiation, *Jpn. Agric. Res. Q.*, 7, 55, 1973.
61. **Schwimmer, S., Burr, H. K., Harrington, W. D., and Seston, W. J.**, γ Irradiation potatoes: effects on sugar content, chip color, germination, greening, and susceptibility to mold, *Am. Potato J.*, 34, 31, 1957.
62. **Ogata, K., Iwata, T., and Chachin, K.**, The effect of gamma radiation on sprout prevention and physiological mechanism of onion bulb, *J. Jpn. Soc. Hortic. Sci.*, 28, 143, 1959.
63. **Dallyn, S. L. and Sawyer, R. L.**, Effect of sprout inhibiting levels of gamma irradiation on the quality of onions, *Proc. Am. Soc. Hortic. Sci.*, 73, 398, 1959.
64. Wholesomeness of Irradiated Food, Rep. Joint FAO/IAEA/WHO Expert Committee, WHO Tech. Rep. Ser. 604, World Health Organization, Geneva, 1977.
65. The research group of the Central Laboratory of the National Defence Establishment (Paris) and of the Conservatome Society (Lyon), Analytical study of potato tubers subjected to nuclear radiation, *Proc. 2nd Int. Conf. Peaceful Uses At. Energy*, 27, 434, 1958.
66. **Fujimaki, M., Tajima, M., and Matsumoto, T.**, Effect of gamma irradiation on the amino acid of potatoes, *Agric. Biol. Chem.*, 32, 1228, 1968.

67. **Varela, G. and Urbano, G.**, Influence of irradiation on the digestibility and nutritional value of potatoes, wheat and hake, in *Proc. of a Symposium, Karlsruhe, 29 June to 1 July, 1971*, Bundesforschungsanstalt für Ernährung, Fed. Rep. Germany Rep. No. BFL/2/1971, 1971, 31.
68. **Mikaelsen, K. and Roer, L.**, Improved storage ability of potatoes exposed to gamma radiation, *Acta Agric. Scand.*, 6, 145, 1956.
69. **Panalaks, T. and Pelletier, O.**, The effect of storage on ascorbic acid content of gamma radiated potatoes, *Food Res.*, 25, 33, 1960.
70. Wholesomeness of irradiated food with special reference to wheat, potatoes and onions, Rep. Joint FAO/IAEA/WHO Expert Committee, WHO Tech. Rep. Ser. 451, World Health Organization, Geneva, 1970.
71. Steering Committee on Food Irradiation Research of Japan, Rep. on sprout inhibition in potatoes by irradiation, 1971.
72. **Pätzold, C. and Weiss, H. M.**, The effect of γ-radiation from radioactive cobalt-60 on potato tubers, *Angew. Bot.*, 31, 93, 1957.
73. **Pätzold, C. and Kolb, W.**, The effect of X-radiation on potatoes *(Solanum tuberosum)* and Jerusalem artichoke *(Helianthus tuberosus), Beitr. Biol. Pflanz.*, 33, 437, 1957.
74. **Takai, Y. and Iwao, H.**, Effects of cobalt-60 gamma rays on potatoes, *Annu. Rep. Natl. Inst. Nutr.*, 1969, 79.
75. **Lee, M. S. and Kim, H. L.**, Effects of ionizing radiation on sprout inhibition and nutritive value of potato tubers, *Korean J. Food Sci. Technol.*, 4, 29, 1972.
76. **Uralvá, M.**, Vyskumny Ustav Hygieny, Ph.D. thesis, Department of Radiation Hygiene, Research Institute for Hygiene and Epidemiology, Prague, 1973.
77. **Chachin, K., Honjo, H., and Ogata, K.**, The relationship between the levels of some reducing compounds in onion bulbs and the effect of sprout-inhibition by γ-irradiation, *Food Irradiat.*, 7, 73, 1972.
78. **Mahmoud, A. A., Kálmán, B., and Farkas, J.**, A study of some chemical changes in onion bulbs and in their inner buds as affected by gamma radiation and storage, paper presented to an IAEA/FAO/WHO Int. Symp. Food Preserv. Irradiat., IAEA-SM-221/22, Wageningen, 1977.
79. **Schwimmer, S. and Weston, W. J.**, Chlorophyll formation in potato tubers as influenced by gamma irradiation and by chemicals, *Am. Potato J.*, 35, 534, 1958.
80. **Ramaswamy, N. K. and Nair, P. M.**, Temperature and light dependency of chlorophyll synthesis in potatoes, *Plant Sci. Lett.*, 2, 249, 1974.
81. **Larsen, E. C.**, Chlorophyll formation in potato tubers as affected by temperature and time, *Science*, 111, 206, 1950.
82. **Hilton, R. J.**, Factors in relation to tuber quality in potatoes. II. Preliminary trials on bitterness in Betted Gem potatoes, *Sci. Agric.*, 31, 61, 1951.
83. **Ramaswamy, N. K., Behere, A. G., and Nair, P. M.**, A novel pathway for the synthesis of solanidene in the isolated chloroplast from greening potatoes, *Eur. J. Biochem.*, 67, 275, 1976.
84. **Sreenivasan, A.**, Compositional and quality changes in some irradiated foods, in *Improvement of Food Quality by Irradiation*, STI/PUB/370, IAEA-PK-561/10, IAEA, Vienna, 1974, 129.
85. **Atomic Energy of Canada Ltd.**, Report on the Results of the Canadian Pilot Scale Potato Irradiation Program, 1961 to 1962 season, Ottawa, 1962.
86. **Mathur, P. B.**, Reversal of γ-ray-induced dormancy of potato tubers by gibberellic acid, *Nature (London)*, 190, 547, 1961.
87. **Ussuf, K. K. and Nair, P. M.**, Effect of gamma irradiation on indole acetic acid synthesizing system in potatoes, *Phytochemistry*, 10, 929, 1971.
88. **Tajima, M., Kida, K., and Fujimaki, M.**, Effect of gamma irradiation on volatile compounds from cooked potato, *Agric. Biol. Chem.*, 31, 935, 1967.
89. Japan Isotope Association, Reports on Food Irradiation Contract Research for Science and Technology Agency, Tokyo, 1967 to 1969 and 1971.
90. **Kawakishi, S., Namiki, K., Nishimura, H., and Namiki, M.**, Effects of γ-irradiation on the enzyme relating to the development of characteristic odor of onions, *J. Agric. Food Chem.*, 19, 166, 1971.
91. **Nishimura, H., Asahi, A., Fujiwara, K., Mizutani, J., and Obata, Y.**, Changes in flavor components of onion by γ-irradiation, *Agric. Biol. Chem.*, 35, 1831, 1971.
92. **Bandyopadhyay, C., Tewari, G. M., and Sreenivasan, A.**, Studies on some chemical aspects of gamma-irradiated onions, in *Radiation Preservation of Food*, Proc. Symp., Bombay, 1972, IAEA, Vienna, 1973, 11.
93. **Umeda, K., Takano, H., and Sato, T.**, Sprout inhibition of potatoes by ionizing radiation. I. Effect of delay between harvest and irradiation on sprouting, *J. Food Sci. Technol.*, 16, 508, 1969.
94. **Umeda, K.**, Sprout inhibition of potatoes by radiation, *New Food Ind.*, 11, 27, 1969.
95. **Hendel, C. E. and Burr, H. K.**, Treatment of potatoes with gamma rays: effects of delay between harvest and irradiation, *Food Technol. (Chicago)*, 15, 218, 1961.

96. **Ogata, K. and Chachin, K.**, Irradiation and biological response of fruits and vegetables, *Kagaku-to-Seibutsu*, 10, 234, 1972.
97. **Takano, H., Umeda, K., and Sato, T.**, Sprout inhibition of onions by ionizing radiation. III. Effect of irradiation dose level on the sprouting of onions var. "Sapporoki", *Rep. Natl. Food Res. Inst.*, 29, 55, 1974.
98. **Mullins, W. R. and Burr, H. K.**, Delay between harvest and irradiation, *Food Technol. (Chicago)*, 15, 178, 1961.
99. **Ojima, T., Hori, S., Kawasaki, T., Toratani, H., Ujihara, A., Fujimoto, H., Kitoh, S., and Sekii, I.**, Gamma irradiation of onion bulbs to inhibit sprouting, *Annu. Rep. Radiat. Cent. Osaka Prefect.*, 4, 112, 1963.
100. **MacQeen, K. F.**, Sprout inhibition of vegetables using gamma radiation, in Radiation Preservation of Foods, National Research Council-National Academy of Sciences, Washington, D.C., 1965, 127.
101. **Kim, H. S., Park, N. P., Cho, H. O., Choi, E. H., Lee, H. S., and Kim, Y. M.**, Studies on the sprout inhibition and biochemical changes of onions by gamma-irradiation, *J. Nucl. Sci.*, 9, 95, 1970.
102. **Mullins, W. R. and Burr, H. K.**, Treatment of onions with gamma rays: effects of delay between harvest and irradiation, *Food Technol. (Chicago)*, 15, 178, 1961.
103. **Grünewald, T.**, Studies about sprout inhibition of onions by irradiation in the Federal Republic of Germany, a paper presented to an IAEA/FAO/WHO Int. Symp. Food Preserv. Irradiat., IAEA-SM-221/17, Wageningen, 1977.
104. **Fischnich, O., Pätzold, C., and Heilinger, F.**, Influence of low doses of irradiation (X-rays and gamma rays of cobalt-60) on potato seed, in *Effects of Ionizing Radiations on Seeds*, IAEA, Vienna, 1961, 553.
105. **Nuttall, V. W., Lyall, L. H., and MacQeen, K. F.**, Some effects of gamma radiation on stored onions, *Can. J. Plant Sci.*, 41, 805, 1961.
106. **Meshitsuka, G., Iba, S., Oka, M., Matsubara, H., Tabei, M., Yamazaki, K., Inokoshi, Y., and Kaneko, S.**, Sprout inhibition of onion bulbs by γ-rays, *Tokyo Metrop. Isot. Cent. Annu. Rep.*, 1, 78, 1962.
107. **Matsuyama, A.**, Recent advances in food irradiation research in Japan, in *Food Irradiation*, IAEA/FAO Symp., Karlsruhe, 1966, IAEA, Vienna, 1966, 767.
108. **Solanas, J. and Darder, A.**, Food irradiation programme in Venezuela, in *Preservation of Fruit and Vegetables by Radiation*, Proc. of a FAO/IAEA Panel, 1966, IAEA, Vienna, 1968, 13.
109. **Temkin-Gorodetski, N., Kahan, R. S., and Padova, R.**, Development of the damage in the buds during storage of irradiated onions, *Can. J. Plant. Sci.*, 52, 817, 1972.
110. **Takano, H., Aoki, S., Umeda, K., and Sato, T.**, Sprout inhibition of onions by ionizing radiation. IV. Effect of post-harvest treatment on extension of irradiation time for storage of onions, *J. Food Sci. Technol.*, 21, 273, 1974.
111. **Furuta, J., Hiraoka, E., Okamoto, S., Fujishiro, M., Kanazawa, I., Ohnishi, T., Tsujii, Y., Tabata, T., Hori, S., and Ojima, T.**, The gamma-ray dose-rate effect in the sprout inhibition of onion bulbs, *Food Irradiat.*, 12, 1, 1977.
112. **Kume, T., Tachibana, H., Aoki, S., Umeda, K., and Sato, T.**, Effects of dose and dose rate of gamma radiation on tbe sprout inhibition of onions, *J. Food Sci. Technol.*, 24, 37, 1977.
113. **Waggoner, P. E.**, Radiation and resistance of tubers to rot, *Am. Potato J.*, 32, 448, 1955.
114. **Gustafson, F. C., Brownell, L. E., and Martens, R. A.**, The influence of gamma irradiation of potato tubers on rate of respiration, *Am. Potato J.*, 34, 177, 1957.
115. **Schreiber, J. S. and Highlands, M. E.**, Commercial storage of irradiated white potatoes, *Maine Exp. Stn. Bull.*, 563, 1957.
116. **Truelsen, T. A.**, Irradiation of potatoes, onions and carrots to inhibit sprouting, Risö Rep. No. 16, 1960, 67.
117. **Skou, J. P.**, Studies on the effects of ionizing radiation for extending the storage lives of onions, Risö Rep. No. 238, 1971.
118. **Mathur, P. B.**, Variety, developmental stage and dose rate in irradiation of the potato, *Nature (London)*, 198, 99, 1963.
119. **Ellis, W., Workman, M., Patterson, M., and Johnson, R.**, A Study of the Feasibility of Utilizing Ionizing Radiation to Increase the Storage Life of White Potatoes, Purdue University, Agricultural Experiment Station, Lafayette, Ind., Rep. 11, (final), Contract DA-129-QM-789, November 1, 1956 to April 30, 1959, Quartermaster Food and Container Institute for the Armed Forces, Chicago.
120. **Sawyer, R. L. and Dallyn, S. L.**, Effect of irradiation on storage quality of potatoes, *Am. Potato J.*, 38, 227, 1961.
121. **Workman, M., Patterson, M. E., Ellis, N. K., and Heiligman, F.**, The utilization of ionizing radiation to increase the storage life of white potatoes, *Food Technol. (Chicago)*, 14, 395, 1960.
122. **Sato, T.**, Sprout control of potatoes and onions by radiation treatment, *Shokuhin Kogyo*, 17(10), 71, 1965; 17(11), 63, 1965.

123. **Mikaelsen, K., Brenna, H. K., and Landfall, R.**, Improved storage quality of potatoes, vegetables and fruit exposed to gamma radiation, *Proc. 2nd Int. Conf. Peaceful Uses At. Energy*, 27, 401, 1958.
124. **Kahan, R. S. and Temkin-Gorodetske, N.**, Storage tests and sprouting control on Up-to-Date variety potatoes and on an experimental onion variety (Bet Alpha), in *Preservation of Fruit and Vegetables by Radiation*, Proc. Panel, Vienna, 1966, IAEA, Vienna, 1968, 29.
125. **Sato, T.**, Present status of food irradiation in Japan, *Peaceful Uses At. Energy*, 12, 325, 1972.
126. **Parks, N. M., Cloutier, J. A. R., and MacQueen, K. F.**, Gamma irradiation of potatoes. Effects on storage properties, sugar, starch and ascorbic acid content, cooking and chipping qualities in Proc. Int. Hortic. Congr., Nice, France, 1958.
127. **Umeda, K., Kawashima, K., Takano, H., and Sato, T.**, Sprout inhibition of potatoes by ionizing radiation. II. Radiation treatment of Norin-ichigo and the effect of irradiation on the crisps made of irradiated potato, *J. Food Sci. Technol.*, 16, 515, 1969.
128. **MacQueen, K. F. and Robb, J. A.**, Report on the results of the Canadian potato irradiation program 1961—62 season, *Gamma Irradiat. Can.*, 3, 42, 1964.
129. **Umeda, K., Takano, H., and Sato, T.**, Sprout inhibition of onions by ionizing radiation. I. Effect of delay between harvest and irradiation and sprouting of var. "Sapporoki" and "Senshuki", *Rep. Natl. Food Res. Inst.*, 25, 24, 1970.
130. **Lewis, N. F. and Mathur, P. B.**, Extension of storage lives of potatoes and onions by cobalt-60 γ-rays, *Int. J. Appl. Radiat. Isot.*, 14, 447, 1963.
131. **Dharkar, S. D.**, Report on BARC programmes in food irradiation: status of project on preservation of potatoes and onions by sprout inhibition, *Food Irradiation*, Proc. Semin., Trombay, 1969, Bhabha Atomic Research Centre, Trombay, India, 1969, 24.
132. **Khan, I. and Wahid, M.**, Feasibility of radiation preservation of potatoes, onions and garlic in Pakistan, paper presented to an IAEA/FAO/WHO Int. Symp. Food Preserv. Irradiat., IAEA-SM-221/48, Wageningen, 1977.
133. **Brownell, L. E., Gustafson, F. G., Nehemias, J. V., Isleib, D. R., and Hooker, W. J.**, Storage properties of gamma-irradiated potatoes, *Food Technol. (Chicago)*, 11, 306, 1957.
134. **Takano, H., Tanaka, Y., Umeda, K., and Sato, T.**, Sprout inhibition of onions by ionizing radiation. II. Effects of radiation dose and storage condition on the sprouting of var. "Senshuki", *Rep. Natl. Food Res. Inst.*, 27, 59, 1972.
135. **Duncan, D. T., Hooker, W. J., and Heiligman, F.**, Storage rot susceptibility of potato tubers exposed to minimum inhibiting levels of ionizing radiation, *Food Technol. (Chicago)*, 13, 159, 1959.
136. **Brownell, L. E., Burns, C. H., and Gustafson, P. G.**, Storage properties of gamma irradiated potatoes, *Food Technol. (Chicago)*, 10, 33, 1956.
137. **Takano, H., Aoki, S., Umeda, K., and Sato, T.**, Sprout inhibition of potatoes by ionizing radiation. IV. Effect of gamma irradiation on the cure of wounded potatoes, *Rep. Natl. Food Res. Inst.*, 29, 48, 1974.
138. **Dallyn, S. L. and Sawyer, R. L.**, Effect of gamma and fast electron irradiation on storage qualities of onions, *Proc. Am. Soc. Hortic. Sci.*, 73, 390, 1959.
139. **Staden, O. L.**, Experiences with the irradiation of vegetables in the Netherlands, in *Food Irradiation*, Proc. IAEA/FAO Symp., Karlsruhe, 1966, IAEA-SM-73/68, IAEA, Vienna, 1966, 609.
140. **Heiligman, F.**, Effects of ionizing radiation on white potatoes, *Am. Potato J.*, 34, 153, 1957.
141. **Sawyer, R. L., Dallyn, S. L., and Cotter, D. J.**, Some physical aspects of irradiated potatoes, in Conf. on Biological, Physical and Industrial Aspects of Potato Irradiation, Brookhaven National Laboratory, Upton, N.Y., 1955, 13.
142. **Gustafson, F. C., Brownell, L. E., and Martens, R. A.**, The influence of gamma irradiation of potato tubers on rate of respiration, *Am. Potato J.*, 34, 177, 1957.
143. **Cotter, D. and Sawyer, R. L.**, The effect of gamma irradiation on the incidence of black spot and ascorbic acid, glutathione and tyrosinase content of potato tubers, *Am. Potato J.*, 38, 58, 1960.
144. **Ojima, T., Hori, S., Toratani, H., Fujimoto, H., Katayama, T., Kitoh, S., Shiomi, N., and Kowyama, Y.**, On radiation-induced browning of vascular region in potato tubers, *Annu. Rep. Radiat. Cent. Osaka Prefect.*, 11, 143, 1970.
145. **Ogata, K., Chachin, K., and Tatsumi, Y.**, Studies on the browning of potatoes by gamma radiation, *Food Irradiat.*, 5, 40, 1970.
146. **Ogawa, M. and Uritani, I.**, Tissue browning of potato tubers induced by gamma irradiation, *Food Irradiat.*, 5, 106, 1970.
147. **Ogata, K., Tatsumi, Y., and Chachin, K.**, Studies on the browning of potato tubers by gamma radiation. I. Histological observation and the effects of the time of irradiation after harvest, low temperature and polyethylene packaging, *J. Food Sci. Technol.*, 17, 298, 1970.
148. Steering Committee on Food Irradiation Research, Japan, Final rep. on potato irradiation, 1971.

149. Tatsumi, Y., Chachin, K., and Ogata, K., Studies on the browning of potato tubers by gamma radiation. II. The relationship between the browning and changes of o-diphenol, ascorbic acid and activities of polyphenol oxidase and peroxidase in irradiated potato tubers, *J. Food Sci. Technol.*, 19, 508, 1972.
150. Tatsumi, Y., Chachin, K., Matsuzuka, M., and Ogata, K., Studies on the browning of potato tubers by gamma radiation. III. Influence of cultural localities on the browning of irradiated potato tubers, *J. Food Sci. Technol.*, 20,132, 1973.
151. Matsuyama, A., Present status of food irradiation research in Japan with special reference to microbiological and entomological aspects, in *Radiation Preservation of Food*, Proc. Symp., Bombay, 1972, IAEA, Vienna, 1973, 261.
152. Smith, O., Effect of cultural and environmental conditions on potatoes for processing, in *Potato Processing*, Talburt, W. F. and Smith, O., Eds., AVI Publishing, Westport, Conn., 1959, 70.
153. Kiermeier, F. and Rickerl, E., Cause of discoloration of steamed potatoes. I. Influence of phenolic materials, *Z. Lebensm. Unters. -Forsch.*, 100, 441, 1955.
154. Kiermeier, F. and Rickerl, E., Cause of discoloration of steamed potatoes. II. Influence of iron, *Z. Lebensm. Unters. -Forsch.*, 102, 330, 1955.
155. Sawyer, R. L., Effect of irradiation on potatoes for processing, Proc. 7th Natl. Potato Utilization Conf., Orono, Maine, 1956, 3.
156. Gardner, D. S. and MacQeen, K. F., The effect of gamma rays on storage life and chipping qualities of Ontario grown Kennebec potatoes, *Potato Chipper*, AECL-2175, 1965.
157. Thomas, P. and Joshi, M. R., Prevention of aftercooking darkening of irradiated potatoes, *Potato Res.*, 20, 77, 1977.
158. Grünewald, T., Experience in irradiating potatoes, in *Aspects of the Introduction of Food Irradiation in Developing Countries*, IAEA, Vienna, 1973, 7.
159. Sparenberg, H., Irradiation of potatoes-after cooking darkening-Fusarium susceptibility during storage, *Food Irradiat. Newsl.*, 2, 8, 1977.
160. Sparenberg, H. and Buitelaar, N., Can reconditioning prevent the after-cooking darkening of irradiated potatoes?, paper presented to an IAEA/FAO/WHO Int. Symp. Food Preserv. Irradiat., Wageningen, 1977.
161. Asselbergs, E. A. M. and Wethington, E., The effect of gamma radiation on the cooking and chipping quality of stored potatoes, in *Gamma Irradiation in Canada*, AECL, Ottawa, 1960, 30.
162. Takano, H., Suzuki, T., and Umeda, K., Sprout inhibition of potatoes by ionizing radiation. V. Effect of irradiation on the reducing sugar and amino acid content, and quality of crisps produced from irradiated potatoes, *J. Food Sci. Technol.*, 21, 484, 1974.
163. Brunelet, L. and Vidal, P., Inhibition de la montée en végétation des bulbes alimentaires, *Revue de la Conserve*, Mars-Avril, Paris, 1960, 3.
164. Hori, S., Kawasaki, T., and Kitoh, E., Gamma irradiation of onion bulbs to inhibit sprouting. II. Browning phenomenon in inner buds of irradiated onion bulbs, and a method of early detection for inhibition of sprouting, *Annu. Rep. Radiat. Cent. Osaka Prefect.*, 5, 90, 1964.
165. Chachin, K. and Ogata, K., Effects of delay between harvest and irradiation and of storage temperatures on the sprout-inhibition of onions by gamma radiation, *J. Food Sci. Technol.*, 18, 378, 1971.
166. Okuneva, K. A., The hygienic characteristics of potatoes irradiated with radioactive cobalt for the arrest of sprouting, Voprossi Pitaniya, 1958, (5).
167. Parks, N. M., MacQueen, K. F., and Cloutier, J. A. R., General summary and conclusions, *Gamma Irradiation in Canada*, AECL, Ottawa, 1960, 32.
168. Josephson, E. S., Heiligman, F., Wadsworth, C., Howker, J. J., and Wierbicki, E., Low dose food irradiation at Natick, a paper presented to the FAO/IAEA Advisory Group Meeting on Low Dose Irradiation of Agriculture Products, Rio de Janeiro, 1975; Tech. Rep. 78/003, U.S. Army Natick Research and Development Command, Natick, Mass., June 1977.
169. Results of the Canadian pilot scale potato irradiation program, *Food Irradiat.*, 3, 2, 1963.
170. Pedersen, S., The effect of ionizing radiations on sprout prevention and chemical composition of potatoes, *Food Technol. (Chicago)*, 10, 532, 1956.
171. Sawyer, R. L. and Dallyn, S. L., Vaporized chemical inhibitors and irradiation, two new methods of sprout control for tuber and bulb crops, *Proc. Am. Congr. Hortic. Sci.*, 67, 514, 1956.
172. Adesuyi, S. A. and Mackenzie, J. A., The inhibition of sprouting in stored yams, *Kioscorea rotundata* poir, by gamma radiation and chemicals, in *Radiation Preservation of Food*, Proc. Symp., Bombay, 1972, IAEA, Vienna, 1973, 127.
173. Lewis, N. F. and Mathur, P. B., Extension of storage lives of potatoes and onions by cobalt-60 γ-rays, *Int. J. Appl. Radiat. Isot.*, 14, 447, 1963.
174. Dharkar, S. D., Report on BARC programmes in food irradiation: status of project on preservation of potatoes and onions by sprout inhibition, in *Food Irradiation*, Proc. Semin., Trombay, 1969, Bhabha Atomic Research Centre, Trombay, India, 1969, 24.

175. **Burton, W. G. and de Jong, W. H.**, The irradiation of ware potatoes, *Int. J. Appl. Radiat. Isot.*, 6, 167, 1959.
176. **Dharkar, S. D. and Sreenivasan, A.**, Irradiation of tropical fruits and vegetables, in *Food Irradiation*, Proc. IAEA/FAO Symp., Karlsruhe, 1966, IAEA-SM-73/64, IAEA, Vienna, 1966, 635.
177. **Kahan, R. S. and Temkin-Gorodetski, N.**, Storage tests and sprouting control of Up-to-Date variety potatoes and on an experimental onion variety (Bet Alpha), in *Preservation of Fruit and Vegetables by Radiation*, Proc. of Panel, FAO/IAEA, IAEA, Vienna, 1968, 29.
178. **Roushdy, H. M., Shukry, K., and Mahmoud, A. A.**, Lower radiation levels for better storage ability of potatoes and onions using certain chemical treatments, in *Radiation Preservation of Food*, Proc. IAEA/FAO Symp., Bombay, 1972, IAEA, Vienna, 1973, 105.
179. **Chowdhury, M. S. U. and Rahman, A.T.M.F.**, Food irradiation research in Bangladesh, in *Radiation Preservation of Food*, STI/PUB/317, IAEA, Vienna, 1973, 637.
180. **Farkas, J.**, Present status and prospects for the commercialization in Hungary of irradiated food items for human consumption, in *Requirements for the Irradiation of Food on a Commercial Scale*, IAEA, Vienna, 1975, 37.
181. **Sekhavat, A., Zare, Z., Kudva, M. G., Chopra, K. L., and Shariat-Panahi, J. M.**, Preservation of potatoes and onions by irradiation ahd chemical treatments, a paper presented to an IAEA/FAO/WHO Int. Symp. Food Preserv. Irradiat., IAEA-SM-221/29, Wageningen, 1977.
182. **Roushdy, H. M., Abd Elhufez, A. A., Sharabash, M. T. M., and Mahmoud, A. A.**, Effect of pre-storage chemical and/or radiation treatments on the storage ability of potatoes and onions, paper presented at an IAEA/FAO/WHO Int. Symp. Food Preserv. Irradiat., IAEA-SM-221/21, Wageningen, 1977.
183. **Salková, Z., Kubin, K., Staněk, M., and Horáček, P.**, Present status of food irradiation in Czechoslovakia, paper presented to an IAEA/FAO/WHO Int. Symp. Food Preserv. Irradiat., IAEA-SM-221/7, Wageningen, 1977.
184. **Farkas, J., Kalman, B., and Kiss, I.**, Hungarian research in the area of the inhibition of sprouting of onions by ionizing radiation, in *Peaceful Uses of Atomic Energy for Scientific and Economic Development*, Proc. 1st Sci. Conf. Iraqi, At. Energy Comm., Baghdad, 1975, Iraqi Atomic Energy Comm., Baghdad, 1975, 129.
185. **Kalman, B., Kiss, I., and Farkas, J.**, Pilot scale studies on the irradiation of onions in Hungary, paper presented to an IAEA/FAO/WHO Int. Symp. Food Preserv. Irradiat., IAEA-SM-221/23, Wageningen, 1977.
186. **Singson, C. C., de Guzman, Z. M., Mendoza, E. B., Lustre, A. O., Roncal, R., Villaruel, F., and Kolendo, A. L.**, The use of gamma irradiation for the extended commercial storage of Philippine onions and other agricultural produce, paper presented to an IAEA/FAO/WHO Int. Symp. Food Preservation by Irradiation, IAEA-SM-221/76, Wageningen, 1977.
187. **Bandyopadhyay, C., Tewari, G. M., and Sreenivasan, A.**, Studies on some chemical aspects of gamma-irradiated onions, in *Radiation Preservation of Food*, Proc. Symp., Bombay, 1972, IAEA, Vienna, 1973, 11.
188. **Sparks, W. C.**, Influence of ventilation and humidity during storage on weight and quality changes of *Russet Burbank* potatoes, *Potato Res.*, 16, 213, 1973.
189. **Rumpf, G.**, Gas chromatographic studies on the suitability of irradiated potatoes for storage and processing, *Potato Res.*, 16, 296, 1973.
190. **Warland, H. M. F. and MacQueen, K. F.**, Canadian food irradiation facilities, in *Food Irradiation*, Proc. IAEA/FAO Symp., Karlsruhe, 1966, IAEA, Vienna, 1966, 715.
191. **Iwata, T. and Ogata, K.**, Studies on the storage of chestnuts treated with gamma radiation, *Bull. Univ. Osaka Prefect., Ser. B*, 9, 59, 1959.
192. **Hayakawa, A., Umeda, K., and Shiroishi, M.**, Preservation of some agricultural products by ionizing radiation, *Nosan Kako Gijutsu Kenkyu Kaishi*, 8, 48, 1961.
193. **Nara, S.**, Changes of sugar contents in gamma irradiated and stored chestnuts, *J. Agric. Chem. Soc. Jpn.*, 35, 415, 1961.
194. **Mather, P. B.**, Extension of storage life of garlic bulbs by γ-irradiation, *Int. J. Appl. Radiat. Isot.*, 14, 625, 1963.
195. **Metlitsky, L. V., et al.**, Rest-period-controlled use of ionizing radiation during storage of onions and garlic, *Biokhim. Plodov. Ovoshehei*, Akad. Nauk U.S.S.R. Inst., Giokhim, 1964, 163.
196. **Watanabe, T. and Tozaki, H.**, The Co-60 irradiation of garlic to prevent sprouting and its influence to allicin-lyase activities, *Food Irradiat.*, 2, 106, 1967.
197. **Mikaelsen, K., Brenna, H., and Roer, L.**, Effects of gamma rays on sprouting and growth during storage in carrots and potatoes, *Proc. Int. Conf. Peaceful Uses At. Energy*, 12, 108, 1956.
198. **Madsen, K. A., Salunkhe, D. K., and Simon, M.**, Certain morphological and biochemical changes in gamma-irradiated carrots *(Daucus carota L.)* and potatoes *(Solanum tuberosum L.) Radiat. Res.*, 10, 48, 1959.

199. **Okanoue, S., Fukutani, M., Sugihara, M., and Hashida, I.,** Studies on the preservation of foods and the changes of their nutrients by gamma-ray irradiation. VI. Effects on the sweet potato, *Radioisotopes*, 14, 103, 1965.
200. **Ogawa, M. and Uritani, I.,** Metabolic changes in sweet potato roots induced by gamma radiation in response to cutting, *Radiat. Res.*, 39, 117, 1969.
201. **Ogawa, M. and Uritani, I.,** Metabolic changes in tissues prepared from irradiated sweet potato roots with regard to ethylene effects, *Radiat. Bot.*, 11, 67, 1971.
202. **Ogawa, M. and Uritani, I.,** Effect of gamma radiation on peroxidase development in sweet potatoes disks, *Radiat. Res.*, 41, 342, 1970.
203. **Wittwer, S. H. and Sharma, R. C.,** The control of storage sprouting in onions by preharvest foliage sprays of maleic hydrazide, *Science*, 112, 597, 1950.
204. **Nishi, Y., Mori, M., and Inui, N.,** Chromosomal abberations induced by maleic hydrazide and related compounds in Chinese hamster cells in vitro, *Mutat. Res.*, 67, 249, 1979.
205. **McCann, J., Choi, E., Yamasaki, E., and Ames, B. N.,** Detection of carcinogens as mutagens in the *Salmonella*/microsome test: assay of 300 chemicals, *Proc. Natl. Acad. Sci. U.S.A.*, 72, 5135, 1975.
206. **Shimomura, H., Kuriyama, E., and Takai, Y.,** A method for early detection of irradiated potato tubers. II. Morphological changes by gibberellin and kinetin on dormant buds, *J. Food Hyg. Soc. Jpn.*, 13, 536, 1972.
207. **Thomas, P., Srirangarajan, A. N., Padwal-Desai, S. R., Ghanekar, A. S., Shirsat, S. G., Pendharkar, M. B., Nair, P. M., and Nadkarni, G. B.,** Feasibility of radiation process for post-harvest storage of potatoes under tropical conditions, paper presented to an IAEA/FAO/WHO Int. Symp. Food Preserv. Irradiat., IAEA-SM 221/25, Wageningen, 1977.
208. **Penner, H. and Fromm, H.,** Studies on the identification of irradiated potatoes. II. Determination of chlorogenic acid, *Z. Lebensm. Unters.-Forsch.*, 150, 84, 1972.
209. **Hori, S., Kawasaki, T., and Kitoh, S.,** Gamma irradiation of onion bulbs to inhibit sprouting. II. Browning phenomenon in inner buds of irradiated onion bulbs, and a method of early detection for inhibition of sprouting, *Annu. Rep. Radiat. Cent. Osaka Prefect.*, 5, 90, 1964.
210. **Münzner, R.,** Nachweis einer Strahlenbehandlung bei Küchenzwiebeln, *Z. Lebensm. Unters.-Forsch.*, 162, 47, 1976.
211. **Diehl, J. F.,** Preparations for marketing irradiated potatoes and onions in the Federal Republic of Germany, in *Requirements for the Irradiation of Food on a Commercial Scale*, IAEA, Vienna, 1975, 31.
212. **de Zeeuw, D.,** Commercialization of irradiated potatoes, mushrooms, onions and spices in The Netherlands, in *Requirements for the Irradiation of Food on a Commercial Scale*, IAEA, Vienna, 1975, 133.
213. **Lapidot, M.,** Status of the food irradiation programme in Israel, in *Requirements for the Irradiation of Food on a Commercial Scale*, IAEA, Vienna, 1975, 179.
214. **Kálmán, B.,** Pilot scale experiments for sprout-inhibiting onions by ionizing radiation, a paper presented to FAO/IAEA Res. Coord. Meet. Technol. Econ. Feasibility Food Irradiat. Accra, Ghana, February 1978.
215. **Baraldi, D.,** Technical and economic study of irradiated foodstuffs wthin a coordinated programme on technological and economic feasibility of Food Preservation by Irradiation, paper presented to FAO/IAEA Res. Coord. Meet. Technol. Econ. Feasibility Food Irradiat., Accra, Ghana, February 1978.
216. **Tsai, C. M.,** Institute of Nuclear Energy Research, Atomic Energy Council, Republic of China, April 1978, private communication.
217. **Shinozaki, Y., Hōgetsu, D., Okuyama, N., Manabe, T., Sasasagawa, T., Minamikawa, T., and Matsuyama, A.,** Mutagenicity studies on alcohol extracts from gamma-irradiated potatoes. Preparation of samples and their chemical analysis, *Radioisotopes*, 30, 655, 1981.
218. **Ishidate, M., Jr., Yoshikawa, K., Sofuni, T., Iwahara, S., Shibuya, T., Muramatsu, S., Kitayama, S., and Matsuyama, A.,** Mutagenicity studies on alcohol extracts from gamma-irradiated potatoes. Tests for biological activities in bacterial and mammalian cell systems, *Radioisotopes*, 30, 662, 1981.
219. **Shibuya, T., Murota, T., Iwahara, S., Hashimoto, K., Minegishi, A., Hōgetsu, D., and Matsuyama, A.,** Mutagenicity studies on alcohol extracts from gamma-irradiated potatoes. Dominant lethal test in mice, *Radioisotopes*, 31, 74, 1982.

Chapter 7

INSECT DISINFESTATION OF GRAIN AND FRUIT

Elvin W. Tilton and Arthur K. Burditt, Jr.

TABLE OF CONTENTS

I. INTRODUCTION

The efficient control of insects and their removal from food commodities has long been the goal of producers and processors. Although many methods of control are now being used, the food industry is continually seeking safer and more economical means. Fumigation, though by far the most effective method of disinfestation in current use, has its own serious limitations. It is not possible for fumigants to penetrate some commodities in sufficient concentration to control some pests, such as the mango seed weevil *Cryptorhynchus mangiferae* (Fabricius). As a result, mangoes cannot be shipped to the U.S. mainland from Hawaii or other infested areas.[1] Also, in some cases, chemical treatments may create problems by injuring the fruit or grain or by leaving undesirable residues. The development of resistance to insecticides by insects is yet another incentive to develop alternate methods for pest control.[2] Irradiation techniques seem to offer solutions that are desirable in many respects. The treatment with radiation takes less time than fumigation, leaves no undesirable residue, and can be as effective in controlling an insect infestation as any procedure now available.

A great deal of information has now been accumulated on radiation effects on insects infesting stored cereal products, fruits, and other commodities. For the most part, however, the data are scattered and incomplete, and little more than qualitative observations. Most data are not strictly comparable because of differences in types of radiation used, exposure rates, and dosimetry. In addition, it is now known that various preradiation and postradiation treatments or conditions, as well as those during treatment, may profoundly affect the results obtained. Also a uniformity in criteria for determining mortality or other radiation effects is needed. Further problems arise because radiosensitivity changes greatly through the course of an insect's life cycle and may change drastically even within a single stage.

The earliest studies of radiation effects on stored-product insects were concerned with the cigarette beetle, *Lasioderma serricorne* (Fabricius), and the rice weevil, *Sitophilus oryzae* (L.). Hunter[3] found no effect on fertility of the rice weevil with the doses of X-rays he used, while Morgan and Runner[4] found that X-ray equipment designed for radiation of cigars did not kill any stage of the cigarette beetle. Runner[5] later showed that eggs, larvae, and adults of cigarette beetle could be killed in cigars by X-rays, but that the doses required would be very large. This work discouraged other actual control attempts for a time, but interest was revived with the availability of radioisotopes in the early 1950s. Many studies have been carried out since that time on the effects of radiation on grain insects. Hilchey[6] compiled an excellent summary of work up to 1956 on radiation effects on storage insects. Balock et al.[7] showed that radiation of eggs or larvae of the oriental fruit fly, *Dacus dorsalis* Hendel, could prevent adult emergence. O'Brien and Wolfe[8] discussed the principles and some of the work relating to radiation disinfestation, and Cornwell[9] edited several original research reports and added editorial comments in a review of grain irradiation.

Two main types of radiation have been proposed. Some workers advocate use of electron accelerators with high grain flow rates and low penetrating powers, while others favor radioisotopic sources such as ^{60}Co or ^{137}Cs for the more penetrating gamma rays.

Machine radiation and gamma radiation of grain each has some inherent advantages. One advantage of machine radiation that has received little attention involves the method of exposing the infested grain. The methods usually suggested for moving large quantities of grain past a machine-produced ionizing beam, involve moving the grain at high speed in an air stream. Proposals for isotopic gamma radiation usually use other grain moving systems. The use of the high-speed air stream and the consequent high impact of the grain at the end of its movement will doubtless destroy all adult moths, the most radioresistant stage of the most radioresistant stored-product insects. All adult and immature insects including beetles (many species) that are living external to individual grains probably would suffer a high degree of mortality. In an experiment to determine the effects of gamma radiation on infested wheat, Cogburn et al.[10] attributed a high degree of control to the movement of the grain in an air stream. The destruction of this portion of the insect population will do much to enhance the radiation treatment. A disadvantage of this type of grain movement is that some kernels are broken because of the high impact and special effort in the engineering design of the equipment is necessary to keep grain breakage to a minimum.

Suggestions have been made for mobile radiation facilities installed in trucks, railroad cars, and ships. By moving such facilities to locations where they are needed, maximum use can be made of the irradiator, and this reduces the unit cost proportionately.

II. DOSIMETRY

For lowest cost and highest efficiency in a food irradiation facility, all of a product should receive the same dose. The consideration of high throughput, however, forces irradiation plant designers to compromise between uniform dose distribution and high throughput. There are limits on the allowable variability of dose distribution, as too low an absorbed dose would fail to bring about the desired effect, and too high a dose could exceed the legally permitted level of treatment. A method of monitoring and recording the absorbed dose in a food product would be a valuable aid for commercial food irradiation facilities. Dosimeters could be included with any food product passing through the radiation plant to measure the actual dose distribution in the treated product. General requirements for a dosimetry system for use in food irradiation are discussed in Chapter 8, Volume 1.

Bulk-grain radiation presents special problems because the dosimeter must withstand being mixed with the grain, have the same flow characteristics as the grain, remain undamaged, and be easily recoverable after treatment. Most dosimeters used in food-irradiation studies have depended on changes in optical density and these, without exception, are useless in bulk-grain irradiation because of the grain's abrasive action. Systems using liquids or glass are unsuitable because of breakage and grain contamination. Large dosimeters are unsuitable because they have flow characteristics unlike those of grain and they may be damaged by grain-handling equipment.

In a series of tests, Tilton et al.[11] found that a lithium fluoride thermoluminescent dosimetry system had unique properties which suited it remarkably well for use with a bulk-grain irradiator. Their tests showed that the flow characteristics of the dosimeters approached those of wheat, and the dosimeters measured the variability of absorbed dose in a mass of wheat. In that series of tests 50 mg of lithium fluoride powder was loaded into each small capsule.

Flow characteristics of various sizes and configurations of dosimeter capsules, with and without coatings to increase friction, were determined by mixing large numbers of capsules with batches of wheat which had been color-marked with various dyes. These colored batches were interspersed between large amounts of normal grain during operation of the bulk-grain irradiator and the distribution of dosimeters was monitored at the exit unit. Dosimeters were recovered from the grain stream by using a specially constructed, enclosed, vibrating screen which allowed the wheat to pass through into the storage bins. The dosimeters were retained above the screen and discharged into a container. Results showed that uncoated cylindrical polyethylene capsules (4 × 17 mm with tight-fitting friction caps) remained with the dyed wheat at both high and low speeds of grain flow through the irradiator. Other types of dosimeter capsules either were subject to greater damage in the conveying system or failed to have the desired flow characteristics.

Lithium fluoride impregnated polyethylene rods were not evaluated, but if rods of the same approximate dimensions and weight are used, they should be satisfactory. In addition, the rods are safer and easier to use.

Fruit being irradiated as a treatment for quarantine purposes would be prepackaged in fiberboard cartons or other containers. Suitable dosimeters could be placed within each container, attached to a drawstring in order to permit retrieval of the dosimeter to verify that the proper dosage had been applied. Also, a radiation reactive label could be applied to the outside of the container, and/or to the fruit itself, to show that it had been irradiated as a quarantine measure.

III. TREATMENT LEVELS

A major problem in disinfestation of commodities by any method is the large number of insect species that may be present. Therefore, an effective dose of radiation must sterilize or kill the most resistant of the species present. At the same time, the lowest effective dose should be chosen because of economic savings. Because, at present, there is no way of predicting the radiosensitivity of a species, the sensitivity of all species commonly found infesting products must be determined empirically. There have been numerous reports by many workers on the sensitivity of various species. The sensitivity of any species can be affected by many factors, including age, sex, strain, food, temperature, type of radiation, and even dose rate. With so many factors affecting sensitivity, together with the variations in methodology of handling insects and criteria used for death or sterility, it is not surprising that there has been some disagreement in the reported sensitivities of species.

At the U.S. Department of Agriculture Laboratories at Savannah, Ga. and Fresno, Calif., over 30 species of insect pests of stored products have been tested for radiosensitivity, using similar handling techniques and criteria for evaluation. A summary of that data is given in Table 1. The tests were performed in a systematic manner so comparisons of sensitivity among species are valid.

Differences among species of *Coleoptera* are substantial. The most resistant beetles are 6 to 7 times more resistant than the most sensitive species; the bruchids appear the most sensitive but curculionids are almost equally sensitive. The cucujids and most of the tenebrionids are intermediate in sensitivity, and the anobiids, dermestids, and ptinids are increasingly resistant. These data indicate that *Palorus subdepressus* (Wollaston), a tenebrionid, may be the most resistant to radiation of all stored-product *Coleoptera*; both males and females have reproduced after exposure to 30 krad.[12]

In general, the females of the species are more sensitive to the sterilizing effects of radiation than are the males of the same species. Thus, in some situations, a dose high enough to sterilize the females might be selected, even if viable males persisted. Also, the dose required to produce 100% mortality in the several metamorphic stages depends almost entirely on age at the time of treatment. For example, the sensitivity of eggs of the yellow mealworm, *Tenebrio molitor* L., decreased 250-fold as the eggs aged from 0.5 to 7.5 days. This extreme variability greatly limits the usefulness of mortality as a means of establishing the relative sensitivity of species.

The *Lepidoptera* as a group are more resistant to radiation than are the *Coleoptera* (Table 1), especially if the dose for sterilization is the basis for comparison. It was found that 100 krad may not sterilize some moth species but that fertility and fecundity were very low after this dose. The almond moth, *Ephestia cautella* (Walker), appears to be somewhat more sensitive than the Angoumois grain moth, *Sitotroga cerealella* (Olivier), or the Indian meal moth, *Plodia interpunctella* (Hübner), but other phycitids have sensitivies similar to that of *E. cautella*. The grain mite, *Acarus siro* L., and other grain mites appear to be intermediate in sensitivity between the *Coleoptera* and the *Lepidoptera*.

Table 1
COMPARATIVE SENSITIVITY OF STORED-PRODUCT INSECTS TO THE STERILIZING EFFECTS OF GAMMA RADIATION (THE SEX IRRADIATED IS SPECIFIED)

Species	Highest dose (krad) permitting reproduction			Sterilizing dose (krad)		
	Male	Female	Both*	Male	Female	Both*
Acarina						
Acarus siro L.	50	25	25			45
Coleoptera						
Attagenus megatoma (F.)	13.2	13.2		17.5	17.5	
Callosobruchus maculatus (F.)	5	5		7	7	
Cathartus quadricollis (Guérin-Méneville)	20	10		30	20	
Gibbium psylloides (Czenpinski)	20	20		30	30	
Gnathocerus maxillosus (F.)	10	10		20	20	
Lasioderma serricorne (F.)	17.5	13.2		25	17.5	25
Latheticus oryzae Waterhouse	10	5		20	10	
Oryzaephilus mercator (Fauvel)			10			20
O. surinamensis (L.)	10	10		20	20	
Palorus subdepressus (Wollaston)	30	30		40	40	
Rhyzopertha dominica (F.)			13.2			17.5
Sitophagus hololeptoides (LaPorte)	5	5		10	10	
Sitophilus granarius (L.)	5	5		10	10	
S. oryzae (L.)	<13.2	<13.2		13.2	13.2	
S. zeamais Motschulsky	5	5		10	10	
Tenebrio molitor (L.)	10	<5		15	5	
T. obscurus (F.)	5	5		10	10	
Tenebroides mauritanicus (L.)	5	<5		10	5	
Tribolium castaneum (Herbst)	10	10		20	20	
T. confusum Jacquelin du Val			13.2			17.5
T. destructor Uyttenboogaart	5	5		10	10	
T. madens (Charpentier)	20	5		30	10	
Trogoderma glabrum (Herbst)	17.5	13.2		25	<13.2	
T. inclusum (LeConte)	20	10		30	20	
T. variabile (Ballion)	20	5		30	10	
Lepidoptera						
Ephestia cautella (Walker)	50	20		100	30	
Plodia interpunctella (Hübner)	100	100		>100	>100	
Sitotroga cerealella (Olivier)	100	100		>100	>100	

* Unsexed irradiated populations left together.

The dose of radiation selected for control of stored-product insects can therefore be low if only the more sensitive beetle species are present; it must be much higher if some of the most resistant beetles, mites, or moths are present. However, if commodities are infested with many species, a dose of 50 krad will control even the most resistant beetle species and the immature stages of moths.[12] Some of the adult moths might remain fertile, but the few progeny they produce would be sterile because of inherited genetic damage.[13]

Brower and Tilton[14] conducted tests to determine the feasibility of gamma radiation for insect control in dried fruits. Raisins, zante currants, prunes, and dried apricots were infested with Indian meal moths or sawtoothed grain beetles, *Oryzaephilus suri-*

namensis (L.), and then radiated. The data indicated that a 40-krad dose, possibly much less, is sufficient to control the saw toothed grain beetle and the Indian meal moth in dried fruits. If packages of dried fruit contain only eggs and/or young larvae, then a dose of 20 krad is adequate for control. Presumably other species of insects that infest dried fruits would be controlled by these levels of treatment, as they exhibit similar sensitivity levels to radiation. Papadopoulou[15] recommended 100 to 200 krad as the level necessary to kill Indian meal moths and saw toothed grain beetles. However, he was apparently trying to achieve death of all stages present within a few days. Although Shchegoleva[16] stated that a dose of 300 krad was necessary to control dermestids, more recent data indicate that 40 krad would be adequate to control most species of dermestids. The development of mite populations would probably be arrested at these levels (40 krad), but the mites present would not be killed.[14]

At the U.S. Department of Agriculture Laboratory in Honolulu, Hawaii, the effects of gamma radiation on three species of fruit flies were studied extensively to determine the potential for its use as a quarantine treatment.[17] In these tests it was found that low doses of radiation (5 to 25 krad) did not result in immediate kill of eggs or larvae but did prevent emergence of adults. Eggs of fruit flies did not hatch when exposed to 5 to 25 krad gamma radiation, if they were less than 50% developed. However, once they had passed this point in development, the eggs continued to develop and hatch even at dosages from 25 to 120 krad. After the eggs hatched, the larvae continued to develop but usually did not reach maturity or form puparia. Fruit flies that were in the larval stage at the time of irradiation continued to feed and eventually reached maturity and formed puparia, but they were unable to complete metamorphosis and become adults when irradiated at doses of from 15 to 25 krad. This research showed that exposure to dosages in excess of 100 krad was required to prevent pupation of mature larvae. However, exposure of fruit fly eggs and larvae to less than 5 krad prevented 95% of the insects from emerging as adult flies.

These studies[17] showed that the melon fly, *Dacus cucurbitae* Coquillet, was more resistant to the effects of gamma radiation than the oriental fruit fly and that the Mediterranean fruit fly, *Ceratitis capitata* (Wiedemann), was most susceptible of the three species. Rate of development may be a factor in radiation effects, as the melon fly develops faster than the oriental fruit fly and the Mediterranean fruit fly is slowest to develop of the three species tested in Hawaii.

Research on immature stages of the Mexican fruit fly, *Anastrepha ludens* (Loew), using a ^{60}Co teletherapy source was conducted by Benschoter and Telich.[18] These studies showed that gamma radiation would prevent hatching of 1-day-old eggs and emergence of adults from mature larvae or 5- to 7-day-old pupae.

Thomas and Rahalkar[19] reported that larvae of oriental fruit flies and melon flies developing from irradiated eggs or larvae were sluggish and had a reduced growth rate when compared to those from untreated eggs or larvae. The larvae were reared on semiripe mangos or pumpkin, respectively.

Macfarlane[20] found that the effect of gamma radiation on immature stages of the Queensland fruit fly, *D. (Strumeta) tryoni* (Froggatt), was similar to that found for the Mediterranean fruit fly by Balock et al.[17] A dose of 5 krad prevented adult emergence from irradiated eggs and larvae. Shipp and Osborn[21] confirmed that nearly all of the adults that might emerge from any pupae treated at this dosage would be sterilized. However, they suggested a dose of 10 krad in order to provide an adequate treatment for this species.

Cavalloro and Delrio[22] determined the effects of gamma radiation on various stages of the olive fly, *D. oleae Gmelin*, as well as the Mediterranean fruit fly. They found that there were no significant differences in effect of radiation on larvae of the olive fly and those of the Mediterranean fruit fly. However, eggs and older pupae of the olive fly were slightly more resistant to radiation than comparable stages of the Mediterranean fruit fly. Three-day-old pupae of the olive fly were slightly more susceptible, possibly due to differences in stage of development, as discussed later. They demonstrated that radiation delayed hatching and reduced emergence, survival, and mating of adults from treated eggs. Growth of irradiated 1-day-old larvae was delayed; and at higher doses the larval weight and pupation were reduced.

Bughio et al.[23] reported on the effects of gamma radiation on eggs of *D. zonatus* (Saunders). They found that 4- to 5-hr-old eggs exposed to 4 kR did not hatch. Hatch of older eggs was reduced but not prevented by 12 kR, the highest dose tested. The duration of the larval period was extended and pupation was reduced. No normal adults emerged after radiation of eggs at 4 kR.

Hossain et al.[24] radiated older larvae of *D. ferrugienus* F. taken from infested fruit and placed in petri dishes containing mango flesh. They noted that no adults emerged from larvae treated at 2 to 5 kR, but adults emerged sooner from larvae treated at 1 kR than from the controls.

Upon reaching maturity, the third instar fruit fly larva generally leaves the fruit to form the puparium. There would be a slight possibility of third instar larvae reaching maturity and pupating after the fruit has been placed in the shipping containers. However, some subtropical and tropical fruit flies contain a fourth instar larval stage within the puparium during the first 2 to 3 days after the larva has formed the puparium.[25] The effect of gamma radiation on this stage was similar to that on the first three instars which were found within the fruit.[17] Limited research in Hawaii[17] demonstrated that radiation of larvae at dosages below 2 krad would permit substantial adult emergence and that these adults would be able to lay viable eggs. However, radiation at dosages above 2 krad resulted in such low adult emergence that it was impossible to ascertain if the adults could reproduce. Also, the presence of carbon dioxide, nitrogen, or water increased the dosage required for the radiation treatment.

The preceding studies[17] involved application of gamma radiation to immature stages of fruit flies under laboratory conditions, using eggs on moist blotting paper, larvae in rearing media or petri dishes, and pupae in paper cups. Results of tests using infested papayas and other fruit were reported by Balock et al.[26] and Seo et al.[27] for the three species of fruit flies found in Hawaii. Fruit used in the former study were radiated in an aluminum cannister, approximately 12.7 cm in diameter, using the center 22.9 cm depth. A dosage of less than 10 krad prevented emergence of most adults from such fruit. However, in order to achieve quarantine security of 99.9968% mortality,[28] a dosage of approximately 18 krad was required for papayas infested by immature oriental fruit flies. In the latter study,[27] fruits were placed in fiberboard cartons and irradiated in either a large research irradiator or a quasicommercial developmental unit. The authors demonstrated that the minimum dosage required for quarantine security must be increased to compensate for absorption owing to the additional mass involved in irradiation of packaged fruit. Thus, the minimum dosage required for infested papayas varied from 20.9 to 29.1 krad for thicknesses of 1.47 to 3.14 g/cm^2.

The mango weevil is found in most mango growing areas of the world except Central America, Mexico, and the mainland U.S. Experiments in Hawaii[29] and South Africa[30] have demonstrated that gamma radiation of infested mangos would prevent eggs or larvae from producing adults or would result in adult weevils that could not reproduce. Thomas[30] reported that treatment at 50 krad gave 100% mortality of larvae and pupae and 75 krad prevented adults from escaping from the seedpod. Seo et al.[29] reported that adult females exposed to 10 or 20 krad died without laying eggs and that the longevity after irradiation was related to dose.

Table 2
AVERAGE PERCENTAGE REDUCTION OF ADULT EMERGENCE FROM WHEAT AFTER GAMMA PLUS INFRARED OR MICROWAVE RADIATION BASED ON PERCENTAGE EMERGENCE FROM CONTROLS

	Actual % reduction due to single treatment		Expected % reduction for combination*	Actual % reduction due to combination	
Species	Gamma	Infrared		Gamma + infrared	Infrared + gamma
Sitotroga cerealella	52	55	79	93	93
Rhyzopertha dominica	54	55	80	99	—
	Gamma	Microwave		Gamma + microwave	Microwave + gamma
S. cerealella	54	74	88	96	97
R. dominica	54	42	72	96	—
Sitophilus oryzae	86	28	90	92	91

* Expected percentage of reduction was calculated for the combination by multiplying the percentage of survival in the first treatment by the percentage of survival in the second treatment and subtracting this figure from 100%.

IV. COMBINING RADIATION WITH OTHER CONTROL METHODS

Efforts have been made by several researchers to reduce radiation disinfestation costs by adding a second treatment that would predispose insects to increased damage when irradiated.

As treatments with infrared and microwave radiation were likely to provide stresses that would be complementary to gamma radiation, microwave, and infrared radiations were tried both before and after gamma radiation of Angoumois grain moths, lesser grain borers, *Rhyzopertha dominica* (F.), and rice weevils in wheat[12] as a part of the U.S. Department of Agriculture studies at the Savannah facility.

Each type of treatment alone was found to produce some mortality. The expected percentage mortality for the combined treatments was calculated by using these mortalities (Table 2). However, actual mortality for the combinations was always greater than that calculated (Table 2), an average 16 and 11% greater for the infrared and microwave treatments, respectively; indeed, differences of as great as 24% were obtained. Also, the results showed that the sequence of the treatments did not alter the percentage mortality.

The dose of gamma radiation can then be reduced without sacrificing a high level of mortality if a supplemental treatment is used, and most such supplemental treatments would greatly reduce the cost of the radiation. For example, if natural gas-fired IR heaters are used, the cost of the combined treatment might be less than the cost of gamma radiation alone. Also, the quantity of grain treated by the irradiator would be increased proportionately.

Other tests were made to determine whether radiation is compatible with currently used chemical methods of control.[12] Since malathion is commonly used to protect grain from insects, tests were made to determine whether gamma radiation adversely affected the residues of malathion. Neither the toxicity nor the degradation of malathion on wheat or on kraft paper was affected by doses of as much as 4.3 Mrad. However, red flour beetles, *Tribolium castaneum* (Herbst), were slightly more susceptible to malathion after they were irradiated with a dose of 10 krad or more. Also, the combination of treatments produced more and earlier mortality than did radiation alone. Erdman[31] also reported that malathion or DDT in combination with γ radiation was more effective than either insecticide alone or γ radiation alone.

In addition, fumigation with methyl bromide and treatment with gamma radiation were found to be compatible if the interval between treatments was short or complementary if the interval was longer. For example, mortality from the combination was greater than from either treatment alone if 1 week or more elapsed between treatments. Thus, disinfestation by radiation also appears compatible with currently used methods of chemical control and might actually be increased by the combination.[12]

Fresh fruit could be placed in cold storage following use of irradiation as a quarantine treatment. A combination of methyl bromide fumigation plus cold storage has been shown to be more effective than fumigation alone;[32] mortalities were independent for each treatment, but were summed. Combination of gamma radiation with subsequent cold storage would probably permit a slight reduction in the dosage required compared to radiation alone.

V. FEEDING BY IRRADIATED INSECTS

Insects in bulk grain[9] or fresh fruit[33] do not have to be killed immediately, as some have suggested. Control can be obtained with lower, more economical doses if the resident pupal or adult populations are sterilized or eggs and larvae continue to develop but do not emerge as adults. Also, Cornwell[9] suggested that the presence of sterile but sexually competitive insects might offer some measure of protection against reinfestation of stored grain. However, gamma sterilization for such purposes allows these insects to continue to feed and further damage the commodity.

Brower and Tilton[34] reported that the wheat consumption by irradiated rice weevils and lesser grain borers during a 5-week period was reduced 90 and 97%, respectively. Also, Rogers and Hilchey[35] reported that although red flour beetles continued to feed after treatment with high-speed electrons, it was at a reduced rate. Cornwell[36] treated granary weevils, *Sitophilus granarius* (L.) with 16 krad of gamma radiation and found that they consumed only half as much food as did untreated adults. Watters and MacQueen[37] found that four stored-product insect species could still damage wheat 14 weeks after irradiation with 6.25 krad, but the amount of damage was greatly reduced.

The cessation of feeding that occurs with the smaller doses of gamma radiation appears to result primarily from damage to the midgut of irradiated insects. For example, the levels required to sterilize both yellow mealworm and Indian meal moth produce considerable damage to the midgut.[12]

As previously noted, emergence of viable adult fruit flies from fruit that has been infested with eggs or larvae of fruit flies can be prevented by exposure of the fruit to less than 25 krad of gamma radiation.[26] This dosage will provide adequate control (probit 9 security) for quarantine purposes. However, it will permit some eggs to hatch and larvae to continue their development and form puparia,[38] although most do not pupate. Quarantine treatments using fumigation, cold storage, or vapor heat had been based on preventing pupation of larvae or hatch of eggs.[39] The dosage of gamma irradiation required to accomplish this was found to be too high to be practical.[17] Therefore, any use of gamma radiation as a quarantine treatment for fruit infested by fruit fly eggs and larvae could permit continued feeding and subsequent pupation of irradiated larvae. Adults that might emerge from irradiated pupae would be sterile.

VI. EFFECTS OF IRRADIATED DIETS ON INSECTS

Tests to control insects in various food commodities necessitated that the insects be irradiated in the commodity under study. Subsequent changes in these insect populations could then result either from the direct effects of radiation on the insects or from radiation-induced changes in the food medium affecting the insects. Several research workers have conducted tests to determine if there is an effect and if so, to determine its significance. Considerable variation in results has been reported and some appear contradictory.

Most earlier studies of insects reared for several generations on irradiated diets indicated that irradiated diets were not harmful to insect development, behavior, or progeny production. For example, Cornwell and Burson[40] reported that the behavior and numbers of progeny of granary weevils and rice weevils were not affected when reared on wheat irradiated at dosages ranging between 0 and 50 krad. Hodges and Guyer[41] found that confused flour beetles, *T. confusum* Jacquelin duVal; granary weevils; and Angoumois grain moths, reared for three generations on wheat irradiated with dosages as high as 1 million rep had at least as many progeny as the control. Confused flour beetles had a higher number of progeny in the irradiated flour than in the nonirradiated flour, but the increases were significant in only two generations at the lowest radiation dosage.

Tilton et al.[42] found no increase or decrease in progeny production by rice weevils reared on irradiated wheat, a finding similar to that of Cornwell and Burson.[40] Similarly, their data for the red flour beetle agreed with that of Hodges and Guyer[41] for the confused flour beetle. They found significantly increased numbers of progeny on the irradiated diets for three dosage levels. Wiendl et al.[43] reported that the maize weevil, *S. zeamais* (Motschulsky) reared on 20 krad irradiated rice had an increase in the longevity and progeny numbers. A decrease in longevity and progeny was observed for rice preirradiated at 100, 500, and 1500 krad.

Bagheri-Zenouz[44] irradiated wheat at 25, 100, and 500 krad, and 1 Mrad and granary weevils were reared on this medium. He found that the average fertility was significantly higher for the insects reared on wheat treated at 100 krad or more. There was no significant difference between control medium and medium treated at 25 krad. A difference was found in the case of adults from the second generation. Longevity was found to have increased more significantly in the males than in the females in rearings on irradiated media. This effect was observed at all dosage levels, but was greatest at 500 krad. When calculated without distinction of sex, there were no significant differences in effects on longevity between irradiated and control media.

In contrast, several recent reports indicate that insects fed irradiated food may be subject to lowered fecundity or to shortened life span. Singh and Liles[45] reported a significant decrease in the survival of lesser grain borers fed on 160- or 320-krad irradiated wheat kernels. They also reported that feeding the beetles on irradiated wheat resulted in a significant suppression of egg production and reduced egg viability, with the effect more pronounced in beetles fed on 320-krad than in beetles fed on 80-krad irradiated wheat. Seugé et al.[46] showed a 32% decrease in fecundity of Indian meal moths when they were reared for a single generation on a diet of pistachio nuts irradiated with 66 krad and 42% reduction in the fecundity of mealy bugs, *Pseudaulacaspis pentagona* (Targioni-Tozzetti) reared on potato tubers irradiated at 33, 66, or 165 krad. The reduction in fecundity at the 66 to 165 krad range was not dose dependent. Brower et al.[47] using irradiated diets reported that four generations of Indian meal moths reared on raisins or nine generations on flour showed no dietary effect on fecundity or sex ratio of progeny. This was true even though very high doses (1000 and 4300 krad) were used. In a follow-up study of the report of Seugé et al.,[46] Brower and Tilton[48] radiated four nutmeats including pistachios, pecans, walnuts, and almonds at dosage levels of 33, 66, and 100 krad. Indian meal moths were reared on each kind of nutmeat from each treatment level for four generations on pistachios and pecans, for three generations on almonds, and for a single generation on walnuts. They reported no decrease in development success or in fecundity of moths reared on any of the diets.

Morère and Seugé[49] continued the work reported by Seugé et al.[46] with both the mealy bug and Indian meal moth. In 1976, they reported that with mealy bugs feeding on irradiated potato tubers there was a slight but not statistically significant reduction in fecundity, up to 5.5 krad. Further increases in dose produced further decreases in fecundity, and after 16 krad, fecundity was reduced by 45%. Beyond 20 krad, the degree of reduction was dose independent. In this study the fecundity of Indian meal moth females reared on whole preirradiated meridic diet or on a meridic diet including irradiated casein or irradiated dextrose was unchanged. The number of emerging moths was lower on the whole irradiated diet and on the diet with only casein irradiated, and the mean duration of larval development was longer when only casein or dextrose was irradiated.

The data from extensive studies of the effects of irradiated diets on *Drosophila* are equally confusing. Hossain et al.[50] reared *D. melanogaster Meigen*, on irradiated banana diets (treated at 35, 45, and 55 krad) for 7 generations. They reported no effects on egg-pupae ratio, percent adult emergence, longevity, egg production, or viability. Redi et al.[51] radiated a basic *Drosophila* medium at 150 and 300 krad and found no effects of the treated medium on recessive sex-linked lethals and translocations or on sex-linked visible mutations and large deletions of *Drosophila*. Rinehart and Ratty[52] cultured *D. melanogaster* on food irradiated at 150, 500, or 3000 rad of X- or γ-rays and tested the offspring for dominant, sex-linked, and F_3 lethality. They found no decrease in F_1 offspring survival associated with the irradiated food. There was an approximate twofold increase in sex-linked recessive lethality associated with the irradiated food that they attributed to an increase in gonial mutants. Aging the food prior to use decreased the numbers of nongonial mutations, but there was still an effect of the irradiated food in increasing lethality above control levels that was associated with gonial mutations.

VII. INSECT RESISTANCE TO RADIATION

The development of resistance to insecticides by species of stored-product insect pests is an increasingly serious problem. Also, cross resistance in resistant populations can make new insecticides obsolete almost as soon as they are introduced. Thus disinfestation by radiation has a large advantage over currently used chemical control in that insecticide resistance is unrelated to radiosensitivity. For example, Brower[53] showed that two extremely malathion-resistant strains of Indian meal moth were just as susceptible to gamma radiation as the control (susceptible) population; Cole et al.[54] found that a highly DDT-resistant strain of the human body louse, *Pediculus humanus humanus* L., was no more resistant to gamma radiation than a DDT-susceptible strain; and Erdman[55] showed that a strain of red flour beetles that was slightly more resistant to DDT than the control strain was more sensitive to X-ray. If the same is true for most insect pests, then disinfestation of grain or other stored commodities by radiation might be economical in areas where there are insecticide-resistant insect populations. Brower[56] reported that a strain of red flour beetles resistant to DDT and malathion did not possess greater resistance to gamma radiation than a susceptible strain.

Because insects can be sterilized with doses of radiation lower than those needed to produce quick mortality, and because these lower doses make disinfestation by radiation less costly, the use of radiation in bulk raw agricultural commodities will probably be limited to sterilizing doses. However, the use of these lower doses might mean that a small proportion of the insects is exposed to a substerilizing dosage. Thus it is important to know whether substerilizing doses applied to several successive generations can cause insects to develop radiation-resistant strains.

Few workers have studied the possibility of inducing resistance to radiation in insects. However, Cornwell and Morris[57] found no increase in the resistance in F_1 and F_2 granary weevils from an irradiated parent population. Indeed, the F_2 generation appeared to be slightly more susceptible than a strain with no past history of radiation treatment. Also, extensive tests have been made at the U.S. Department of Agriculture's Savannah laboratory on four diverse species of stored-product insects exposed to substerilizing doses. These strains were then analyzed periodically for changes in longevity, fertility, fecundity, and resistance to acute exposures. No evidence of any pronounced increase in resistance to radiation has been found in any of the species, even after the prolonged periods of selection. However, population "fitness" tended to be reduced in the selected strains (higher percentages of sterility, lower fecundities, and decreased lifespans).[12,58-61] Similar results were reported for the rice weevil by Matin.[62] Therefore, the possibility of accidental exposure of some insects to substerilizing doses during commercial disinfestation does not appear likely to cause the development of radiation-resistant populations.

VIII. USE OF STERILE-MALE TECHNIQUE WITH STORED-GRAIN INSECTS

The sterile-male technique shows little promise as a means of controlling grain-infesting stored-product *Coleoptera.* Cornwell[9] explored the possibilities of this technique and concluded it was not practical. The picture is somewhat different with *Lepidoptera,*[63] and several research organizations are presently working on modifications of the sterile-release technique for stored-product insects of this order. Some characteristics of the biology of these stored-product moths that make them more promising subjects are (1) adults are nonfeeding, (2) infestations of single moth species occur, (3) most species are primarily surface feeders, (4) and the adults of most species emerge from the product before mating. In addition, genetic damage in *Lepidoptera* can be transmitted through several generations so effects of the treatment are often intensified in the subsequent generations.[13] A more extensive discussion of the sterile insect release technique is given in Chapter 7, Volume 2.

REFERENCES

1. **Balock, J. W. and Kozuma, T. T.,** Notes on the biology and economic importance of the mango weevil, *Sternochetus mangiferae* (Fabricius), in Hawaii (*Coleoptera:* Curculionidae), *Proc. Hawaii. Entomol. Soc.,* 18, 353, 1964.
2. **Champ, B. R. and Dyte, C. E.,** Report of the FAO Global Survey of Pesticide Susceptibility of Stored Grain Pests, Food and Agriculture Organization (FAO), Rome, 1976, 298.
3. **Hunter, W. D.,** Results of experiments to determine the effects of Röentgen rays upon insects, *J. Econ. Entomol.,* 5, 188, 1912.
4. **Morgan, A. C. and Runner, G. A.,** Some experiments with Röentgen rays upon the cigarette beetle, *Lasioderma serricorne* Fabr., *J. Econ. Entomol.,* 6, 226, 1913.
5. **Runner, G. A.,** Effects of Röentgen rays on the tobacco, or cigarette beetle and the results of experiments with a new form of Röentgen tube, *J. Agric. Res.,* 6, 383, 1916.
6. **Hilchey, J. D.,** Action of ionizing radiations on insects, in *Radiation Preservation of Food,* Bailey, S. D., Davies, J. M., Morgan, B. H., Pomerantz, R., Siu, R. G. H., and Tischer, R. G., Eds., U.S. Government Printing Office, Washington, D.C., 1957, chap. 25.
7. **Balock, J. W., Christenson, L. D., and Burr, G. O.,** Effect of gamma rays on immature stages of the oriental fruit fly, (*Dacus dorsalis* Hendel) and possible application to commodity treatment problems, *Proc. Hawaii. Acad. Sci.,* 31, 18, 1956.

8. O'Brien, R. D. and Wolfe, L. S., *Radiation, Radioactivity and Insects*, Academic Press, New York, 1964, 211.
9. **Cornwell, P. B.**, *The Entomology of Radiation Disinfestation of Grain*, Pergamon Press, Elmsford, N.Y., 1966, 236.
10. **Cogburn, R. R., Tilton, E. W., and Brower, J. H.**, Bulk grain irradiation for control of insects infesting wheat, *J. Econ. Entomol.*, 65, 818, 1972.
11. **Tilton, E. W., Brower, J. H., and Cogburn, R. R.**, A method of dosimetry for a bulk-grain irradiator, *Int. J. Appl. Radiat. Isot.*, 22, 577, 1971.
12. **Tilton, E. W. and Brower, J. H.**, Status of U.S. Department of Agriculture research on irradiation disinfestation of grain and grain products, in *Radiation Preservation of Food*, Conf. Proc., Bombay, International Atomic Energy Agency (IAEA), Vienna, 1973, 295.
13. **Ashrafi, S. H., Tilton, E. W., and Brower, J. H.**, Inheritance of radiation-induced partial sterility in the Indian meal moth *Plodia interpunctella*, *J. Econ. Entomol.*, 65, 1265, 1972.
14. **Brower, J. H. and Tilton, E. W.**, Insect disinfestation of dried fruit by using gamma radiation, *Food Irradiat.*, 11(1—2), 10, 1970.
15. **Papadopoulou, C. P.**, Disinfestation of dried figs by gamma radiation, in *Proc. Symp. Radiation and Radioisotopes Applied to Insects of Agricultural Importance*, Athens, 1963, IAEA, Vienna, 1964, 485.
16. **Shchegoleva, G. I.**, Disinfestation of dried fruits (in Russian), *Zashch. Rast.*, 19, 14, 1963; Abstr. only seen, in *Radioisotopes and Ionizing Radiation in Entomology III* (1964—1965), Bibliogr. Ser. 24, IAEA, Vienna, 1967, 267.
17. **Balock, J. W., Burditt, A. K., Jr., and Christenson, L. D.**, Effects of gamma radiation on various stages of three fruit fly species, *J. Econ. Entomol.*, 56, 42, 1963.
18. **Benschoter, C. A. and Telick, C., Jr.**, Effect of gamma rays on immature stages of the Mexican fruit fly, *J. Econ. Entomol.*, 57, 690, 1964.
19. **Thomas, P. and Rahalkar, G. W.**, Disinfestation of fruit flies in mango by gamma irradiation, *Curr. Sci.*, 44, 775, 1975.
20. **Macfarlane, J. J.**, Control of the Queensland fruit fly by gamma irradiation, *J. Econ. Entomol.*, 59, 884, 1966.
21. **Shipp, E. and Osborn, A. W.**, Irradiation of Queensland fruit fly pupae to meet quarantine requirements, *J. Econ. Entomol.*, 61, 1721, 1968.
22. **Cavalloro, R. and Delrio, G.**, Biological effects of gamma radiation on the life stages of two fruit fly species, *Radionuclides in Ecosystems*, Vol. 2, Proc. 3rd Natl. Conf. Radioecology, Oak Ridge, Tenn., 1971, 1197.
23. **Bughio, A. R., Qureshi, Z. A., and Mecci, A. K.**, Effects of gamma-radiation on eggs of *Dacus zonatus* (Saunders), *Int. J. Radiat. Biol.*, 16, 183, 1969.
24. **Hossain, M. M., Hamid, M. A., Siddiqui, A. K., and Matin, M. A.**, Irradiation preservation of certain tropical fruits, in A Comprehensive Rep. on IAEA/FAO Research Contract No. 361/R1/RB on Banana and Mango, AEDC/RB-9, IAEA, Vienna, 1967, 267.
25. **Christenson, L. D. and Foote, R. H.**, Biology of Fruit Flies, *Annu. Rev. Entomol.*, 5, 171, 1960.
26. **Balock, J. W., Burditt, A. K., Jr., Seo, S. T., and Akamine, E. K.**, Gamma radiation as a quarantine treatment for Hawaiian fruit flies, *J. Econ. Entomol.*, 59, 202, 1966.
27. **Seo, S. T., Kobayashi, R. M., Chambers, D. L., Dollar, A. M., and Hanaoka, M.**, Hawaiian fruit flies in papaya, bell pepper, and egg-plant: quarantine treatment with gamma irradiation, *J. Econ. Entomol.*, 66, 637, 1973.
28. **Baker, A. C.**, The basis for treatment of products where fruit flies are involved as a condition for entry into the United States, *U.S. Dep. Agric. Circ.*, 551, 1939, 8.
29. **Seo, S. T., Kobayashi, R. M., Chambers, D. L., Steiner, L. F., Lee, C. Y. L., and Komura, M.**, Mango weevil: cobalt 60 gamma irradiation of packaged mangos, *J. Econ. Entomol.*, 67, 504, 1974.
30. **Thomas, A. C.**, The Application of Ionising Radiation to the Shelf Life Extension of Mangoes in South Africa, Atomic Energy Board, Pelindaba, Pretoria, Rep. South Africa, PEL-qcc, 1975, 15.
31. **Erdman, H. E.**, Modification of fitness in species and strains of flour beetles due to X-ray and DDT, *Ecology*, 47, 1066, 1966.
32. **Seo, S. T., Kobayashi, R. M., Chambers, D. L., Steiner, L. F., Balock, J. W., Komura, M., and Lee, C. Y. L.**, Fumigation with methyl bromide plus refrigeration to control infestations of fruit flies in agricultural commodities, *J. Econ. Entomol.*, 64, 1270, 1971.
33. **Barnes, B. N.**, Irradiation-disinfestation of export fruit as a quarantine treatment for proclaimed insects: the potentialities and problems, in *Proc. Int. Congr. Entomol. Soc. South Africa*, Stellenbosch, S. A., Ed., Entomological Society of South Africa, Pretoria, 1975, 163.
34. **Brower, J. H. and Tilton, E. W.**, Weight loss of wheat infested with gamma-irradiated *Sitophilus oryzae* (L.) and *Rhyzopertha dominica* (F.), *J. Stored Prod. Res.*, 9, 37, 1973.
35. **Rogers, W. L. and Hilchey, J. D.**, Studies on the post-irradiation feeding activity of *Tribolium castaneum* (Tenbrionidae: *Coleoptera*), *Ann. Entomol. Soc. Am.*, 53, 4, 1960.

36. **Cornwell, P. B.**, Insect control, in *Massive Radiation Techniques*, Jefferson, S., Ed., George Newnes, London, 1964, 141.
37. **Watters, F. L. and MacQueen, K. P.**, Effectiveness of gamma irradiation for control of five species of stored-product insects, *J. Stored Prod. Res.*, 3, 223, 1967.
38. **Burditt, A. K., Jr. and Seo, S. T.**, Dose requirements for quarantine treatments of fruit flies with gamma irradiation, in *Disinfestation of Fruit by Irradiation*, Proc. 1970 Panel, Honolulu, Hawaii, IAEA, Vienna, 1971, 33.
39. **Burditt, A. K., Jr., Seo, S. T., and Balock, J. W.**, Basis for developing treatments for fruit flies, in *Disinfestation of Fruit by Irradiation*, Proc. 1970 Panel, Honolulu, Hawaii, IAEA, Vienna, 1971, 27.
40. **Cornwell, P. B. and Burson, D. M.**, Grain weevils, *Calandra granaria* L. and *C. oryzae* L. reared on irradiated wheat, *Nature (London)*, 181, 1947, 1958.
41. **Hodges, R. and Guyer, G.**, The effect of an irradiated wheat diet on the confused flour beetle, granary weevil and the Angoumois grain moth, *J. Econ. Entomol.*, 51, 674, 1958.
42. **Tilton, E. W., Brower, J. H., and Cogburn, R. R.**, Progeny production by *Sitophilus oryzae* and *Tribolium castaneum* reared for several generations on irradiated diets, *J. Ga. Entomol. Soc.*, 8, 168, 1973.
43. **Weindl, F. M., Arthur, V., Sgrillo, R. B., Domarco, P. E., Tornisielo, V. L., Pacheco, J. M., and Walder, J. M. M.**, Mortalidade e reproducao de *Sitophilus zeamais* Mots. em anoz. irradiadado, in *Boletim Cientifico BO-015*, Centro de Energia Nuclear na Agricultura, Escola Superior de Agricultura "Luis de Queiroz," Piracicaba, S. P., Brasil, 1974, 9.
44. **Bagheri-Zenouz, E.**, Influence de l'irradiation des grains de sur le development et la reproduction de Sitophilus granarius, paper IAEA-SM-138132, in *Proc. Symp. Sterility Principle for Insect Control or Eradication*, Athens, 1970, IAEA, Vienna, 1971, 23.
45. **Singh, H. and Liles, J. N.**, Effect of irradiated food on the adult survival and reproduction of *Rhyzopertha dominica* (F.) (*Coleoptera*, Bostrichidae), *J. Stored Prod. Res.*, 8, 155, 1972.
46. **Seugé, J., Morère, J. L., and Ferradini, C.**, Effect of food preirradiation on the fecundity of two insects: mealy bugs *(Pseulaulacaspis pentagona* Targ.) and Indian meal moths (*Plodia interpunctella* Hübner), *Radiat. Res.*, 67, 120, 1976.
47. **Brower, J. H., Tilton, E. W., and Cogburn, R. R.**, Effects of irradiated diets on production of progeny by several successive generations of the Indian meal moth, *Plodia interpunctella* (Hubner), *Radiat. Res.*, 48, 283, 1971.
48. **Brower, J. H. and Tilton, E. W.**, Development and fecundity of the Indian meal moth (*Plodia interpunctella* Hübner) reared on diet of irradiated nutmeats, *Int. J. Appl. Radiat. Isot.*, 24, 327, 1973.
49. **Morère, J. L. and Seugé, J.**, Effect of natural or artificial preirradiated food on the life history of two unirradiated insects: mealy bugs (*Pseudaulacaspis pentogona* Targ.) and Indian meal moths (*Plodia interpunctella* Hübner), *Radiat. Res.*, 67, 120, 1976.
50. **Hossain, M. M., Mollah, S. A., and Mallik, M. U.**, Studies on the experimental population of *Drosophila melanogaster* raised on irradiated banana diet, *Food Irradiat.*, 7, 45, 1967.
51. **Redi, O. S., Reddy, G.M., Rao, J. J., Ebenezer, D. N., and Rao, M. S.**, Lack of mutagenic effect of irradiated *Drosophila* medium, *Nature (London)*, 208, 702, 1965.
52. **Rinehart, R. R. and Ratty, F. J.**, Mutation in *Drosophila melanogaster* cultured on irradiated food, *Genetics*, 52, 119, 1965.
53. **Brower, J. H.**, Gamma radiation susceptibility of insecticide-resistant strains of the Indian meal moth, *Plodia interpunctella, J. Econ.Entomol.*, 66, 461, 1973.
54. **Cole, M. M., LeBreque, G. C., Burden, G. S.**, Effects of gamma radiation on some insects affecting man, *J. Econ. Entomol.*, 52, 448, 1966.
55. **Erdman, H. E.**, Modification of fitness in species and strains of flour beetles due to X-ray and DDT, *Ecology*, 47, 1066, 1966.
56. **Brower, J. H.**, Radio-sensitivity of an insecticide-resistant strain of *Tribolium castaneum* (Herbst), *J. Stored Prod. Res.*, 10, 129, 1974.
57. **Cornwell, P. B. and Morris, J. A.**, Susceptibility of the grain and rice weevils *Calandra granaris* and *C. oryzae* to gamma radiation, in *Large Radiation Sources in Industry*, Vol. 2, Conf. Proc. I., Warsaw, IAEA, Vienna, 1960, 291.
58. **Brower, J. H.**, Radioresistance of the red flour beetle, *Tribolium castaneum* (*Coleoptera:* Tenebrionidae), exposed to sublethal doses of gamma radiation for 25 generations, *Can. Entomol.*, 106, 241, 1974.
59. **Brower, J. H.**, Lack of radioresistance in *Plodia interpunctella* (*Lepidoptera:* Phycitidae) exposed to sublethal gamma radiation for 30 generations, *Radiat. Res.*, 57, 73, 1974.
60. **Brower, J. H.**, Inability of populations of *Callosobruchus maculatus* to develop tolerance to exposures of acute gamma radiation, *Ann. Entomol. Soc. Am.*, 67, 287, 1974.
61. **Brower, J. H.**, Fitness and radioresistance of the rice weevil, *Sitophilus oryzae* (*Coleoptera:* Curculionidae), after 25 generations of substerilizing irradiations, *J. Kan. Entomol. Soc.*, 47, 437, 1974.

62. **Matin, A. S. M. A.**, On the possible development of radioresistance in *Sitophilus oryzae* (L.) following exposures of successive generations to ionizing radiation, *Int. J. Appl. Radiat. Isot.*, 26, 625, 1975.

63. **Brower, J. H. and Tilton, E. W.**, Potential for control of *Cadra cautella* (Walker) by release of fully or partially sterile males, *J. Appl. Radiat. Isot.*, 26, 720, 1975.

Chapter 8

RADAPPERTIZATION OF MEAT, POULTRY, FINFISH, SHELLFISH, AND SPECIAL DIETS*

E. S. Josephson

TABLE OF CONTENTS

* Publ. No. 3995 of Department of Nutrition and Food Science, Massachusetts Institute of Technology, Cambridge, Mass., 02139.

I. INTRODUCTION

Radappertization requires exposing food in sealed containers to ionizing radiation at absorbed doses high enough (25 to 70 kGy) to kill all organisms of food spoilage and public health significance to achieve "commercial sterility". Radappertization, therefore, is analogous to thermal canning in achieving shelf stability (long-term storage without refrigeration). Except for dry products in which autolysis is negligible, the radappertization process also requires that the food be heated to an internal temperature of 70 to 80°C (bacon to 53°C) to inactivate autolytic enzymes which catalyze spoilage during storage without refrigeration.

It took more than a century following Nicolas Appert's successful demonstration in 1809 of thermal canning for this process to become established on a commercial scale with an outstanding record of reliability and safety to the customer. For radappertization to become established on a commercial scale, it must match or better the reliability and safety record achieved by thermal canning. The high standards for radappertization which have been set by the health authorities at national levels and by international bodies[1] have posed an unprecedented challenge to the scientists and engineers working in this field. This challenge has been met.

To minimize the occurrence of irradiation induced off-flavors and odors, undesirable color changes, and textural and nutritional losses from exposure to the high doses required for radappertization, the foods are vacuum sealed and irradiated while frozen (−40 to −20°C).

Foods which are eaten raw (e.g., fresh fruits and vegetables) or foods which deteriorate because of freezing and thawing (e.g., fresh fruits and vegetables, whole milk) are not good candidates for radappertization. On the other hand, radappertized meats, poultry, fin fish, shellfish, nutritionally complete meals for humans, and diets for germ-free (gnotobiotic) animals have excellent organoleptic and nutritional quality.

Because of the high cost to develop the technology and prove safety for consumption (wholesomeness) of radappertized foods, almost all of the research and development has been sponsored by the U.S. government, mostly by the Army's laboratories at the Quartermaster Food and Container Institute in Chicago, Ill. (1953 to 1962) and at Natick, Mass. (1962 to 1980). During this period, the U.S. Army Medical Department has been responsible for the toxicological and nutritional aspects of radappertization and for assurance of no measurable induced radioactivity.

Two bibliographies dealing primarily with radappertization have been published under U.S. Army auspices.[2,3] Publications on the wholesomeness aspects of radappertization are encompassed in a comprehensive bibliography prepared by Barna.[4]

II. OBJECTIVES

The objective of radappertization is to provide foods, food products, meals, and diets for humans and animals which, even when kept for several years without refrigeration, are highly palatable, nutritious, and wholesome. Radappertization must be competitive with other food preservation methods not only in taste but also in price, energy requirements, and impact on the environment.

Facets unique to radappertization should also be considered in overall comparisons with competing processes. These facets include the ability to produce shelf-stable foods which (1) on the plate of the consumer are almost indistinguishable from food similarly prepared for consumption, (2) cannot be produced by any other known process or with the same degree of palatability (e.g., whole roasts, luncheon meats), and (3) reduce or eliminate the use of chemical preservatives such as ethylene oxide and sodium nitrite which, per se or by combining with secondary amines in the food or in the digestive tract to form nitrosamines, are carcinogenic or toxic.

Table 1
AUTOLYTIC ENZYME INACTIVATION PRIOR TO IRRADIATION[a]

Product	Method	External temperature (°C)	Time (min)	Final internal temperature (°C)
Rolls (beef, pork, lamb, chicken)	Cookhouse	60° start, 88° end	540—660	70—75
Roasts (beef, pork, lamb)	Oven	177	120—180	70—75
Ham	Smokehouse	Commercial practice	540—600	68—75
Corned beef (4 cm thick)	Boiling water	100	60	70—75
	Simmering	75—80	50	75—80
Beef patties	Oven	204	20—25	72—75
Bacon	Smokehouse	Commercial practice[b]	1320	54
Chicken parts	Oven	177	ca. 120	75—80

[a] All meats were refrigerated (4—6°C) before enzyme inactivation.
[b] Stepwide process, starting with 38°C and ending with 65°C smokehouse temperature. Smoking "to brown" 20—45 min after first 2 hr processing.

From Wierbicki, E., U.S. Army Natick Research and Development Laboratories, Natick, Mass., personal communication, July 31, 1980.

III. TECHNOLOGY

The major causes of food spoilage encountered in radappertization are autolytic enzymes, oxygen, and food spoilage and pathogenic organisms. In addressing these factors, it is important to prevent or minimize creation of new problems such as measurable induced radioactivity, loss of organoleptic and nutritional quality, and formation of radiolysis products of possible toxicological significance.

A. Inactivating Autolytic Enzymes

The highest radiation doses to which foods have been exposed without serious loss in palatability (about 70 kGy) result in only 25 to 50% loss in autolytic enzyme activity. If autolytic enzymes are not inactivated, postirradiation storage of food without refrigeration limits shelf life because of gradual loss in texture and palatability. Application of heat to inactivate the autolytic enzymes has been the only known safe and reliable method to minimize the occurrence of autolysis. Chemical inhibitors tested cannot be used because they are toxic and/or unreliable. The conditions for inactivating autolytic enzymes while maintaining the fresh-like quality of some typical radappertized foods prepared for eating have been summarized by Wierbicki[6] in Table 1.

A rapid method to measure total tissue proteolytic activity with a hemoglobin substrate labeled with $K^{14}CNO$ has been used to assess the effectiveness of inhibiting autolysis.[7,8] This permits rapid assessment of the efficacy of the procedures used to inactivate proteases and obviates the need to wait for several months of postirradiation storage before visible signs of autolysis appear.

Enzyme inactivation of raw meats can lead to a 35% loss of natural juices and water-soluble nutrients. This results in drier products than the nonirradiated counterparts, particularly when the food is reheated for serving. Such products, however, score well within the acceptable range when assessed by expert and consumer panels. Loss of juiciness can be limited to 10 to 15% with concomitant improvement in organoleptic quality in ground beef and beef, poultry, pork, and lamb rolls by pretreatment with 0.75% sodium chloride (below the threshold for salty taste) and 0.25 to 0.5% food grade phosphates such as sodium tripolyphosphate. Shults et al.[9] reported on the effect of the combination of condensed phosphates and salt on the water-holding capacity of beef. Heating to inactivate autolytic enzymes provides an additional benefit to the radappertization process because it also inactivates most of the food-borne viruses and several species of vegetative bacteria (e.g., *Moraxella-Acinetobacter*) which are radiation resistant at the subfreezing temperatures at which the food is held during irradiation.

B. Preventing Oxidative and Other Undesirable Physical and Chemical Changes

Undesirable changes in color, taste, odor, texture, and nutrients are minimized by excluding atmospheric oxygen from the package in which exposure of the food to irradiation occurs and by using sealed packaging materials providing an impermeable barrier to light, moisture, oxygen, and microorganisms. Although oxygen can be excluded from the package by flushing with nitrogen gas, evacuation of the air has been the method generally used. Evacuation makes it possible to eliminate by visual inspection flexible containers of foods where oxygen removal has been incomplete or sealing of the package or the packaging material itself has been defective. When rigid metal cans are used, the degree of evacuation can be measured nondestructively by the simple flip test to assure that the internal pressure specified in the can containing the radappertized food has been met. The packaging materials used are critical to the radappertization process and are discussed in Chapter 8, Volume 2.

Undesirable physical and chemical changes in the food are also minimized by exposing the food while frozen to irradiation. The beneficial effect by lowering the temperature of the food during irradiation is offset in part by the increased resistance of *Clostridium botulinum* to irradiation and to the added cost for freezing. Taking into account the energy cost for low temperature, the required radiation dose for radappertization, the organoleptic and nutritional quality of the food, and the availability and capability of mechanical refrigeration, the optimum temperature of the food at the beginning of irradiation has been determined to be −40°C. If gamma radiation sources are used, the food packages during exposure to irradiation can be perfused with liquid nitrogen, pumped in from an external storage tank to minimize the temperature rise from the room temperature environment and dissipate the heat formed because of absorption of the ionizing energy. Since electron sources have much higher dose rates than isotope sources, irradiation exposure time at the room-temperature environment is so limited that it is not necessary to cool the packages during the exposure period. However, the packages are protected from the room-temperature environment by irradiating them in insulated Styrofoam® boxes. In practice the temperature of the food at the end of irradiation should not exceed −20°C.

C. Eliminating Organisms Causing Spoilage and Health Hazards

The *raison d'etre* for radappertization is to destroy in the prepackaged autolytic enzyme inactivated food all organisms of food spoilage and public health significance. As in the case of thermal canning, the dose required for radappertization is geared to reducing to statistically negligible numbers the population of the most radiation-resistant spoilage or pathogenic organism. Since spores of *C. botulinum* are the most radiation-resistant organisms of concern in radappertization,[1,5,8,10-15] the entire process has been centered around the dose needed to reduce the numbers of viable spores from 10^{12} to 10^{0} (the 12-D concept where D is the dose required for 90% population reduction) recommended by a panel of experts convened by the Food and Agriculture Organization (FAO), International Atomic Energy Agency (IAEA), and World Health Organization (WHO).[1]

The 12-D concept is based on the studies by Esty and Meyer[16] in 1922 on the heat

Table 2
TYPICAL INOCULATED PACK PROCEDURE

Prototype food	40—45 g/Container
Container	211 × 101.5 Cylindrical metal can (epoxy enamel liner)
Containers/dose	100 Replicates
Vacuum seal	125 mm Mercury pressure
Spore inoculum	10^6/Strain; 10^7/can
Clostridium botulinum strains	33A, 36A, 62A, 77A, 12885A, 41B, 53B, 67B, 9B, 40B
Radiation temperature	−30±10°C
Doses (kGy)	14—50 in 4 kGy Increments
Incubation	6 Months at 30 ±2°C
Analyses	Swelling, botulinal toxin, recoverable *C. botulinum*

From Rowley, D. B., personal communication, 1980.

resistance of spores of *C. botulinum.* This concept served as the basis for the time and temperature needed to establish the safety and efficacy of thermal canning. To assure that radappertization should provide the same margin of safety as that of thermal canning, the 12-D concept proposed by Schmidt[15] was adopted by the FAO/IAEA/WHO panel of experts[1] and has received acceptance world-wide.

Of 102 strains of *C. botulinum* irradiated in neutral phosphate buffer for comparative radiation resistance, 33A, 36A, 40B, 41B, and 53B which had D values ranging from 3.17 to 3.36 kGy, were the most resistant.[17] Strains 62A, 77A, 12885A, and 9B with intermediate resistance exhibited D values of 2.24 to 2.53 kGy.[17] The D value for the most sensitive strain, 51B, was 1.29 kGy.[17]

Since the D value varies with the food medium, this value must be determined experimentally for each food as it would be processed commercially by radappertization. The procedure for conducting a typical inoculated pack is shown in Table 2.[12] The 12-D doses for typical radappertized foods appear in Table 3.[18]

Nonspore-forming members of the *Moraxella-Acinetobacter* group of bacteria and viruses are sometimes more radiation resistant than *C. botulinum* spores.[19,20] Since these organisms are heat sensitive, their threat is minimized or eliminated during autolytic enzyme inactivation. Larkin[21] reviewed the literature on thermal inactivation of viruses in food and concluded that they can be reduced by a factor of 1×10^3 in 1 min at 71°C or in 6 sec at 75°C. A culture of *Moraxella-Acinetobacter* in minced beef could be reduced by one log cycle at 70°C in 10.7 min and at 75°C in 2.0 min.[22] Thermal injury resulting during inactivation of autolytic enzymes reduces the radiation resistance of *Moraxella-Acinetobacter* organisms.[23] The combination of heat for inactivating autolytic enzymes in food and subsequent exposure of the food to a 12-D dose of radiation based on the radiation resistance of *C. botulinum* should minimize or eliminate the potential hazard of food-borne asporogenous microorganisms such as viruses or members of the *Moraxella-Acinetobacter* group of bacteria.

The addition of sodium chloride and sodium tripolyphosphate for pretreating food to reduce loss in juiciness during thermal inactivation of autolytic enzymes results in a significant lowering of the 12-D irradiation dose on the order of 7 to 20 kGy for radappertization. On the other hand, lowering the temperature of beef slices inoculated with *C. botulinum* strain 53B from +20 to −40°C increased the 12-D dose by 7.4 kGy.[24] The 12-D dose for *C. botulinum* strain 33A in cooked ground beef increased by 5.9 kGy when the dose was lowered from +25 to −25°C.[25]

Based on all the available evidence and more than 2 decades of experience with radappertization using the 12-D concept for microbial safety, it is concluded that foods processed by radappertization are as free of microbial spoilage and food-borne pathogens during an extended shelf life without refrigeration as thermally processed foods.

Table 3
THE 12-D DOSES FOR VARIOUS FOODS IRRADIATED AT −30±10°C

Food	12-D dose[a] (kGy) Extreme value[b]
Beef[c]	41.2[d]
Chicken[c]	42.7
Ham[c,e]	31.4
Pork[c]	43.7
Codfish cake[f]	31.7
Corned beef[f]	26.9
Pork sausage[f]	25.5
Bacon[g]	25.2[h]

[a] Based on recoverable botulinal cells and an assumed one most resistant strain/can.
[b] Based on an assumed exponential spore death rate with an initial shoulder.
[c] Used a mixture of 10 strains (5 type A and 5 type B) of 10^6 botulinal spores/strain or 10^7 spores/can.
[d] Using the binomial confidence limit procedure for handling cases with one partial-spoilage point.
[e] The NO_2/NO_3 levels in the ham were reduced from 156/700 μg/g to 25/100 μg/g.
[f] Screened each of 10 strains of botulinal spores for highest resistance in the food ("screening" pack of 20 cans/dose) followed by use of the one most resistant strain ("clearance" pack of 100 cans/dose).
[g] Irradiated at 5 to 25°C.
[h] Spearman-Karber method for estimating 12-D dose, based on an assumed exponential spore death rate without an initial shoulder.

Rowley, D. B. and Brynjolfsson, A., *Food Technol. (Chicago)*, 34, 75, 1980.

D. Packaging

Durable reliable packaging is needed to protect the radappertized foods after irradiation and long-term nonrefrigerated storage from microbial recontamination and the deleterious effects of light, oxygen, and moisture. The containers must be capable of withstanding low-temperature radiation sterilization and rough handling, yet retain protective qualities during storage without any adverse effects on the food contained therein. There is no problem in irradiation of tinplate containers at doses up to 75 kGy at temperatures as low as −90°C, provided the can enamels used are of the epoxyphenolic or phenolic types and the end-sealing compounds are blends of cured and uncured butyl elastomers, neoprene and butadiene-styrene elastomers, and neoprene and uncured butyl elastomers.[26]

The U.S. Food and Drug Administration (FDA) has approved several plastic films as food contactants for radappertized foods (60 kGy maximum dose). These are listed in Volume 1, Chapter 1, Table 2. Additional flexible materials satisfactory for packaging radappertized foods are described in Volume 2, Chapter 8. These films are used as the food contactant layer in a laminated structure with aluminum foil (middle layer) as a moisture, light, and oxygen barrier, and either polyethylene terephthalate or

Table 4
ACCEPTANCE OF RADAPPERTIZED MEATS

Product	kGy at −30°C	Recipe	Number raters	Number tests	Average acceptance rating[a]
Beef	33—47	Onion gravy	33	2	6.4
	33—47	Roast au jus	89	4	6.2
	33—47	Brown gravy	85	4	6.5
Beef steak[b]	37—43	Fried[c]	64	2	7.0
Ham	37—44	Grilled	32	2	8.1
Ham[d]	37—44	Baked	79	3	7.7
Ham[b]	37—43	Baked[c]	64	2	7.4
Corned beef[b]	25—29	Boiled[c]	64	2	7.0
Turkey slices[b]	37—43	Roasted[c]	64	2	6.4
Pork sausage	27—33	Fried	91	4	7.7
Chicken	45—54	Breaded, fried	79	2	7.0
Cooked salami	25—29	Cold	64	2	6.4
Shrimp	38—49	Cocktail	115	4	7.2

[a] 9-point hedonic scale where 9 is "like extremely"; 5 is "neither like nor dislike"; 1 is "dislike extremely".
[b] Apollo-Soyuz Test Program.
[c] Eaten at cabin environmental temperature.
[d] Apollo 17.

From Rice, L. J. and Mason, V. C., U.S. Army Natick Research and Development Laboratories, Natick, Mass., 1975.

polyaminocaproyl as the outside layer to give strength to the laminate in the form of pouches. Both metal containers and flexible packages must be sealed under vacuum to prevent rancidity of the lipids in the foods packaged for radappertization.

E. Organoleptic Quality

Table 4 shows the organoleptic quality of several typical radappertized meats using the 9-point hedonic scale for preference.[27] In the case of meat and poultry products, a rating of 5 ("neither like nor dislike") is considered to be the threshold of acceptability. A rating of 7 or above indicates a highly acceptable product. The baked ham shown with footnote d in Table 4 was from the same batch as that used by the astronauts of the Apollo 17 flight to the moon in December 1972. The beefsteak, ham, corned beef, and turkey slices (footnote b in Table 4 and Plate 1, Volume 1) were from the same batches as those eaten by American astronauts participating in the joint flight in space with Soviet cosmonauts in 1975.

All irradiated foods eaten on the Apollo 17 and Apollo-Soyuz Test flight were considered to be highly acceptable.[28] For detailed comments consult Reference 28 and Chapter 1, Volume 1.

Advances in technology for producing irradiation sterilized foods have made it possible to produce a broad spectrum of meat, poultry, and seafood items which score well in the acceptable range by both expert and consumer panels. The remaining major scientific hurdle is to provide health officials at national levels with the data demonstrating that radappertized foods are wholesome, i.e., safe to eat, and their nutritional quality has not been significantly impaired.

F. Summary of Technology for Radappertization

In the radappertization process it is important to start with quality food which has been handled under good sanitary conditions. Radiation cannot reverse deterioration and spoilage of food once it has begun; it can only arrest or prevent these conditions. Nor should radiation be used as an excuse for poor sanitation. Radappertization depends upon prepackaging in an impermeable barrier to protect the food from recontamination during post irradiation storage without refrigeration and from undesirable chemical changes due to oxidation and change in moisture content because of diffusion of gases and vapors. Detailed processing conditions for radappertized beef rolls have been described elsewhere.[29-33]

The first step in the process is to remove as much of the inedible material as possible by deboning, trimming off gristle and excess fat. The next step is to inactivate the proteolytic enzymes in these foods. This is done by heating (blanching) to an internal temperature between 68 and 80°C (Table 1). The foods, while still hot, are vacuum packed and sealed in rigid metal cans or flexible packaging materials. The foods are frozen without delay in a blast freezer or by cold nitrogen gas to a temperature of −40°C and are exposed while held frozen (−40°C at the start and −20°C maximum at the end of irradiation) either to gamma rays from cobalt-60 or cesium-137, X-rays, or electrons. Irradiation in the frozen state minimizes adverse chemical and physical changes which may occur so that the quality (taste, color, odor, texture, and nutrients) of the product is maintained.

Under the processing conditions just described the main drawback is that the final product appears to have a tendency to be somewhat drier to the consumer. This condition has been largely eliminated for beef, chicken, lamb, and pork through pretreatment of the raw material with 0.25 to 0.5% sodium tripolyphosphate (TPP) and 0.5 to 1.5% sodium chloride. The dry salt and TPP are added to chunks of meat weighing between approximately 250 and 750 g. Crushed ice is added (3 to 5%) to facilitate solubilizing the additives. Distribution is accomplished by continuous mixing for 10 to 15 min. The material is then stuffed into casing of a diameter which will permit a close fit when the product is placed at a later stage in the final cylindrical container. If the configuration of the final package is square or rectangular the stuffed casing is compressed in a wire metal cage to the desired dimensions.

After holding for a minimum of 2 hr at 1 to 4°C, the material in the casing is placed in a cookhouse for inactivation of proteolytic enzymes. During the first 3 hr, the temperature of the cookhouse is held at 60°C. The cookhouse temperature is then allowed to rise slowly to 88°C until the internal temperature of the meat roll reaches 70 to 75°C. The salt and TPP increase the water-holding capacity of the meat and the natural juice content. The additives distribute within the system and solubilize some of the muscle proteins which are drawn to the surface during the first 3 hr in the cookhouse and then form a skin when denatured at the higher temperatures. Total time in the cookhouse is approximately 12 hr. Cooking losses of juice can be kept to 10 to 20% by the addition of salt and TPP. After enzyme inactivation is completed, the casing is removed and the meat is fitted into cylindrical metal cans of the same diameter as the casing or in rectangular cans or pouches. The geometric configuration of the rolls permits more effective packaging by reducing head spaces in the containers and providing better means for proper dosimetry during irradiation.

G. Reducing Nitrite in Cured Meats

Irradiation can reduce dependence upon some of the chemical food additives which have been indicted or convicted as carcinogens or which possess properties injurious to health. The use of nitrite to impart the characteristic flavor and color in bacon, ham, corned beef, and other cured meats is of concern to health authorities because of the possibility that the residual nitrite in the meat forms nitrosamines (e.g., nitrosopyrrolidine during frying of bacon) and because nitrite itself has been indicted as a carcinogen.[34] Therefore, a reduction in the use of nitrite in bacon and other cured meats is being considered in the U.S.

Table 5
MINIMAL ADDITIONS OF NITRITE TO IRRADIATED MEATS

Product	Minimum addition (ppm $NaNO_2$)	Product quality
Bacon	None	Slightly different color and flavor
	20	Color, flavor, and taste like in normal commercial bacon
Ham	25	Color fading
	25/25[a]	Color stabilized
	None	In research state (texture excellent)
Corned Beef	50[b]	Regular quality product
	None	No color, otherwise acceptable
Frankfurters	50	Good quality product
	None	Acceptable, different flavor and color

[a] 25 ppm $NaNO_3$ addition is needed to prevent fading of the color.
[b] 25 ppm $NaNO_2$ addition is sufficient if $NaNO_2$ is uniformly distributed during pumping.

From Wierbicki, E. and Brynjolfsson, A., presented at 25th Eur. Meet. Meat Res. Workers, Budapest, August 27—31, 1979.

One of the purposes of nitrite in cured meats is to protect consumers against botulism. Because irradiation destroys *C. botulinum*, the nitrite addition to cured meats can be reduced and, in the case of bacon, can be eliminated entirely.[35-37] The objective then becomes one of reducing the amount of nitrite to the minimum needed for maintaining the characteristic color and flavor.

Table 5 indicates the extent to which nitrite can be reduced in several typical cured meats from the maximum of 156 ppm $NaNO_2$ allowed by the U.S. Department of Agriculture for cured meats (120 ppm $NaNO_2$ for bacon in combination with 550 ppm ascorbate). The color and general appearance of "no-nitrite" and "low-nitrite" (20 ppm) radappertized bacon are shown in Plate 4.* No nitrosamines were detected in radappertized (30 kGy at −30°C) "no-nitrite" bacon even after frying[36] and only a trace (1 ppb nitroso-pyrrolidine) was found in radappertized (30 kGy at −30°C) prefried bacon cured with 20 ppm sodium nitrite.[36] Irradiation also destroys residual nitrite in commercial bacon,[36] thus decreasing the possibility of forming nitrosamines during frying. However, after frying, bacon without added nitrite has a darker appearance than bacon to which nitrite has been added. In some radappertized cured meats (bacon, corned beef, frankfurters) the use of nitrite can be eliminated. The resulting products are organoleptically acceptable although slightly different in flavor and color from the nitrite-cured counterparts.[36]

IV. SAFETY FOR CONSUMPTION (WHOLESOMENESS)

The four criteria which must be met to assure safety for consumption (wholesomeness) of radappertized foods are

1. The absence of measurable induced radioactivity
2. Freedom from viable pathogens and their toxins
3. Avoidance of excessive loss in nutritive value
4. The absence of toxic, mutagenic, or carcinogenic radiolysis products.

* Plate 4 appears after page 144.

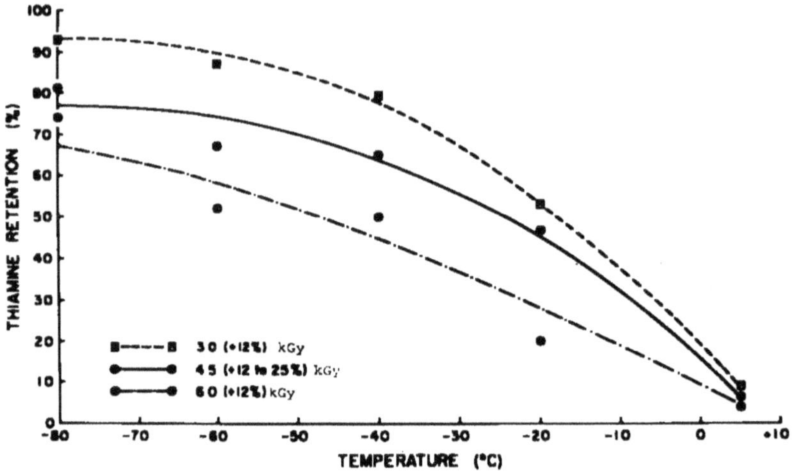

FIGURE 1. Thiamin retention (%) in irradiation-sterilized canned ham as a function of irradiation dose and temperature of the ham during exposure to ionizing radiation. (Data provided by Mrs. Miriam H. Thomas, U.S. Army Natick Research and Development Laboratories, Natick, Mass., 1971.)

A. Assurance of No Measurable Induced Radioactivity

It is now generally accepted by the scientific community that, by limiting gamma radiation sources to ^{60}Co and ^{137}Cs, the maximum energies of the gamma rays are below the threshold needed to activate elements in food.[1] Similarly, by restricting the maximum energies of electron and X-ray sources to 10 and 5 MeV, respectively, no measurable induced radioactivity in the food will occur.[1]

B. Absence of Pathogens

Pathogens have been discussed earlier in this chapter and in Volume 2, Chapters 4 and 6. For all practical purposes they should not present a hazard provided that the minimum radiation dose adheres to the 12-D concept and the recommended temperature for heat inactivation of autolytic enzymes is used.

C. Nutritional Adequacy

When foods are exposed to ionizing radiation under conditions envisaged for radappertization on a commercial scale, no significant impairment in the nutritional quality of the protein, lipid, and carbohydrate constituents was observed.[38-40] Because of the protective qualities inherent in foods, the sensitivity of nutritional components is less than that of the same nutrients irradiated in pure form or in artificial solutions and mixtures.[41,42]

Protection of nutrients is improved if the food is held at low temperatures in an oxygen-free medium during irradiation.[42-45] The radicals produced by irradiation in the water phase of foods react with other ingredients and lead to changes. Irradiating frozen food minimizes these changes because, according to Brynjolfsson,[46] "the ice has a much lower shielding capability than water and, consequently, electrons knocked out by the radiation can find their way back to their place of origin." ... and "the radicals formed in the ice are largely immobilized and will build up and react with each other rather than the surrounding foodstuffs." The beneficial effect on thiamin retention by low temperature radappertization is shown in Figure 1.

1. Composite Diets

Radappertization (25 kGy) of feed and protein concentrates for germ-free rat and mouse colonies maintained for many years has been used with excellent results with little or no effect on nutritional quality or palatability.[47,48] A variety of foods and whole frozen meals (25 kGy at −15°C) have been radappertized for patients whose natural resistance to infections has been impaired or suppressed (e.g., organ transplant recipients). Among the radappertized foods served these patients are meats, fish, vegetables, potato, rice crispies, corn flakes, bread, jam, marmalade, sugar, tea, instant coffee, biscuits, and cake.[48] Others have reported that the biological value of the proteins and metabolizable energy value of composite rodent diets are unaltered by radappertization at 56 kGy.[49-51]

2. Carbohydrates

The main effects of radiation upon carbohydrates are those of hydrolysis (depolymerization) and oxidative degradation. Radappertization (55.8 kGy) did not affect the availability of the carbohydrates of eight vegetables, fruits, and cereal products.[50]

3. Lipids

Radappertization of lipids results in oxidation, degradation, and decarboxylation. Unsaturated fatty acids are more readily oxidized than the saturated acids. In human test subjects the apparent digestibility of the fat in nonirradiated and radappertized (28 kGy) pork stored for 1 year at room temperature is nearly identical.[52] However, lard irradiated to 55.8 kGy was absorbed more slowly than nonirradiated lard in the dog owing to delayed emptying of the stomach contents.[53] Nevertheless, overall digestibility was unaffected.

4. Proteins and Amino Acids

Although very high levels of irradiation have marked effects upon proteins and amino acids, doses envisioned for use in radappertization show minimal deleterious effects. Radappertization (70 kGy) of protein supplied in a rat diet in the form of soya, meat and bone meal, and fishmeal had no significant effect on digestibility, biological value, net protein utilization, and amino acid composition.[54] Based upon the amino acid analyses for 18 amino acids and total amino acids performed by a contractor for the U.S. Army Medical Department, the protein in beef radappertized (47 to 71 kGy) under conditions proposed for commercial processing showed no significant loss in nutritional quality when compared with beef preserved by freezing (−18°C) or heat sterilization (F_o = 5.8).[39]

5. Vitamins

The curves in Figure 1 show dramatically the effects of radappertization dose and temperature on the percent retention of thiamin in ham, indicating that irradiation of ham while frozen is mandatory for thiamic conservation. Results with pork loin (Table 6), ham, and beef (Table 7) show that when irradiation is performed at low product temperature, it is no more destructive to thiamine, riboflavin, niacin, and pyridoxine than is preservation by thermal processing.

Information is sparse on the effects of radappertization upon the so-called lesser-known vitamins. Although vitamin B_{12}, paraamino-benzoic acid, pantothenic acid, and folacin are all radiosensitive at room temperature irradiation in aqueous solution, their radiosensitivity is reduced considerably in food.[41,42] A considerable reduction in radiosensitivity of vitamin B_{12} in whole milk was observed.[55] Radappertization (55.8 kGy) resulted in less than 10% destruction of pantothenic acid and no destruction of folacin in ground pork.[55] There was no significant decrease in folacin activity in radappertized diets fed to chickens.[56]

Table 6
EFFECT OF DIFFERENT PRESERVATION METHODS ON THE VITAMIN CONTENT OF SHELF-STABLE CANNED PORK LOIN

Vitamin	Treatment	Mg/100G[a]	Retention (%)
Thiamin	Control	3.69 ± 0.22[b]	
	45 kGy at −80 ± 5°C	3.14 ± 0.25	85
	Thermally processed	0.76 ± 0.08	20
Riboflavin	Control	1.02 ± 0.28	
	45 kGy at −80 ± 5°C	0.79 ± 0.06	78
	Thermally processed	0.82 ± 0.02	81
Niacin	Control	20.3 ± 5.1	
	45 kGy at −80 ± 5°C	15.9 ± 2.6	78
	Thermally processed	13.2 ± 1.8	65
Pyridoxine	Control	0.76 ± 0.05	
	45 kGy at −80 C ± 5°C	0.75 ± 0.07	98
	Thermally processed	0.63 ± 0.07	84

[a] Moisture, fat, salt-free basis.
[b] Mean ± SD; three samples per treatment.

From Thomas, M. H. and Josephson, E. S., *Sci. Teach.*, 37, 59, 1970.

Table 7
EFFECT OF DIFFERENT PROCESSING METHODS UPON THIAMINE, RIBOFLAVIN, NIACIN, AND VITAMIN B_6 CONTENT OF ENZYME-INACTIVATED BEEF (mg/kg)

		Treatment			
	Storage (months)	Frozen	Heat sterilized	^{60}Cobalt (47—71 kGy)[a]	Electron (47—71 kGy)[a]
Thiamin	0	0.97	0.63	0.83	0.77
	15	0.68	0.14	0.21	0.26
Riboflavin	0	2.80	2.63	2.83	2.60
	15	1.69	2.60	2.60	1.46
Niacin	0	48.6	48.1	48.8	46.8
	15	57.2	54.9	50.1	44.5
Vitamin B_6	0	2.50	2.13	3.93	5.20
	15	0.97	0.57	0.35	0.42

[a] Beef temperature between −40 and −20°C during irradiation.

From office of the U.S. Army Surgeon General (Industrial Bio-Test, Inc., Contractor), Internal Rep., Col. Roger W. Baker, Project Officer, 1976.

Less information is available about the effects of irradiation on fat-soluble vitamins than there is in regard to water-soluble vitamins. Carotene in solution was found to be radiosensitive to cathode rays.[57] Carotene destruction during irradiation can, however, be minimized by the addition of ascorbic acid. For food in which ascorbic acid is present in significant amounts, most of the carotene is retained even at doses of 48 kGy (Table 8).[58] Irradiation of fluid whole milk with 4 kGy resulted in destruction of 40% of the carotenoids, 70% of vitamin A, 60% of the tocopherols, and 61% of vitamin E.[59]

Table 8
ASCORBIC ACID AND CAROTENE RETENTION IN RADAPPERTIZED VEGETABLES (48 kGy)

Product	Percent Retention	
	Ascorbic acid	Carotene
Beans, green	73	169
Carrots	78	87
Maize	71	56

Calculated from Thomas, W. H. and Calloway, D. H., *J. Am. Dietet. Assoc.*, 39, 105, 1961.

Biological evidence indicates that vitamin D activity for chicks is decreased by gamma radiation of the total diet (not frozen) with 27.9 kGy.[55]

Alfalfa leaf meal and fresh spinach radappertized (27.9 kGy) at ambient temperature showed no loss in vitamin K content.[60] No destruction of vitamins K_1, K_3, or K_5 was apparent in irradiated semisynthetic diets when they were fed to chickens.

6. Summary and Conclusion

The data in this section have been selected from the vast literature on the basis of relevancy to the nutrition of the consumer. For this reason, less attention was devoted to the effects upon nutrient constituents irradiated in pure form or in artificial solutions and mixtures, although there could be applications for intravenous use if radappertized. It is concluded that when food is preserved by radappertization under processing conditions proposed for commercial production, nutrient destruction is no greater than that which occurs when food is preserved by more conventional methods. This conclusion is based upon data derived from in vivo feeding studies and physical and chemical means.

D. Toxicological Studies

Until recently the approach to "wholesomeness" testing was largely based upon complicated chronic toxicity feeding studies using rodents and dogs. Control and test animals are challenged prenatally and during the entire test period until sacrifice. The tests with rodents involve four generations; for dogs the test period is generally 3 years or production of 3 litters, whichever occurs first. Details for wholesomeness testing of radappertized beef have been provided elsewhere.[28]

Parameters evaluated in animal feeding studies of radappertized beef are growth and body weights, food consumption, reproduction performance, longevity, pathology (gross and microscopic), ophthalmoscope examination, urinalysis (dogs), and semen examination (dogs). It is obvious that 3 to 5 years, much money, and large numbers of animals are involved. Liver function tests and tests for teratogenicity and mutagenicity are also conducted.

In recent years, increasing emphasis is being given to the direct examination of the radiolysis products formed in the food and their toxicological significance. A Select Committee on Health Aspects of Irradiated Beef, organized by the Federation of American Societies for Experimental Biology (FASEB), reviewed the possible toxicity to man of the volatile compounds detected in radappertized (41 to 71 kGy) beef. Sixty-five compounds extracted from the beef were identified as saturated and unsaturated aliphatic compounds containing 2 to 27 carbon atoms and certain of their alcohol, aldehyde, and ketone derivatives; 3 aromatic hydrocarbons; and some sulfur, nitrogen, and chlorine containing compounds. The chlorine compounds were found in both irradiated and unirradiated beef samples and were demonstrated not to be radiolysis products. The concentrations of the individual components ranged from 1 to 100 μg/kg of beef with a total concentration of 9.4 mg/kg. The committee concluded that "there were no grounds to suspect that the radiolytic compounds evaluated in the report would constitute any hazard to health to persons consuming reasonable quantities of beef irradiated in the described manner."[61]

A considerable body of information on radiolytic products has been developed in recent years including two symposia devoted to this subject.[8,62-66]

It is now possible to assess wholesomeness with a high degree of certainty by extrapolating chemical data from one food to another generically related food using the principle of chemiclearance. Sufficient data are now available to conclude that all major classes of foods irradiated up to an average dose of 10 kGy pose no toxicological problems and should be permitted for use. According to Taub et al.[67] four important principles underlie this conclusion:

1. Irradiation is a process in which chemical changes are uniform and predictable.
2. The radiolysis products pose no toxicological (short- or long-term) hazards as evidenced by the feeding studies and genotoxicity data.
3. Nutrition is not compromised.
4. The doses used in specific applications would be consistent with the objective for irradiation and the public health requirements set by the licensing authorities.

Expressed in another way, the major radiolysis compounds and their approximate yields can be predicted on the basis of the chemical composition of the food components, their weight fractions in the food, and the irradiation conditions.[67,68] Foods similar in composition and in radiolytic response for a basic set of irradiation conditions should be taken as a generic group and viewed as one test substance in wholesomeness studies. Health authorities should be able to grant clearances for generic foods or to extend an existing clearance on one member of the group to other foods in the group — on the basis of chemical data and related toxicological and microbiological tests, without requiring extensive animal feeding tests.[67] In the case of radappertized foods, if the data from studies in progress on chicken indicate this food to be wholesome, ham, pork, beef, and other foods of animal origin should be deemed wholesome according to the chemiclearance principle.[68]

The foregoing principles were adopted by the Joint FAO/IAEA/WHO Expert Committee on Wholesomeness of Food Irradiation during its October 27 to November 3, 1980 meeting in Geneva, Switzerland. The committee concluded that "the irradiation of any food commodity up to an overall average dose of 10 kGy presents no toxicological hazard; hence, toxicological testing of foods so treated is no longer required."[69] In the same report the "Committee recognized that higher doses of radiation were needed for the treatment of certain foods but did not consider the toxicological evaluation and wholesomeness assessment of foods so treated because the available data are insufficient for this purpose. Further studies in this area are therefore needed."[69]

The FDA, on the basis of a study by an internal Irradiated Food Committee established within its Bureau of Foods, has also recognized the foregoing principles for evaluating and approving irradiated foods. For further details please refer to references 28c and 28d of Volume 1, Chapter 1 describing FDA's proposal published in the Federal Register on March 27, 1981 to adopt the policy that a food class comprising only a minor portion of the diet (e.g., spices) and irradiated at a dose of 50 kGy or less may be considered safe for human consumption based upon minimum biological testing.

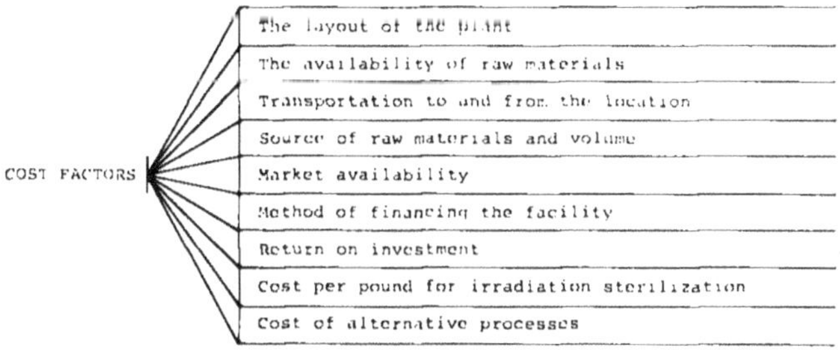

FIGURE 2. Factors influencing costs of food irradiation. (Prepared by Dr. Martin S. Peterson from an unpublished report by Mr. Jerome Deitch.)

Once the wholesomeness data for radappertized foods become available, it is expected that the FAO/IAEA/WHO Joint Expert Committee on the Wholesomeness of Irradiated Foods will apply these same principles at doses in excess of the overall average dose of 10 kGy. Preliminary data suggest that irradiation in the 10 to 71 kGy dose range of frozen foods (−40 to −20°C) results in radiolysis products which are qualitatively not different from those found in the same foods irradiated unfrozen in the 1 to 10 kGy dose range and quantitatively in lesser concentrations.

With adoption of these principles by the international community, approvals of irradiated foods by health authorities should quickly follow. With the uniformity of the chemistry and predictability of radiolysis products demonstrated, it would be scientifically valid to extrapolate wholesomeness clearances from one food to another and from one radiation dose to another. The consumer can rest assured that the absence of toxic products resulting from radiation in the food eaten has been established on scientifically firmer ground than for food processed by any other method. To reiterate, the chemistry studies coupled with animal feeding studies and tests for mutagenicity[70] and teratogenicity conducted during the past 27 years have revealed no confirmed evidence derogatory to the safety for consumption of radappertized foods.

V. COST AND ENERGY CONSIDERATIONS

A. Introduction

Legal approvals by the health officials to produce, market, and eat radappertized foods do not guarantee that these foods will be produced commercially. Radappertization must serve a useful purpose which justifies the cost (cost vs. benefit) and the cost must be competitive with alternate processes or be reasonable for the product involved.

B. Cost

There have been several reports on the costs projected for food irradiation in anticipation that this process will become a commercial reality[71-75] and the published proceedings of a symposium dealing with this subject.[76] An updated report will be published in *CRC Critical Reviews in Food Science and Nutrition*, CRC Press, Inc., Boca Raton, Fla., 1982. The reader is urged to refer to these publications to understand the factors involved in developing costs, such as product characteristics (size, shape, density), facility characteristics (geographic location, plant capacity and utilization, degree of specialization of integration), source characteristics (type, installed cost, efficiency of source utilization, source reliability, source maintenance and replenishment, method of depreciation), and methods of financing and return on investment. These factors are graphically presented in Figure 2.

Table 9
COST OF RADAPPERTIZATION OF BACON

Source	5-Year plant depreciation ¢/lb	Operation ¢/lb	Total ¢/lb
^{60}Co Isotope	2.03	1.20	3.23
^{137}Cs Isotope	2.03	0.35	2.38
10 MeV accelerator	0.49	0.43	0.92
4 MeV accelerator	0.36	0.40	0.76

Note: Dose: 25 kGy at refrigerated temperature. Plant size: 100 million lb/year.

From Brynjolfsson, A., Tech. Rep. FEL-89, U.S. Army Natick Research and Development Command, Natick, Mass., March 1979.

Table 10
COMPARATIVE ENERGY USE IN PROCESSING CHICKEN[a]

Process	kJ/kg
Frozen, raw, cut up	46,600
Frozen, cooked rolls	27,500[b]
Thermal treated rolls	20,180
Radappertized, cooked rolls	14,260

[a] Based on 3.5 weeks storage.
[b] 5150 kJ/kg for each 3.5 weeks.

From Brynjolfsson, A., Tech. Rep. FEL-89, U.S. Army Natick Research and Development Command, Natick, Mass., March, 1979.

Table 9 gives projected costs ranging from 0.76 to 3.23 ¢/lb for irradiation sterilization (25 kGy) of refrigerated, not frozen bacon in a plant with an annual throughput of 100 million lb.[77] If this product were irradiated while frozen (−40 to −20°C) and the price of the liquid nitrogen is assumed to be 1.5 ¢/lb (1967 price), these refrigeration costs would be 10 ¢/lb at 1 million lb annual throughput and 1.9 ¢/lb at 30 million lb annual throughput.[72]

C. Energy

The energy used in food systems in the U.S. amounts to about 16.5% of the total energy used in that country. Brynjolfsson[78] found that radappertization requires 157 kJ of energy per kilogram of food processed, which is less energy than that required for competing processes. The comparative data extracted from his report are shown in Table 10.

VI. APPLICATIONS

The major alternative processes to radappertization to make foods shelf stable are heating, freezing, and drying. Currently used thermal processing of shelf-stable non-acid foods has severe limitations owing to a high can purge (loss of natural juices from heating), wet packaging of the containers to assure better heat penetration, use of only small containers because of poor heat conductivity, very poor (soft) texture, and altered flavor. Frozen storage has the drawback of high energy cost, the need to thaw before use, vulnerability to sudden loss of power, and uselessness in regions where electricity is not available. Drying, including freeze drying and compression with freeze drying, is limited to powdered products or thin slices. Energy costs for drying, particularly freeze drying, are high.

Radappertization offers a most promising method for producing shelf-stable products which, on the plate of the consumer, closely resemble fresh food. The unique characteristics of radappertization are dry pack of the containers, no can purge during processing, use of a wide range in sizes of containers, portion control packaging, and assurance of absence of pathogens.

A survey in the U.S. of the "grocery-line" executives among selected meat packers, seafood processors, retailers, as well as trade organizations, was conducted in 1972 to determine their interest in radappertization of foods.[79] The representatives in the grocery line of the companies surveyed were definitely interested in radappertized ham, corned beef (no thermally processed counterpart exists), pork sausage, pork, lamb, luncheon meats, chicken, turkey slices, and shrimp. Radappertized beef was considered highly acceptable for its texture, dry pack, and portion control possibilities, but further improvements in flavor and taste needed for the product's unlimited marketing and distribution (vintage 1972) were desirable. Great improvements have been achieved during the period between 1972 and 1980. Shelf-stable beef roll products in individual consumer sizes and precooked shelf-stable roast beef for home and restaurant use were considered attractive from the marketing point of view.

The grocery line representatives visualized the opening of a sizeable new market for shelf-stable meats and a considerable movement of meats in the supermarket from the refrigerator or freezer to the open shelf. The limitations of the heat sterilization process to produce shelf-stable canned meats have resulted in very low volume of product output and dollar sales because of low or marginal acceptance by consumers. In the frozen food category, according to Chain Store Age/Supermarkets,[80] "energy is the root of most of the retailers' worries about the (frozen food) department. Display cases for frozen food use 29% of a super's entire power requirement. Faced with steadily rising utility bills, supermarket owners know they must cut costs and increase profitability."

It is not expected that the advent of radappertization will cause abrupt changes in our society's methods for food preservation, distribution, and marketing. An evolutionary process, depending upon many factors (e.g., organoleptic, psychological, legal, political, and economic) with competition in the marketplace, will determine what foods and how much of these foods will be processed and marketed by any particular method.

Radappertization's versatility will make it possible to produce shelf-stable products not on the market today (e.g., frankfurters, bologna, luncheon meats, salami, sausage, fish fillets, etc.). The magnitude of dollar sales for these products could be considerable.

Because of the variety, palatability, and nutritional quality of radappertized foods, whole meals and their components have been approved in the U.K., the Netherlands, and the Federal Republic of Germany to feed hospital patients whose immune systems have been suppressed.

Because radappertized foods have been preheated to inactivate autolytic enzymes, they can be eaten either with or without further preparation. Since they can be prepackaged in containers in a variety of sizes and shapes, they can be sold in packages for individual servings or for families. They are well adapted for use by campers and other vacationers, hikers, fishermen, mountain climbers, truck drivers on long runs, farmers in the field, and astronauts. Their use in institutional and restaurant feeding, fast food chains, and airline catering services would have a high market potential. They can be stockpiled for use in case of both natural disasters (e.g., earthquakes, floods, pestilence, volcanic eruptions) and those that are manmade (e.g., wars).

VII. CONCLUSION

1. Radappertization opens a new era for food preservation — the means to extend without refrigeration the shelf life of foods, particularly meat, poultry, fin fish, shellfish, food products, meals, and, for humans and animals, diets which are highly palatable, nutritious, and wholesome. These foods have the characteristics of fresh foods prepared for eating. By reducing spoilage, this process can contribute to closing the world's nutrition gap, particularly in animal protein, in the face of a rapidly increasing world population.
2. Radiation sterilization lends itself readily to the concept of convenience. Irradiation-sterilized meats can be offered as components of meals or as snacks in ready-to-eat form, in the form of slices or sandwiches, or as warm-and-serve or cook-and-serve foods. Radappertized foods require no preparation or simple preparation with reduction of labor in the kitchen.
3. Radappertization can provide high quality, nutritious, disease-free, wholesome foods to those living in large cities and remote villages far from the place of production.
4. Radappertization can substitute for some chemical food additives, such as ethylene oxide to sterilize spices and nitrites in cured meats, which are either toxic, carcinogenic, mutagenic, or teratogenic.
5. Radappertization is competitive in cost and demand for energy with the well-established methods to make foods shelf stable.
6. Radappertization can provide a number of shelf-stable products (luncheon meats) or free products (spices) of food spoilage or pathogenic organisms which cannot be accomplished by any of today's established food processing methods.
7. In 27 years of testing for "wholesomeness" of radappertized foods, no confirmed evidence has been obtained of any adverse effects of radappertization on the wholesomeness characteristics of these foods.

ACKNOWLEDGMENT

The help of the staff members of the U.S. Army Natick Research and Development Laboratories in providing unpublished data and a review of this chapter is gratefully acknowledged.

REFERENCES

1. The *Technical Basis for Legislation on Irradiated Food, A Report,* Joint FAO/IAEA/WHO Expert Committee, Rome, 1964, FAO Atomic Energy Series No. 6, World Health Organization Tech. Rep. Ser. No. 316, Food and Agriculture Organization, Rome, Italy, 1965.
2. **Cohen, J. S. and Mason, V. B.,** Radappertization (radiation sterilization) of foods, Bibliography of Technical Publications and Papers, Tech. Rep. Natick/TR-77/009, Food Engineering Laboratory, FEL-62, December 1976.

3. **Reber, E. F., Raheja, K., and Davis, D.,** Wholesomeness of irradiated foods — an annotated bibliography, *Fed. Proc.*, 25, 1530, 1966.
4. **Barna, J.,** Compilation of bioassay data on the wholesomeness of irradiated food items, *Acta Aliment.*, 8(3), 205, 1979.
5. **Rowley, D. W., Sullivan, R., and Josephson, E. S.,** Indicators of viruses in foods preserved by ionizing radiation, in *Indicators of Viruses in Water and Food*, Berg, G., Ed., Ann Arbor Science, Ann Arbor, Mich., 1978, chap. 14.
6. **Wierbicki, E.,** U.S. Army Natick Research and Development Laboratories, Natick, Mass., personal communication, July 31, 1980.
7. **Roth, J. S., Losty, T., and Wierbicki, E.,** Assay of proteolytic enzyme activity using a ^{14}C-labelled hemoglobin, *Anal. Biochem.*, 42, 214, 1971.
8. **Josephson, E. S., Brynjolfsson, A., Wierbicki, E., Rowley, D. B., Merritt, C., Jr., Baker, R. W., Killoran, J. J., and Thomas, M. H.,** Radappertization of meat, meat products and poultry, in *Radiation Preservation of Food*, STI/PUB/317, International Atomic Energy Agency (IAEA), Vienna, 1973, 471.
9. **Shults, G. W., Russell, D. R., and Wierbicki, E.,** Effect of condensed phosphates on pH, swelling, and water-holding capacity of beef, *J. Food Sci.*, 37, 860, 1972.
10. **Ingram, M. and Farkas, J.,** Microbiology of foods pasteurised by ionising radiation, *Acta Aliment.*, 6, 123, 1977.
11. **Rowley, D. B. and Brynjolfsson, A.,** The potential utilization of radiation in the processing of food, *Food Technol. (Chicago)*, 34, 75, 1980.
12. **Rowley, D. B.,** personal communication, 1980.
13. **Anellis, A., Rowley, D. B., and Ross, E. W., Jr.,** Microbiological safety of radappertized beef, *J. Food Prot.*, 42, 927, 1979.
14. **Schmidt, C. F.,** Appendix II. Dose requirements for the radiation sterilization of food, *Int. J. Appl. Radiat. Isot.*, 14, 19, 1963.
15. **Schmidt, C. F.,** Dose requirements for the radiation sterilization of food, in *Report of the European Meeting on the Microbiology of Irradiated Foods*, Paris, Food and Agriculture Organization in Collaboration with the International Association of Microbiological Societies, 1960, Rome, 1961, 2; Appendix II, 1.
16. **Esty, J. R. and Meyer, K. F.,** The heat resistance of spores of *B. botulinus* and allied anaerobes, *J. Infect. Dis.*, 31, 650, 1922.
17. **Anellis, A. and Koch, R. B.,** Comparative resistance of strains of *Clostridium botulinum* to gamma rays, *Appl. Microbiol.*, 10, 326, 1962.
18. **Rowley, D. B. and Brynjolfsson, A.,** The potential utilization of radiation in the processing of food, *Food Technol. (Chicago)*, 34, 75, 1980.
19. **Welch, A. B. and Maxcy, R. B.,** Characterization of radiation-resistant vegetative bacteria in beef, *Appl. Microbiol.*, 30, 242, 1975.
20. **Sullivan, R., Fassolitis, A. C., Larkin, E. P., Read, R. B., Jr., and Peeler, J. T.,** Inactivation of thirty viruses by gamma irradiation, *Appl. Microbiol.*, 22, 61, 1971.
21. **Larkin, E. P.,** Thermal inactivation of viruses, Tech. Rep. 78/002, U.S. Army Natick Research and Development Command, Natick, Mass., 1977.
22. **Firstenberg-Eden, R., Rowley, D. B., and Shattuck, E. G.,** Factors affecting the inactivation of *Moraxella-Acinetobacter* cells in an irradiation process, *Appl. Environ. Microbiol.*, 40, 480, 1980.
23. **Firstenberg-Eden, R., Rowley, D. B., and Shattuck, E. G.,** Thermal inactivation and injury of *Moraxella-Acinetobacter* cells in ground beef, *Appl. Environ. Microbiol.*, 39, 159, 1980.
24. **El-Bisi, H. M., Snyder, O. P., and Levin, R. E.,** Radiation death kinetics of *Clostridium botulinum* spores at cryogenic temperatures, in *Botulism*, Ingram, M. and Roberts, T. A., Eds., Chapman and Hall, London, 1966.
25. **Grecz, N., Upadhyay, J., and Tang, T.,** Effect of temperature on radiation resistance of spores of *Clostridium botulinum* 33A, *Can. J. Microbiol.*, 13, 287, 1967.
26. **Killoran, J. J.,** Chemical and physical changes in food packaging materials exposed to ionizing radiation, *Radiat. Res. Rev.*, 3, 369, 1972.
27. **Peryam, D. R. and Pilgrim, F. J.,** Hedonic scale method of measuring food preferences, *Food Technol. (Chicago)*, 11(Suppl. 9), 9, 1957.
28. **Josephson, E. S., Brynjolfsson, A., and Johnson, H. C.,** Food irradiation in the United States — current status and future plans, in *Proc. 12th Japan Conf. Radioisot.*, Japan Atomic Industrial Forum, Tokyo, 1976, 176.
29. **Heiligman, F., Wierbicki, E., Cohen, J. S., and Mason, V. C.,** Industrial productions and quality of whole carcass beef rolls used in the wholesomeness testing of radappertized beef, in *Digest of Papers, First Int. Cong. Eng. Food*, Boston, American Society of Agricultural Engineers, St. Joseph, Mich., 1976, 74.

30. **Killoran, J. J., Howker, J. J., and Wierbicki, E.**, Reliability of the tinplate can for packaging of radappertized beef under production conditions, *J. Food Process. Preserv.*, 3, 11, 1979.
31. **Killoran, J. J., Cohen, J. S., and Wierbicki, E.**, Reliability of flexible packaging of radappertized beef under production conditions, *J. Food Process. Preserv.*, 3, 25, 1979.
32. **MacDonald, B. A.**, Gamma ray sterilization of meat products, in *Digest of Papers, First Int. Cong. Eng. Food,* Boston, American Society of Agricultural Engineers, St. Joseph, Mich., 1976, 79.
33. **Rees, C. W. and Caspersen, J. M.**, Electron irradiation in the sterilization of meat, in *Digest of Papers, First Int. Cong. Eng. Food,* Boston, American Society of Agricultural Engineers, St. Joseph, Mich., 1976, 80.
34. **Newberne, P. M.**, Nitrite promotes lymphoma incidence in rats, *Science,* 204, 1079, 1979.
35. **Wierbicki, E.**, The importance and feasibility of low nitrite meat products, *Act. Rep. Res. Dev. Assoc.*, 31(2), 70, 1979.
36. **Wierbicki, E. and Brynjolfsson, A.**, The use of irradiation to reduce or eliminate nitrite in cured meats, presented at 25th Eur. Meet. Meat Res. Workers, Budapest, August 27—31, 1979.
37. **Wierbicki, E. and Heiligman, F.**, Irradiated bacon without and with reduced addition of nitrite, presented at Int. Meat Res. Cong. (26th Eur. Meet. Meat Res. Workers), Colorado Springs, August 31 to September 5, 1980.
38. **Josephson, E. S., Thomas, M. H., and Calhoun, W. K.**, Nutritional aspects of food irradiation: an overview, *J. Food Process. Preserv.*, 2, 299, 1978.
39. **Josephson, E. S., Thomas, M. H., and Calhoun, W. K.**, in *Nutritional Evaluation of Food Processing,* 2nd ed., Harris, R. S. and Karmas, E., Eds., AVI Publishing, Westport, Conn., 1975, chap. 14.
40. **Josephson, E. S., Thomas, M. H., and Calhoun, W. K.**, Effects of ionizing radiation upon food nutrients — a review and update, in *Proc. IV. Int. Cong. Food Sci. Technol.*, Vol. 1, Selegraf, Valencia, Spain, 1974, 613.
41. **Bregvadze, U. D. and Bokeriya, N. M.**, Effort of γ-irradiation on amino acid composition in wines (Russian), *Tr. Gruz. Nauch-Issled. Inst. Pishch Prom.*, 4, 90, 1971.
42. **Metlitsky, L. V., Rogachev, V. N., and Krushchev, V. G.**, Radiation processing of food products, in Status of the Food Irradiation Program, Hearings before the Subcommittee on Research, Development, and Radiation of the Joint Committee on Atomic Energy, Congress of the United States, U.S. Government Printing Office, Washington, D.C., July 18 and 30, 1968, appendix 9.
43. **Thomas, M. H. and Josephson, E. S.**, Radiation preservation of foods and its effects on nutrients, *Sci. Teach.*, 37, 59, 1970.
44. **Kharlamov, V. T. and Shubnyakova, L. P.**, Radioactive decomposition of methionine by gamma-radiation (Russian), *Zh. Prikl. Khim. (Leningrad),* 37, 1714, 1964.
45. **Southern, E. M. and Rhodes, D. N.**, Radiation chemistry of polyamino acids in aqueous solutions, in *Radiation Preservation of Foods,* American Chemical Society, Washington, D.C., 1967, 58.
46. **Brynjolfsson, A.**, Food irradiation and nutrition, *Prof. Nutr.*, 11, 7, 1979.
47. **Ley, F. J.**, The use of irradiation for the treatment of various animal feed products, *Food Irradiat. Inf.*, 1, 8, 1972.
48. **Ley, F. J.**, Radiation sterilization — an industrial process, in *Radiation Research, Biomedical, Chemical, and Physical Perspectives,* Proc. 5th Int. Cong. Radiat. Res., Seattle, 1974, Nygaard, O. F., Adler, H. I., and Sinclair, W. K., Eds., Academic Press, New York, 1975, 118.
49. **Raica, N., Jr. and Howie, D. L.**, Review of the United States Army wholesomeness of irradiated food program (1955 to 1966), in *Food Irradiat.*, Proc. Symp., Karlsruhe, 1966, IAEA, Vienna, 1966, 119.
50. **Read, M. S., Kraybill, H. F., Worth, W. S., Thompson, S. W., II, Isaac, G. J., and Witt, N. F.**, Successive generation rat feeding studies with a composite diet of gamma-irradiated foods, *Toxicol. Appl. Pharmacol.*, 3, 153, 1961.
51. **Kraybill, H. F.**, The wholesomeness of irradiated foods, in The National Food Irradiation Research Program, Hearings before the Joint Committee on Atomic Energy, Congress of the United States, Part 1, U.S. Government Printing Office, Washington, D.C., January 14—15, 1960, 308.
52. **Plough, I. C.**, An evaluation in human beings of the acceptability, digestibility, and toxicity of pork sterilized by gamma radiation and stored at room temperature, Rep. No. 204, U.S. Army Med. Nutr. Lab., Denver, Colo., 1957.
53. **Schreiber, M. and Nasset, E. S.**, Digestion of irradiated fat *in vivo, J. Appl. Physiol.*, 14, 639, 1959.
54. **Ley, F. J., Bleby, J., Coates, M. E., and Patterson, J. S.**, Sterilization of laboratory animal diets using gamma radiation, *Lab. Anim.*, 3, 221, 1969.
55. **Sheffner, A. L. and Spector, H.**, Action of ionizing radiations on vitamins, steroids, hormones, and other physiologically active compounds, in Radiation Preservation of Food, United States Army Quartermaster Corps., U.S. Government Printing Office, Washington, D.C., 1957, chap. 18.

56. **Richardson, L. R.**, A long range investigation of the nutritional properties of irradiated food, Prog. Rep. III, Texas Agric. Exp. Stn., College Park, Contract DA49-007-MD-582, Defense Documentation Center, Alexandria, Va., September 1, 1954 to July 1, 1955.
57. **Goldblith, S. A. and Proctor, B. E.**, Effect of high-voltage x-rays and cathode rays on vitamins (riboflavin and carotene), *Nucleonics*, 5, 50, 1949.
58. **Thomas, M. H. and Calloway, D. H.**, Nutritional value of dehydrated foods, *J. Am. Dietet. Assoc.*, 39, 105, 1961.
59. **Kung, H., Gaden, E. L., and King, C. G.**, Vitamins and enzymes in milk. Effect of gamma radiation on activity, *J. Agric. Food Chem.*, 1, 142, 1953.
60. **Richardson, L. R., Woodworth, P., and Coleman, S.**, Effect of ionizing radiations on vitamin K, *Fed. Proc.*, 15, 924, 1956.
61. Select Committee on Health Aspects of Irradiated Beef, Herman I. Chinn, Chairman, Evaluation of the Health Aspects of Certain Compounds Found in Irradiated Beef, prepared for the U.S. Army Medical Research and Development Command under Contract No. DAMD-17-76-C-6055 by the Life Sciences Research Office, Fed. Am. Soc. Exp. Biol., Bethesda, Md., August 1977, Suppl. I and II, March 1979.
62. Symposium on the recent advances in the chemistry of food irradiation, *Radiat. Res. Rev.*, 3(4), 1, 1972.
63. Symposium on current studies on the chemistry of food irradiation, First Chemical Congress of the North American Continent, Mexico City, December 1975, *J. Agric. Food Chem.*, 26, 1, 1978.
64. **Merritt, C., Jr., Angelini, P., and Nawar, W.**, Chemical analysis of radiolysis products relating to the wholesomeness of irradiated food, in *Food Preservation by Irradiation*, Vol. 2, STI/PUB/470, IAEA, Vienna, 1978, 97.
65. **Merritt, C., Jr.**, Chemical changes induced by irradiation in meats and meat components, in *Food Irradiation*, STI/PUB/127, IAEA, Vienna, 1977, 197.
66. **Taub, I. A., Robbins, F. M., Simic, M. G., Walker, J. E., and Wierbicki, E.**, Effect of irradiation on meat proteins, *Food Technol. (Chicago)*, 33, 184, 1979.
67. **Taub, I. A., Kaprielian, R. A., Halliday, J. W., Walker, J. E., Angelini, P., and Merritt, C., Jr.**, Factors affecting radiolytic effects in food, *Radiat. Phys. Chem.*, 14, 639, 1979.
68. **Taub, I. A., Halliday, J. W., Walker, J. E., Angelini, P., Vadji, M., and Merritt, C., Jr.**, Chemiclearance: principle and application to irradiated meats, presented at Int. Meat Res. Cong. (26th Eur. Meet. Meat Res. Workers), Colorado Springs, August 31 to September 5, 1980.
69. Rep. Joint FAO/IAEA/WHO Expert Committee on Wholesomeness of Irradiated Food, World Health Organization Tech. Rep. Ser. 659, World Health Organization, Geneva, 1981.
70. **Mittler, S.**, Failure of irradiated beef and ham to induce genetic changes in Drosophila, *Int. J. Radiat. Biol.*, 35, 583, 1979.
71. **Barnes, J. M., Boyd, D. M., Fitzpatrick, G. R., Needels, T. S., O'Dowd, T. A., and Tiller, R. E.**, Radiation processed foods as component of the armed forces feeding system, Operations Division Staff Paper ORO-SP-174, Operations Research Office, Johns Hopkins University, Bethesda, Md., August, 1961.
72. **Josephson, E. S., Brynjolfsson, A., and Wierbicki, E.**, Engineering and economics of food irradiation, *Trans. N.Y. Acad. Sci.*, (Series II) 30, 600, 1968.
73. **Brynjolfsson, A.**, Machine irradiation sources and irradiation technology, in *Chemical and Food Applications of Radiation, Nuclear Engineering, Part XIX*, Brownell, L. E., Josephson, E. S., Manowitz, B., and Harmer, D. E., Eds., American Institute of Chemical Engineers, New York, 1968, 71.
74. **Deitch, J., Osburn, J. W., Jr., and Ketchum, H. W.**, Cost-benefits analysis. Potential radiation sterilized military subsistence items, U.S. Department of Commerce Publ. BDC-788-09-71-002, U.S. Government Printing Office, Washington, D.C., March 1972.
75. **Brynjolfsson, A.**, Factors influencing economic evaluation of irradiation processing, in *Factors Influencing the Economical Application of Food Irradiation*, STI/PUB/331, Proc. Panel Series, IAEA, Vienna, 1973, 13.
76. Factors influencing the economical application of food irradiation, Proc. Panel, June 14—18, 1971, STI/PUB/331, IAEA, Vienna, 1973.
77. **Brynjolfsson, A.**, Cost of irradiating bacon and the associated energy savings, Tech. Rep. FEL-89, U.S. Army Natick Research and Development Command, Natick, Mass., March 1979.
78. **Brynjolfsson, A.**, Energy and food irradiation, in *Food Preservation by Irradiation*, Vol. 2, STI/PUB/470, IAEA, Vienna, 1978, 285.
79. **Josephson, E. S. and Wierbicki, E.**, Radiation preservation of food: past, present, and future, *Act. Rep. Res. Dev. Assoc.*, 25(1), 48, 1973.
80. **Anon.**, Frozen food sales thaw out, *Chain Store Age/Supermarkets*, 53, 291, 1977.

Appendix

GAMMA RADIATION PROCESSING EQUIPMENT AND ASSOCIATED ENERGY REQUIREMENTS IN FOOD IRRADIATION

Frank M. Fraser

Atomic Energy of Canada — Commercial Products (AECL-CP) is one of several agencies worldwide that pioneered in the applications of gamma radiation to food preservation.* More than 25 years of research in this area now appears to be coming to fruition. The barriers to the commercial development are slowly but methodically being overcome.

Irradiation design and operation is a well-developed and advanced technology which does not in any way present a barrier to the future application of food irradiation. Presented here is an overview of experience in the design and operation of gamma irradiators, their licensing, and associated personnel training.

AECLs food irradiator technology began in 1961 with the design, manufacture, and commissioning of the world's first mobile cobalt-60 irradiator. It was built after the food and drug clearance for irradiated potatoes had been obtained in Canada. Its main purpose in the first season of operation was to bridge the gap between laboratory experiment and full-scale commercial operation. The mobile cobalt-60 irradiator was a most useful tool for the food irradiation program. First and foremost, it was able to take cobalt-60 irradiation right into the world of future users and demonstrated that radiation is a safe, reliable source of energy which is simple to use. The mobile irradiator traveled extensively throughout Canada and the U.S. where it was used for the irradiation of a wide variety of fruits and vegetables.

Gamma irradiators generally fall into four categories; namely, *conveyor bed, carrier, bulk,* and *pallet box* systems. Of these, the pallet box irradiator is the most suitable for food processing.

The *conveyor bed irradiators* which utilize tote boxes to contain the product for irradiation can be efficiently used for certain food products but are not ideally suited to many conventional methods of food handling. The standard tote box is relatively small and inflexible, resulting in increased labor and poor product packing efficiency both of which can significantly increase operating costs. However, conveyor bed systems are excellent for efficiently sterilizing certain medical products and this may be a serious consideration in plant selection when, due to the seasonal aspect of food irradiation, there may be an advantage to be gained in having one irradiator capable of more than one commercial application (Figure 1).

Carrier type sterilizers are designed specifically to meet the high throughput, efficiency, dose uniformity, and flexibility demanded by certain medical disposables manufacturers and are not readily adaptable to food irradiation (Figure 2).

Bulk irradiators have been designed for disinfestation of maize, wheat, and grain products and will play an important role in bulk product processing.

The *pallet box irradiator* is designed for the processing of product on pallets. These pallets are loaded into carriers suspended from overhead rails. Each carrier holds two pallets (Figure 3).

The source pass mechanism contains four carriers which are uniquely indexed about the cobalt-60 source such that each face of the pallet is selectively exposed to the source twice before completing its cycle. This, coupled with strategically located attenuators, results in excellent product uniformity for such a large unit.

* The editors are much indebted to the author of this excellent description of gamma radiation processing equipment, but in fairness to other manufacturers of gamma irradiators and machine sources of irradiation, they cannot exclusively endorse the AECL-CP irradiators over those of other commercial agencies.

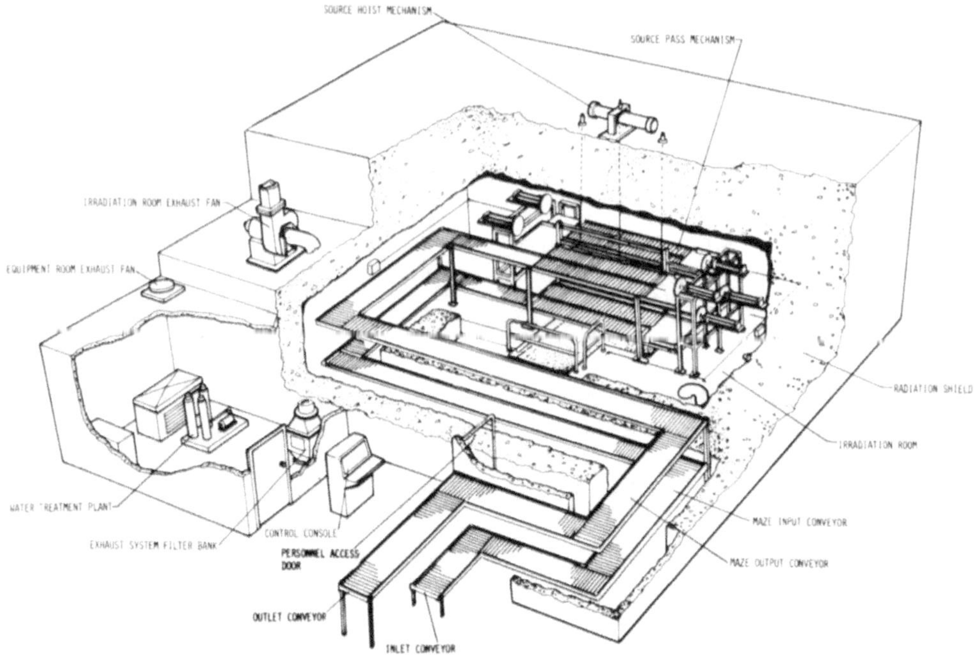

FIGURE 1. JS8500-Irradiator (Tote Box Concept).

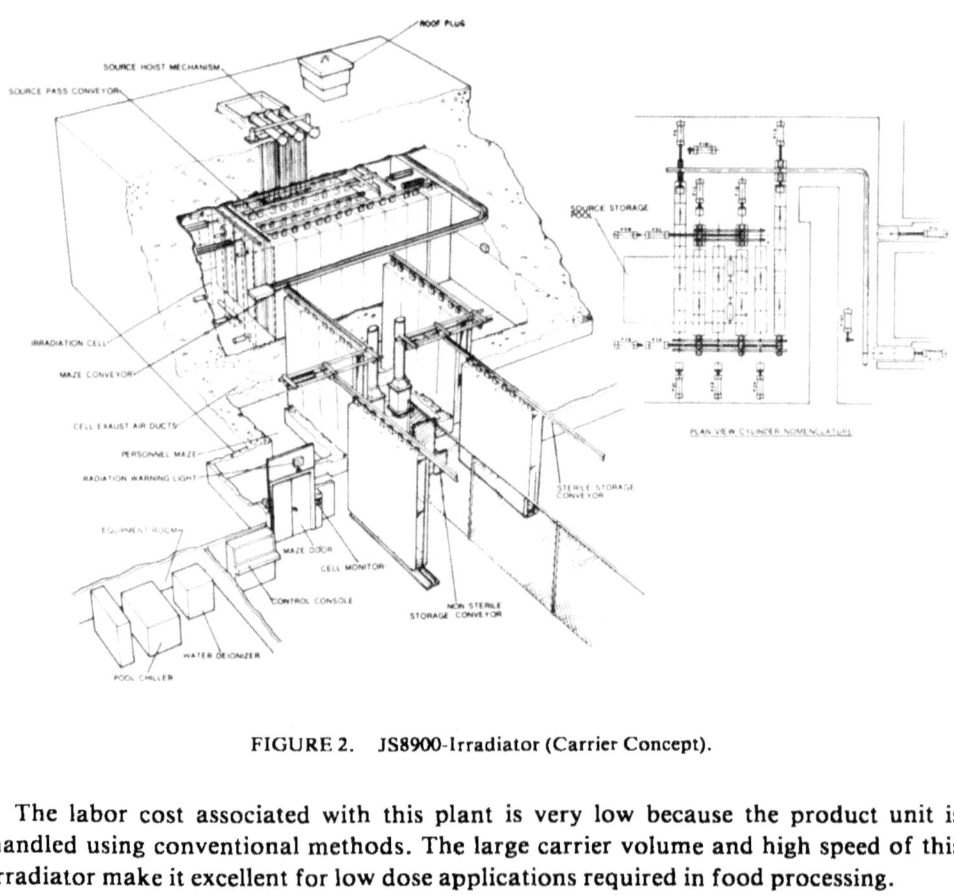

FIGURE 2. JS8900-Irradiator (Carrier Concept).

The labor cost associated with this plant is very low because the product unit is handled using conventional methods. The large carrier volume and high speed of this irradiator make it excellent for low dose applications required in food processing.

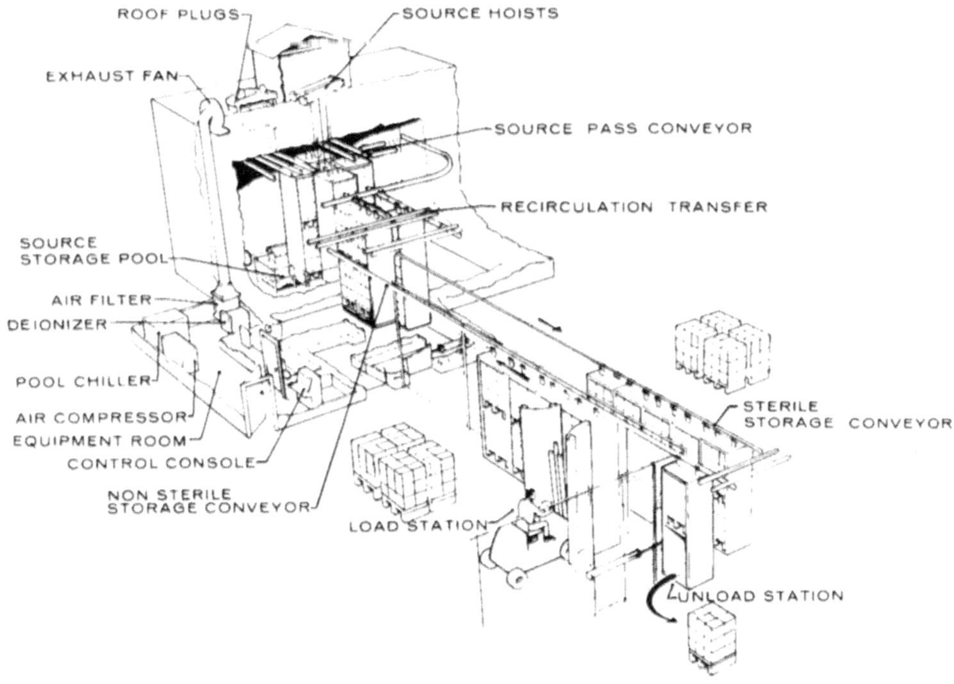

FIGURE 3. JS9000 Pallet Irradiator.

A recirculating loop can be provided to make the irradiator suitable for incremental dose application. This permits operation of the irradiator with products requiring different doses.

A typical pallet box irradiator was recently purchased for a service irradiation facility in The Netherlands. Generally, the specifications were as follows:

For medical products:	
Pallet box size:	100 cm × 120 cm × 195 cm high
Capacity:	~30,000 m³/year
Dose:	2.5 Mrad minimum
Product density:	0.20 g/cm³
Dose uniformity ratio:	1.38
Storage capacity:	8 hr
Annual hours of operation:	8000
For nonmedical products:	
Product density:	0.4 to 0.8 g/cm³
Dose uniformity:	1.58 to 2.14
Minimum dose:	5000 rad

One can readily see from the processing requirements the flexibility in terms of absorbed dose this plant offers. It is perfectly suitable for both food irradiation and for medical products sterilization when pallet box handling is indicated.

The purchase of an irradiator could include building-layout drawings, licensing documentation, dosimetry, and operation and maintenance manuals. Also included could be the training of operators and maintenance personnel during the irradiator installation and commissioning. Building construction drawings and certain building construction materials can also be provided.

At the heart of AECL's industrial irradiators is the gamma radiation emitting isotope cobalt-60. This isotope, produced by the nuclear activation of cobalt-59, has a half-life of 5.3 years, an entirely predictable emission rate, a constant energy and penetration all of which make cobalt-60 ideal for continuous production processes such as food irradiation.

The cobalt-59 is activated in the Canadian Candu nuclear reactors in the form of either ¼ in diameter by 1-in long slugs or 1 mm diameter 1 mm pellets. These forms of cobalt permit fabrication of a wide variety of source configurations required for specific industrial and therapeutic applications. The high thermal neutron flux in the Candu reactors enable the production of high specific activity cobalt-60 in a short activation time. These reactors which are operated by Ontario Hydro, one of the largest public utilities in the world, ensure a reliable and steady supply of cobalt-60 in the foreseeable future.

The economics associated with the operation of radiation processing plants vary considerably but are generally governed by the following basic processing parameters:

1. Product box sizes
2. Product densities
3. Weekly and annual product throughput
4. Minimum sterilization dose
5. Maximum dose which product can tolerate without significant radiation degradation effects
6. Plant operation with regards to shifts/day; days/week, and weeks/year
7. Anticipated annual growth in irradiation requirements
8. Feasibility of standardizing to one box size
9. Storage requirements for unattended operation
10. Location of gamma sterilizer — area available, adjacent to or within an existing building.

These parameters determine the size and geometry of the source, the design of the biological shield, the size and maximum speed of the internal conveyor, the size of the external storage conveyor system, and the requirements for space and equipment for loading and unloading the product.

Food irradiation costs can vary from $0.0015/kg for the disinfestation of maize to $0.06/kg for the irradiation of poultry to eliminate salmonella.

Given the processing parameters and capital write-off guidelines, the vendor of the irradiation source and accessories can provide the customer with an economic analysis of several plants to assist in the selection of the optimum irradiator design.

Let us consider the energy used in the operation of the pallet box irradiator previously discussed. Electrical power is required for the following: (1) Control and Interlock System, (2) Water Deionizing System, (3) Water Cooling System, (4) Ventilation and Exhaust System, (5) Compressed Air, (6) Carrier Monorail Loop, and (7) Lighting.

The electrical power requirement for the above at maximum capacity (30 carriers per hour) is 50 kVA or 50 kW. Based on an annual operation of 8000 hr the power utilized is 400,000 kWh (50 kW × 8000 hr).

The cobalt-60 source strength is 2 million Ci. Over a 1-year period 246 kCi or 12.3% of the energy source strength is lost due to the natural decay of cobalt-60 (cobalt-60 has a half-life of 5.37 years resulting in an annual decay of 12.3%). Therefore the source will decay to 1.754×10^5 Ci, a reduction of 246,000 Ci. The electrical power equivalent generated by the 246 kCi of cobalt-60 is 32,324 kWh (246 kCi × 15 W/kCi × 8760 hr/year). The total annual energy expended thus is 412,324 kWh. This energy will be consumed in the processing of 30,000 m^3 of medical products to a dose of 2.5 Mrad at a product density of 0.20g/cm^3. On a unit basis, the energy used is 179 kJ/kg.

$$\frac{432{,}324 \text{ kWh/year} \times 3.6 \times 10^3 \text{ kJ/kWh}}{30{,}000 \text{ m}^3\text{/year} \times 290 \text{ kg/m}^3} = 179 \text{ kJ/kg}$$

This number compares favorably with A. Brynjolfsson's, August 1978 technical report on energy and food irradiation.*

For most processing applications the dose is several orders of magnitude less than that required for medical products sterilization. As a matter of fact the maximum throughput rate for processing medical products based on the parameters discussed is 1.087 tons/hr as compared to 60 tons/hr for a food product with a density of approximately 0.5 g/cm^3 with a required dose of 20 krad. Assuming the plant would be utilized 2000 hr/year for food processing at a rate of 60 tons/hr, the energy used on a unit basis is 4.0 kJ/kg.

To summarize, Atomic Energy of Canada — Commercial Products' experience in food irradiation dates back to 1956. Irradiator design and operation is a well-developed technology which does not present a barrier to the future application of food irradiation. The newly designed pallet box irradiator is most suitable for food processing. The energy used for gamma radiation processing of food can be as low as 4.0 kJ/kg.

* Technical Report, Natick/TR -78032, Energy and Food Irradiation, A. Brynjolfsson, Food Engineering Laboratory, U.S. Army Natick Research and Development Laboratories, Natick, Mass., 01760.

INDEX

A

B

D

E

I

J

K

L

M

P

Q

R

S

T

U

V